# Radar Meteorology

For a complete listing of the *Artech House Radar Library*,
turn to the back of this book . . .

# Radar Meteorology

**Henri Sauvageot**

Artech House
Boston • London

**Library of Congress Cataloging-in-Publication Data**

Sauvageot, Henri.
[Radarmétéorologie. English]
Radar meteorology / by Henri Sauvageot.
p. cm.
Rev. translation of: Radarmétéorologie.
Includes bibliographical references and index.
ISBN 0-89006-318-4
1. Radar meteorlogy. I. Title.
QC973.5.S2813 1991 91-29276
551.6'353–dc20 CIP

**British Library Cataloguing in Publication Data**

Sauvageot, Henri
Radar meteorology.
I. Title II. [Radarmeterologie. English]
551.6353

ISBN 0-89006-318-4

This work is the translation of a French book by Henri Sauvageot entitled "Radarmétéorologie" published by EYROLLES of Paris.


**685 Canton Street**
**Norwood, MA 02062**



**International Standard Book Number: 0-89006-318-4**
**Library of Congress Catalog Card Number: 91-29276**

**10 9 8 7 6 5 4 3 2 1**

*to Hélène, Laurent, Antoine, and Marguerite*

# Contents

# *Preface*

During the last few decades, the area of radar application has broadened considerably and the technologies contributing to it have advanced considerably as has our understanding of signal processing. Among the conquests of radar, the meteorological observation of the atmosphere is certainly one of the most remarkable: a continuous flow of results has, little by little, established its efficiency in this field. With improving quality and accuracy of the measurements, it has been possible to increase the number of phenomena and the variables that can be accessed by it.

Radar has established itself as the perfect instrument for remote sensing of the atmosphere: after the detection of precipitation, which allowed analysis of the structure of the perturbations, came the measurement of velocity fields by doppler radars, which are now used in operational meteorological arrays, followed by clear air detection with (ST) radars, which have considerable potential in both practical applications and research and are also beginning to be used in operational arrays. Even more recently, polarimetric radar, after a false start, has returned to the fore, promising potential observation of the microstructure of hydrometeors.

Despite all these remarkable results, objective evidence suggests that the best of meteorological radar is still to come, in particular, with the development of sensors implementing multiparameter techniques, the development of dense and homogeneous radar arrays, more complete processing of the signals due to the increased efficiency of the associated data processing systems, satellite-borne systems, improved incorporation of results in the forecasting models, improved dissemination of radar information to users, *et cetera*.

However, the possibilities offered by "meteorological radar" remain inadequately known, even among specialists in the atmospheric sciences or users of the atmosphere. This undoubtedly is due to the rapid advances in this specialty and insufficient teaching of the relevant subjects.

This book has been written to present a compact and accessible introduction to meteorological measurement and observation of atmospheric processes by radar. It is aimed at all those whose activities are related to a knowledge of the local state of the atmosphere: meteorologists; hydrologists; pilots and air traffic controllers; sailors and oceanographers, aeronautical, telecommunication, and electronic engineers interested in the effects of the atmosphere on the propagation of microwaves, and also entomologists and ornithologists for whom radar is a valuable tool, although little used as yet. The book is also intended as an aid to teaching the specialized subjects corresponding to these activities as well as a reference text research workers in the atmospheric sciences. The author therefore assumes that the reader interested in this book possesses the basic knowledge in physics and mathematics at the level of the first two years of higher education in the sciences as well as elementary notions of meteorology. However, the book can be read at several levels and a nonspecialist reader with a good scientific background will be able to understand the main points and to learn something of this subject.

To preserve the review nature of this book and to limit its scope, general technological aspects well covered in other works (see Appendix 1: Bibliographical Note) have been deliberately reduced to the necessary minimum. This also applies to historical aspects. Only the specific problems of the physics of meteorological measurements have been treated. Furthermore, a large number of references are scattered throughout the text making it possible to trace the "bibliographic thread" for all the subjects considered and thus to obtain a more detailed knowledge.

This book is based on the author's experience teaching in various French universities, in particular, the Paul Sabatier University of Toulouse, and also on a first edition, which appeared in French in 1982 published by Eyrolles (Paris). The contents of this first edition have been completely revised and substantially expanded.

The text is divided into five chapters. The first chapter presents the basic concepts that underlie the application of radar to meteorological targets; that is, distributed targets occupying a large atmospheric volume as compared to the dimensions of the beam. The second chapter deals with the propagation of microwaves in clouds and precipitation, in particular, attenuation and scattering by hydrometeors and hydrological measurements. It includes a brief recapitulation of the physics of clouds and precipitations. Chapter 3 is concerned with measurement by scattering the fluctuations in the refractive index of air, for the sake of convenience, incorrectly called measurements in clear air. Chapter 4 classifies the measurements based on the Doppler effect. The fifth and last chapter is an introduction (not exhaustive!) to the observation of some meteorological structures or systems by radar. Additional numerical data, useful for performing calculations on concrete cases, are given in the appendices. However, no function tables are given; most of them can be obtained with scientific pocket calculators.

# Chapter 1
# Basic Concepts of Radar

## 1.1 INTRODUCTION

The propagation of electromagnetic radiation with very short wavelengths in the atmosphere is influenced by the physical properties of the medium that it traverses. These, for the most part, hinder current applications of radar detection and attempts are made to avoid or minimize them. However, with meteorological radar, these effects are harnessed for observing the air space in a number of different ways.

Heinrich Hertz, by experimentally discovering electromagnetic waves at the end of the 19th century and by showing, around 1886, that they are reflected by certain objects, paved the way to radar. In 1904, following experiments on an obstacle detector using the reflection of these waves, a patent was filed by a German engineer. It was quickly followed by many others, which did not enable any convincing technical realization and therefore fell into oblivion. Around 1922, important studies on the propagation of radio waves led Guglielmo Marconi to a more advanced formulation of the principles of electromagnetic detection. Developments in this area have hardly stopped since then: *continuous-wave* (CW) devices detecting interference between the incident wave and the reflected wave with the help of a transmitter and a separate receiver (1922); the first pulsed system making possible the measurement of distance from the target (successfully used for studying the ionosphere in 1925); the first unintentional detection from an airplane (1930) and the intentional detection (1932).

Industrial research and development programs were then systematically undertaken, and around 1938, these resulted in the first version of an antiaircraft system for detection and localization called RADAR (contraction of *radio detection and ranging*). World War II witnessed considerable efforts in the research and development of radar; during this time, it became a mature technology.

The improvement in the performance of the equipment and its diversification, which were required during this first developmental phase, led to a reduction in the wavelengths used, with the aim of improving the operational quality of the radars (directivity of the antennas and thus mobility and accuracy in ranging) and increasing the sensitivity of the equipment in the detection of small objects (for example, submarine periscopes above the waters). First metric waves were used, then decimetric, and finally centimetric. With the enhancement in quality of the radars, precipitating zones started to appear on the screens to the consternation of the engineers who saw this as a serious limitation, capable of reducing the efficiency of surveillance of a certain space to zero (natural echo zones could serve as a refuge for the aggressor!). However, meteorologists and atmospheric physicists were quick to take advantage of the new facilities that radar was now offering. The first meteorological observations exploiting radar date from 1941; while some were trying to find procedures for eliminating echos due to natural precipitation, others were pioneering equipment better adapted to analyzing them. The "meteorological" function of radar therefore appeared from the very beginning. At first, research and application in this area were essentially descriptive and then, with the first attempts at hydrological measurement and speed measurements, they entered a quantitative phase around 1960. (See the historical review in Atlas 1990.)

Not until the early 1970s did the doppler radar, whose principles have long been known, emerge as the major technique providing meteorological research with full efficiency. This happened at the right time, when the meteorological community, which invested much in the laboratory study of the microphysical processes, became aware of the necessity, for progress, of a better understanding of the air movements at cloud scale. This field then became the first concern. With the doppler radars, used separately or in a network, in wide experimentation campaigns, the dynamic study of the clouds and precipitation as well as the mesometeorology could develop. In the 1970s, the "clear air detection" techniques appeared thanks to the recovery by the radar of the VHF and UHF fields, the potentialities of which had been somewhat underestimated. The VHF-UHF systems proved almost perfect for the vertical soundings of winds. Their progressive integration in the form of networks, both in the research and operational meteorology, undoubtedly is one of the major areas of progress in the past 20 years. Because of the VHF-UHF techniques, radar no longer is limited to the precipitating zones; the meteorological atmosphere can be observed permanently and totally.

An inadequacy remains, however, concerning the identification of the nature of the observed hydrometeors. This problem, still partially unsolved, has progressed sensibly with the development of measurements with several wavelengths and measurements in polarization diversity. These methods, interest in which began in the early days of radar, progressed only slowly because of the technical difficulties of implementation. Nowadays they are in full evolution, and the results obtained in recent years are very promising.

The progress of meteorological radar also is linked closely to the techniques of signal acquisition and data processing. In the beginning, the main part of the radar was the captor (transmitter-receiver-antenna), and its exploitation was limited to a basic presentation of the rough signal on a cathode-ray screen, possibly recorded on photographic aids. Then the size of the processors grew. They first were analogical and only processed reflectivity information. The introduction of digital processors in the 1970s, stimulated by the development of doppler measurements, was a small revolution. The quantity of data processed increased considerably, as did measurement precision. In the early 1980s, the cost of data acquisition, preprocessing, and visualization systems almost equaled that of the captor in efficient meteorological radars. The radar used for research could obtain numerical bands which were then processed on general-use computers to obtain the physical value fields of meteorological interest. The present evolution follows from the outstanding progress of the data processing: increase in computer speed and power, decrease in their volume and price; the tendency to return to real-time data processing with the almost exhaustive use of the information contained in the signal, and the representation of three-dimensional fields with high spatial and temporal resolution. The modern meteorological radar is a captor plus a computer, the screen of which is the visualization aid.

Correlatively to this evolution, the use of the radar for meteorological observation and measurement became widespread: creation of wide networks in developed countries; installation of it aboard planes and ships; use of it in all meteorological research; plans for putting it into orbit, *et cetera*. A progressive, although still insufficient, popularization of the use of this technique can be observed, preventing it from remaining the sole prerogative of the specialists, even when the applications are relatively complex.

The application area of radar first of all is to describe the present (detection of rain, identification of structures and their danger, *et cetera*) with the aim of detailed short-term prediction. Only afterward can it be considered for the systematic measurement of useful physical quantities for climatology and the preparation of longer-term forecasts. Finally it can be used to acquire new knowledge, making it possible to develop conceptual models for atmospheric structures and phenomena for constructing and checking numerical models that can be used to simulate behavior on a computer by applying the equations of hydrodynamics and microphysics.

For meteorological radar, the target consists of the atmosphere itself, and more precisely the cloud and precipitation particles, the variations of the air refractive index, and also sometimes insects and birds. Therefore, it does not involve static objects but a mobile medium subject to a permanent evolution under the effect of complex mechanical and thermodynamic processes, intricately intertwined.

The target is a scattering region, the dimensions of which are large compared to those of the beam. Observation at all ranges in the line of sight combined with the movements of the antenna makes volume scanning possible. Signals received by the radar carry information relating to the properties of the medium at each observed point (reflectivity, average speed, turbulence, *et cetera*) and the space-time distribution of the measured parameter (geometry of the volume, structure, relative displacements, *et cetera*).

Compared to other atmospheric observation methods, which are essentially the airplane laboratory and probes transported by different means (balloons, rockets) or passive methods (photographic and radiometric), radar sounding offers the following principal advantages:

- It "sees" through optically opaque zones;
- It makes it possible to carry out active remote detection of a target (the observation is made without having to displace the instrument and the sounded volume can be chosen with precision and in an objective manner within a larger system);
- The signal depends neither on the sun as a light source nor on the emission of the target;
- Scanning can be carried out in three dimensions, as a function of time, and over very large volumes;
- The resolution and the continuity of the observations in time and space are satisfactory for a large number of applications;
- The observation does not disturb the medium in which it is made;
- The data are immediately available.

Radar also has some disadvantages, among which should be mentioned the ambiguity of certain measurements and the limits to resolution.

## 1.2 THE SENSOR

### 1.2.1 Noncoherent Pulse Radar

Conventional radar transmits powerful, very brief pulses of electromagnetic energy with high frequency into the atmosphere at equally spaced intervals of time. The energy is concentrated in a beam with small solid angle by a directive antenna. All types of targets present in the beam intercept some of the incident energy, which they absorb and radiate in different directions. The fraction returned to the radar is the useful signal. If we assume as a first approximation that the electromagnetic waves used are propagated in a straight line at constant velocity ($c \approx 3 \times 10^8$ ms$^{-1}$) over distances not exceeding a few hundred kilometers, the orientation of the antenna and the time elapsed between the transmission of the pulse and the reception of the signal make it possible to locate the scattering region in terms of direction and distance (Figure 1.1).

Figure 1.2 shows a simplified block diagram of a noncoherent pulse radar. Transmission is produced by a power oscillator. The choice of the transmitted frequency $f_0$ results from a compromise, taking into account the conditions of propagation in the atmosphere (diffusion by atmospheric targets, attenuation along the radar target path) and the technological requirements involved in implementing the equipment (these different aspects will be discussed later).

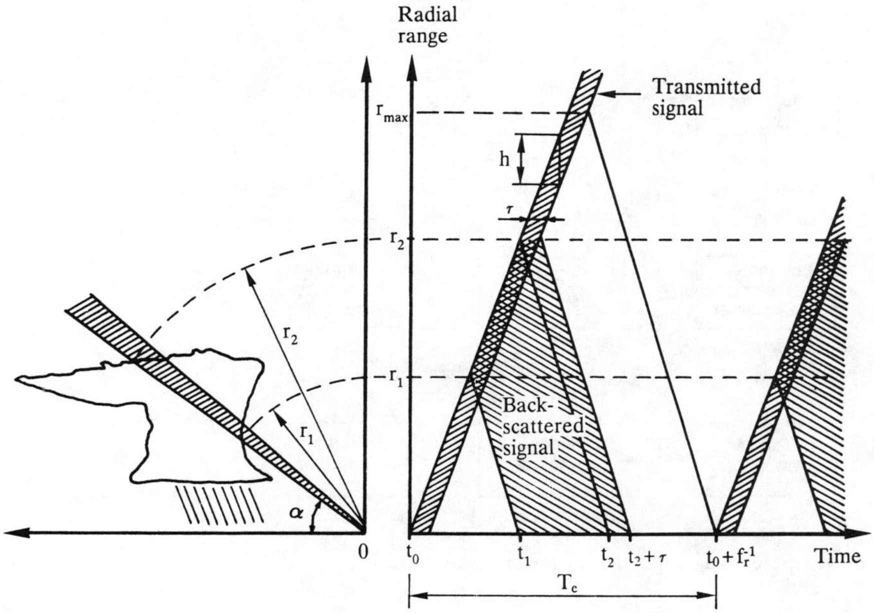


**Figure 1.1** Operating principle for the pulse radar applied to volume sounding. The left-hand side of the diagram represents the beam in the vertical plane of a radar situated at 0, intercepting a cloud between the radial ranges $r_1$ and $r_2$. The right-hand side is a combined range and time representation of the signals transmitted by the radar and scattered by the cloud between two successive pulses. The terms $\tau$, $h$, and $f_r$ are the duration, width, and the repetition frequency of the pulses.

For hydrometeor observations the most common frequencies lie between 3 and 35 gigahertz (GHz) and therefore belong to the microwave region.[1] The corresponding wavelengths ($\lambda = c/f_0$) range from 10 cm to 0.8 cm. For the observation of the clear air, lowest frequencies of the VHF and UHF bands are used (cf. Chapter 4) (the designation of the frequency bands is given in Appendix 4). The power oscillator is a magnetron (self-oscillator) or a power amplifier of the klystron type combined with a master oscillator (klystrons have a better spectral purity than magnetrons). The modulator determines the pulse repetition frequency $f_r$ (often denoted *PRF*) and the pulse duration $\tau$ (or temporal "width"). The value of $f_r$ ranges from a few hundred hertz to a few kilohertz. Its reciprocal is the repetition period of the pulses (or sampling period), $T_S = f_r^{-1}$. The duration $\tau$ lies between a fraction of a microsecond and a few microseconds; it fixes the spatial length ($h = c\tau$) of the microwave train making

[1]In this region, circuits with localized constants are replaced by circuits with distributed constants, such as waveguides or resonating cavities.

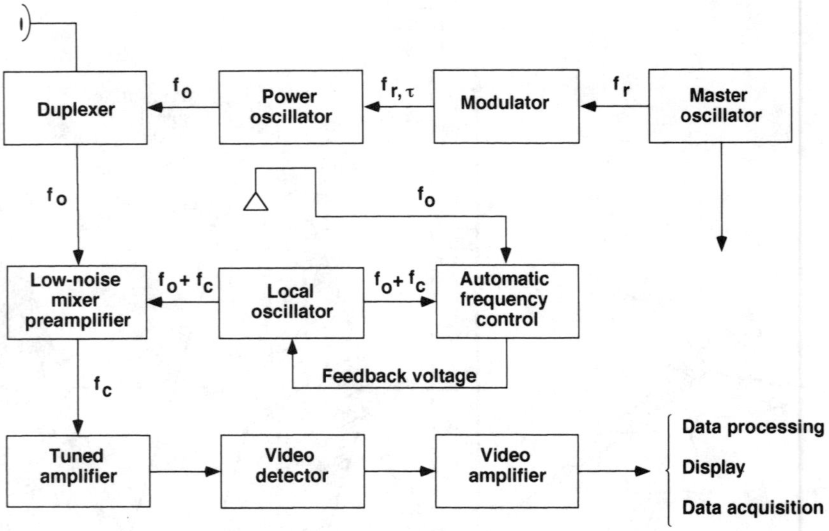


**Figure 1.2** Block diagram of a noncoherent pulse radar.

up the pulse. The transmitted pulse consists of a few thousands of oscillations with frequency $f_0$. The "peak" power $P_t$ of the transmission ranges from a few tens to a few hundred kilowatts. For a rectangular pulse, the average power transmitted is thus

$$P_{\mathrm{av}} = P_t \tau / T_S$$

The ratio $\tau/T_S$ is called the duty factor of the transmitter, and its value is around $10^{-3}$. A switch (or duplexer) makes it possible to use the same antenna for transmission as for reception while protecting the receiver during transmission; nevertheless, sometimes different antennas are used for transmission and reception.

The emitted wave is

$$S(t) = A_t \cos\omega_0 t$$

where $A_t$ is the amplitude, $\omega_0 = 2\pi f_0$ is the angular frequency, and $t$ is the time. For a point target at distance $r$, the instantaneous voltage of signal received by the radar can be written

$$e(t) = a \cos[\omega_0 t + \varphi(t)] \tag{1.1}$$

$a$ is the amplitude and $\varphi(t) = 2\pi(2r/\lambda)$ is the phase.

In the receiver, the principle of which is similar to that of a classical superheterodyne radio receiver, the signal pulsed at frequency $f_r$, is mixed with the continuous wave of a microwave local oscillator of low power. The local oscillator frequency has a constant shift $f_c = \omega_c/2\pi$ composed by using the frequency $f_0$ of the transmitter with respect to what it is fed back by the automatic frequency control system. The local oscillator therefore is of the form $k \cos(\omega_0 - \omega_c)t$. The value of $f_c$, called the intermediate frequency, is on the order of 10 MHz (current values: 30 or 60 MHz). The mixer multiplies the two waves so that a voltage of the form

$$\begin{aligned} e_m(t) &= a \cos[\omega_0 t + \varphi(t)]\, k \cos(\omega_0 - \omega_c)t \\ &= \frac{ak}{2} \{\cos[(2\,\omega_0 - \omega_c)t + \varphi(t)] + \cos[\omega_c t + \varphi(t)]\} \end{aligned} \tag{1.2}$$

appears at the output, consisting of two terms the frequencies of which are equal to the sum and the difference of the frequencies of the mixed signals. Only the low frequency ($f_c$) is preserved. Therefore, mixing (or heterodyning) the signal with the local oscillator amounts to shifting its spectrum by $\pm(f_0 + f_c)$ and multiplying its amplitude by $k/2$. We thus produce a change in frequency without loss of information. One advantage to the intermediate frequency is that it can be handled in ordinary circuits (with localized constants). The signal is then amplified in a low-noise preamplifier (sometimes, preliminary amplification is carried out in the microwave region before mixing), and it is directed to a tuned amplifier that limits the width of the passband to a few megahertz around $f_c$ with the aim of reducing the noise level (see Section 1.2.4). It then passes through the detector (or rectifier) and the video frequency amplifier[2] which together form an "envelope detector" whose function is to extract the modulation and reject the carrier frequency. This modulation is the information required. If the detector is linear, its amplitude is proportionate to the amplitude of the signal at the entrance. However, quadratic detectors are used most often, so that the modulation is proportionate to the square amplitude at the entrance, that is to say, to the power of the signal. The modulation can be used for direct display in real time for all or part of the information it contains with or without preprocessing; it can also be sent to a recording system to save the data for subsequent processing. A clock (or master oscillator) provides the repetition pulses required for the various units of the radar to synchronize the overall system, in particular the simultaneous triggering of the modulator, the display screen sweeps, and the data acquisition. The time interval separating two successive pulses $T_S$ determines the maximum unambiguous observation distance (or range) (Figure 1.1):

$$r_{\max} = \frac{1}{2} cT_S \tag{1.3}$$

[2]Video frequencies are those used for displaying information on the screen.

In fact, echoes originating from a distance $r > r_{\max}$ (or greater than $nr_{\max}$ with $n$ an integer $> 1$) will appear as "second trip" signals at a distance $r - r_{\max}$ (or $r - nr_{\max}$).

A radar designed in this way therefore makes it possible to measure the power (or intensity) of the signal and its fluctuations over time but does not process the phase information that explains the name, noncoherent radar.

### 1.2.2 Pulsed Doppler Radar

*Principle*

In fact, atmospheric targets usually are not static. Let us consider a radar with wavelength $\lambda$ observing a point target at range $r$. The signal from the scattered field received by the antenna is a pulse of duration $\tau$ represented by equation (1.1). The phase of the received signal is

$$\varphi(t) = -\frac{4\pi r(t)}{\lambda} \qquad (1.4)$$

If the target is moving with respect to the radar with a radial velocity $v_r$, the phase of the signal varies and we have

$$\frac{d\varphi}{dt} = -\frac{4\pi}{\lambda}\frac{dr}{dt} = -\frac{4\pi}{\lambda}v_r \qquad (1.5)$$

Thus, the angular frequency of the echo has a shift due to the Doppler effect.[3] When the target moves away from the radar ($V_r > 0$), the doppler shift $d\varphi/dt$ is negative. It is the opposite when the target draws closer. We can write

$$\frac{d\varphi}{dt} = \omega_d = 2\pi f_d \qquad (1.6)$$

that is,

$$f_d = -\frac{2v_r}{\lambda} \qquad (1.7)$$

where $f_d$ is the doppler frequency. Formula (1.7) is a valid approximation for the (nonrelativistic) motion of atmospheric targets.

[3]This effect was discovered by the Austrian, Johann Christian Doppler, in 1842 in the case of sound waves.

The doppler frequencies of atmospheric targets do not exceed a few kilohertz; therefore they are very small with respect to the transmitted frequencies and within the passband of the receiver. To measure them, the local oscillator must retain the memory of the phase of the transmitted wave for comparison with the received signal to extract the information relating to the Doppler effect. Such a measurement is simply carried out in systems with continuous wave transmission and fixed frequency, although with loss of information on distance and poor sensitivity. In this type of system called homodyne, the received signal is made to oscillate with the reference wave supplied by the continuous oscillation of a low-power stable transmitter. The oscillation frequency gives the value of $f_d$ directly. This is the principle of movement detectors and remote measurements of speed (roadside velocimeter, laser anemometer, *et cetera*).

To preserve the range and speed information, systems of the type shown in Figure 1.3 are used (there are several different versions based on the same principle). A local high-stability oscillator, the STALO (stable local oscillator), supplies a microwave signal with frequency $f_0$ mixed with the wave of a coherent reference oscillator, also very stable, the COHO (coherent oscillator) with intermediate frequency $f_c$. The resulting wave of frequency $f_0 + f_c$ (or $f_0 - f_c$) determines, after amplification, the characteristics of the wave transmitted by an amplifier tube (of the klystron type). The received signal is mixed with the STALO waveform to produce a waveform with intermediate frequency having the phase characteristics of the back scattered signal. This waveform is compared with the waveform produced by the reference oscillator, which has retained the phase characteristics of the transmitted wave. Their mixing produces a signal with the doppler frequency.

This system uses a tuned transmitter tube. Other methods also can be used to obtain a coherent system; in fact, the feedback can be applied to the local oscillator (which in this case no longer is a STALO) based on the mixture of stable waves produced by the transmitter and the COHO or back to the COHO with the aid of the reference wave produced by mixing the transmitter wave and that of the stable local oscillator. In the last two cases, the transmitter may consist of either a stable continuous oscillator (STALO) acting as the master oscillator combined with a pulse modulated amplifier tube (klystron) or a magnetron. When the transmitter is a magnetron, the phase relationship is not preserved from one pulse to the next; the coherent reference therefore must be adjusted to the phase of the transmitted signal at the beginning of each firing. For this, a fraction of the transmitter energy is removed to produce a feedback pulse that determines the phase of the servo element (local or COHO oscillator), which must then oscillate independently until the next firing without drifting to ensure coherent reception.

We can see by analogy with equation (1.2) that the signal resulting from mixing the backscattered signal and the reference wave of the COHO has a sinusoidal form and therefore carries no information on the sign of the doppler frequency. To overcome this difficulty, the phase demodulator is used, in which the phase of the signal

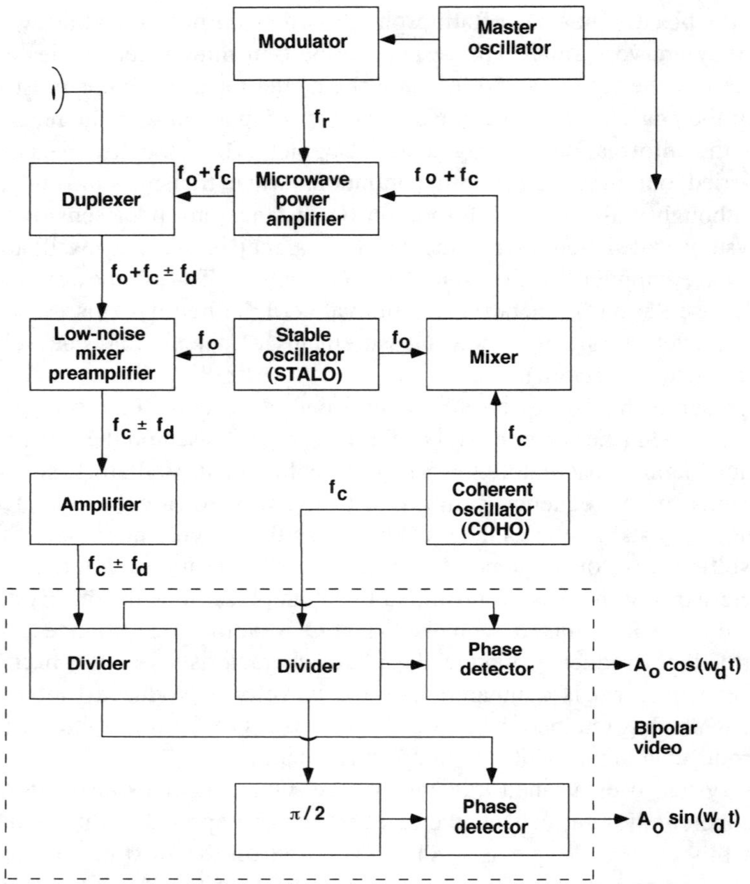


**Figure 1.3** Block diagram of a coherent pulse radar.

is compared to that of the reference wave in two balanced mixers at the same time. On one of the two channels, the phase of the sinusoidal reference wave is advanced by $\pi/2$ rad before comparison. In this way, at the output of the demodulator, two signals are in phase and in quadrature, at the doppler frequency, forming the "bipolar video" (because the amplitudes take positive and negative values); that is,

$$I(t) = A_0 \cos(\omega_d t)$$
$$Q(t) = A_0 \sin(\omega_d t) \tag{1.8}$$

If $f_d > 0$, that is, movement is toward the radar, the phase of $Q(t)$ has a delay of $\pi/2$ with respect to that of $I(t)$. If $f_d < 0$, movement is away from the radar, it is the opposite.

The two components $I(t)$ and $Q(t)$ may be represented by the real and imaginary parts of the complex amplitude:

$$y(t) = A_0 e^{i\omega_d t} \tag{1.9}$$

For a doppler radar observing a meteorological target, the values of $f_d$ are very weak, so that it is not possible to measure it for the duration of a pulse. For example for $v_r = 20$ m s$^{-1}$, $\lambda = 10$ cm, we have $f_d = 400$ Hz and the phase change for the duration of a pulse, $\omega_d \tau$, has a value for $\tau = 1$ $\mu$s of only 0.14 d°; the output of the phase detector has an almost constant phase for the duration of a pulse. Under such conditions, to measure $\omega_d$, the phase change between two successive impulses is taken into account. For the pulse $n$, the complex amplitude is

$$y_n = A_0 e^{i\varphi_{r,n}} \tag{1.10}$$

with $\varphi_r = \arctan Q/I$.

For two samples $n$ and $n - 1$ corresponding to two successive pulses, we have

$$\frac{\Delta\varphi_r}{\Delta t} = \frac{\varphi_{r,n} - \varphi_{r,n-1}}{T_S} = \omega_d \tag{1.11}$$

*Ambiguities*

If the radial velocity of the target is such that the phase difference between one pulse and the next is greater than $\pi$ rads, there is an ambiguity concerning the measurement of the phase and, therefore, the measurement of the speed. This phenomenon is intrinsic to observations made by sampling at constant time intervals. It is met in particular (where it is exploited) in stroboscopy. Let us consider, for example, a sinusoidal signal $s(t)$ sampled in such a manner that the time interval between two consecutive measurements is $T_S$ (with $T_S = f_r^{-1}$). At least two samples per cycle are necessary to define a frequency component of $s(t)$ (Figure 1.4(a)). Consequently, the highest frequency that can be measured by sampling is equal to half the sampling frequency; that is, $f_r/2$. The higher frequencies will be "folded up" in the interval lying between 0 and $f_r/2$ and will be merged with the lower frequency components in this interval (Figure 1.4(b)). This ambiguity, called *aliasing,* is clearly revealed in equation (1.4), taking into account the discontinuous nature of the received signal. The limiting frequency $f_r/2$ is called the *Nyquist frequency* [see Bendat and Piersol, 1971]. The maximum doppler frequency that can be measured therefore is equal to half the repetition frequency; that is,

$$f_{d\max.} = \frac{f_r}{2} \tag{1.12}$$

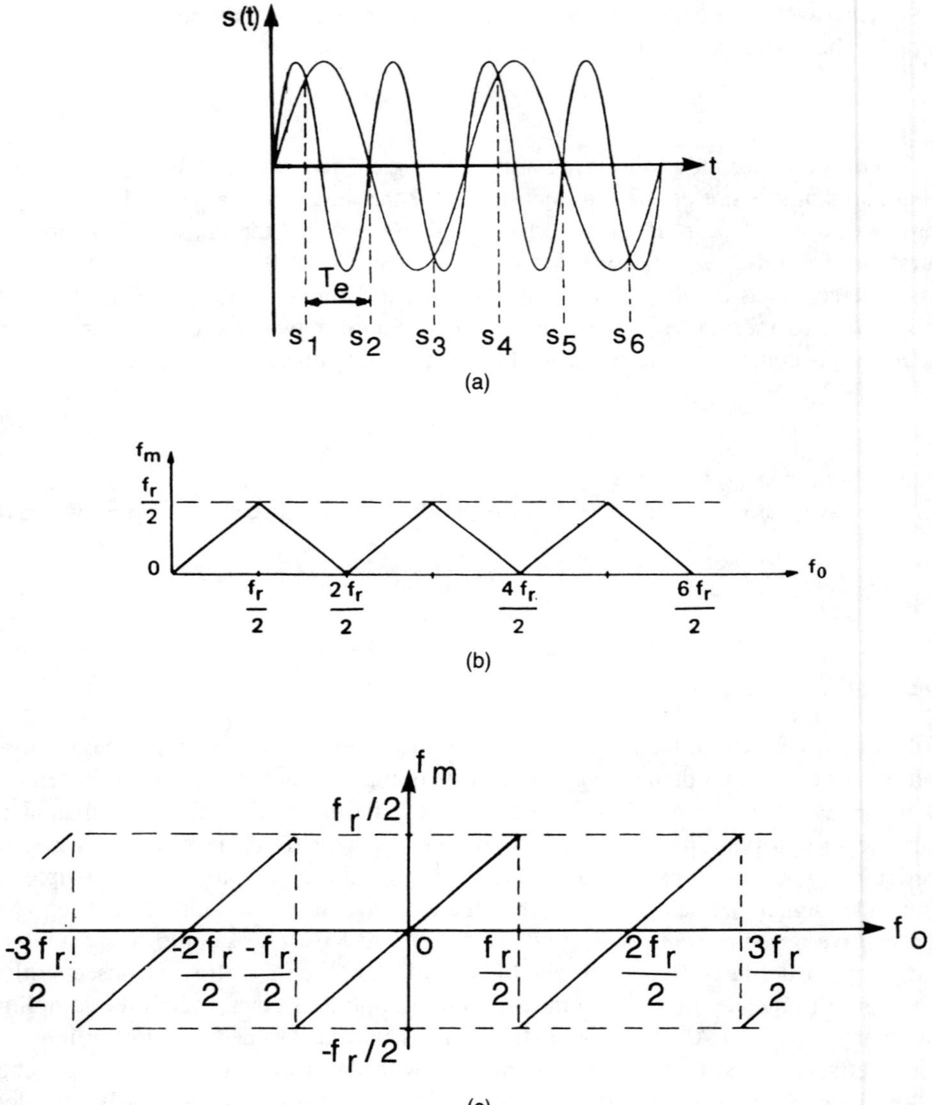


**Figure 1.4** Sinusoidal signal measurements: (a) sampling, at frequency $f_r = 1/T_e$, of two functions; (b) correspondence between the measured value $f_m$ and an original frequency $f_0$, sampling at frequency $f_r$; (c) correspondence between the measured value $f_m$ and an original frequency $f_0$, sampling at frequency $f_r$ combined with sign detection; (d) top—correspondence between the measured values $f_{m1}$ and $f_{m2}$ and an original frequency $f_0$, sampling at two repetition frequencies $f_{r1}$ (solid line) and $f_{r2}$ (broken line), bottom—the difference $f_m = f_{m2} - f_{m1}$ characterizing the ranges of variation of $f_0$ and allowing removal of the ambiguity on the value of $f_m$.

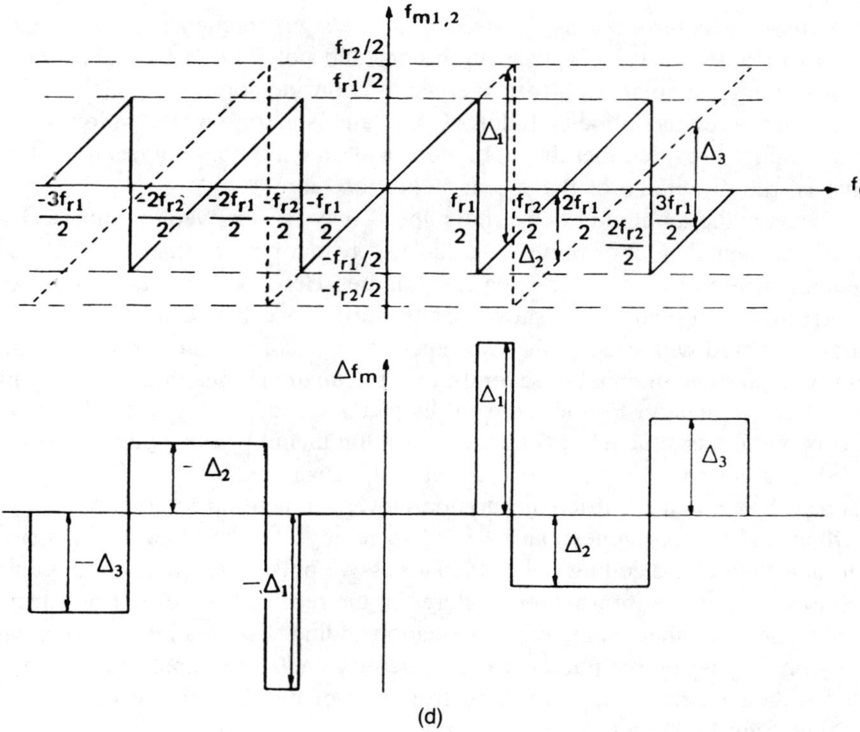


**Figure 1.4** (continued)

According to (1.7) and (1.12), the maximum radial velocity that can be measured is

$$V_{\text{max.}} = f_r \frac{\lambda}{4} \tag{1.13}$$

which means that the target must not travel a distance greater than one-quarter radar wavelength between two successive pulses.

The problem of speed ambiguities thus is related to the technology for measuring the phase difference; if $f_d$ and $\tau$ are sufficiently large that $f_d$ can be measured over a single pulse (for example, if $1/f_d < \tau$), and if a suitable technology is used for measuring the phase difference, the problem of ambiguity disappears (as in the case of missiles for which $f_d$ is large or ionosphere soundings by diffusion where $\tau$ often is equal to a few hundred microseconds).

In fact, the detection of the sign of the doppler frequency by the phase demodulator extends the unambiguous range to an interval of $\pm f_r/2$, and therefore with width equal to $f_r$. This range may be shifted by "heterodyning" the signal. Where the sampling is accompanied by a detection of the sign, frequencies higher than $f_r/2$ in terms of absolute value appear at shifted (and not folded) frequencies of $nf_r$, where $n$ is an integer (Figure 1.4(c)). Frequently, considerations of spatial continuity of the properties of the velocity fields of the "atmospheric targets" allow removal of the ambiguity provided that the ambiguous and unambiguous signals do not overlap; that is, the dispersion of the speeds is not too large.

However, the ambiguity concerning speed may be removed systematically by "coding" the signal. For example, we can use two (or more than two) repetition frequencies, shifted by a few tens to a few hundred Hertz with respect to each other (Figure 1.4(d)). We can easily show that the difference between the doppler frequencies measured with each of the two repetition frequencies allows removal of the ambiguity in an interval equal to several times the unambiguous interval of the highest repetition frequency. Furthermore, a doppler radar can carry out velocity measurements on targets situated beyond the maximum unambiguous range of the radar, provided that the phase coherence between the transmitted and received signals is maintained throughout a sufficient duration (several repetition intervals).

Other coding techniques can be used to remove the ambiguities in meteorological radars. In phase coding radar, the successive pulses are transmitted with random phases. The phase information is stored in the receiver and used to identify the received pulses. Another example is polarization coding. Commuting the polarization ($H = V$) on the successive pulses, we can identify and discriminate the second-trip echoes, as their polarization is different from that of the first-trip echoes.

Combining (1.3) and (1.13), we have

$$v_{\max}\, r_{\max} = \frac{\lambda c}{8} = \text{constant} \qquad (1.14)$$

The product of the maximum values of the speed and range is proportional to the wavelength.

For example, a radar with $\lambda = 5$ cm and $f_r = 4000$ Hz can measure doppler frequencies less than 2000 Hz and speeds of $\pm 50$ m s$^{-1}$ at a range not exceeding 37.5 km. Therefore, unless special technical procedures are used, allowing the ambiguities to be removed in part, the choice of $r_{\max}$ and $v_{\max}$ necessitates a difficult compromise. Furthermore, the smaller the wavelength, the more restricted is the range of the measurable values [*see,* among others, Rihaczek, 1969, on ambiguity problems]. Various techniques also were developed to improve the resolution and the signal to noise ratio of the conventional or doppler radars. Their principle is presented in Section 1.5.5.

### 1.2.3 Display of the Data

The problems associated with displaying meteorological echoes primarily originate from

1. The fact that these usually are not isolated point data but multidimensional value fields for which we wish to display the space-time variations.
2. The dynamics of the signals (or the relationship between the weakest and strongest signals), which reaches a considerable value on the order of 80 dB.[4]

In standard applications of radar, the data are displayed in real time. The simplest display technique uses an oscilloscope to represent the amplitude of the echoes as a function of range (two-variable display). The sweep of the oscilloscope is triggered by a pulse supplied by the master oscillator at the time of transmission, and the video signal is displayed perpendicular to the sweep direction (Figure 1.5). The display of a sequence of successive firings of the radar shows that the meteorological echoes fluctuate rapidly on the screen in a similar manner to a noise, whereas fixed echoes (such as certain ground echoes) or those produced by isolated point targets are comparatively more stable. The interest of this kind of presentation is to show in detail the signal fluctuation. For example, it is useful to locate a particular target for calibration.

To display the signals observed when the radar beam scans a planar or conical surface, a representation in polar coordinates is used (see examples in Chapter 5). In the azimuthal scanning movement of the antenna at constant elevation angle, the origin of coordinates usually is the center of the screen and we obtain a presentation of the *plan position indicator* (PPI) type or a panoramic presentation. When we have constant azimuthal scanning and variable elevation angle, so that the scanning takes place in a vertical plane passing through the radar, the origin of the coordinates is displaced toward the edge of the screen, and we obtain a presentation of *range-height indicator* (RHI) type. When the direction of the beam is left fixed, usually vertical,

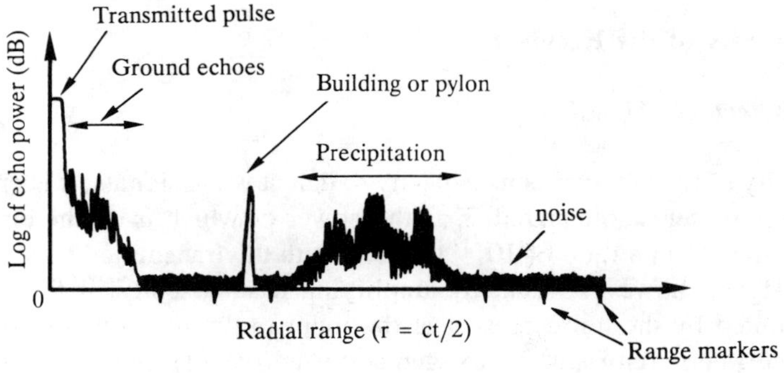


**Figure 1.5** "Type A" representation of echoes.

[4]The decibel (dB) is a logarithmic unit commonly used to express the multiplication by 10 of the logarithm to the base 10 of the ratio of two powers. If the reference level used for the comparison is 1 mW, the result is given in dBm or decibels with respect to the milliwatt.

the signals are presented in height (or range)-time coordinates; this is called *height-time indicator* (HTI). These displays also are provided with grids and markers produced by pulse bursts supplied by the master oscillator that allow the ranges and directions to be conveniently referenced. Various procedures are available for increasing the efficiency of the display; they usually are aimed at revealing variations in the average signal by displaying areas in which the parameters have equal values (for example, isoreflectivity areas). These different types of presentation, which will be discussed in the following pages, give rise to a large number of different versions aimed at satisfying a particular need or correcting various distortions.

For a long time, cathode ray tubes with a monochromatic circular screen were used exclusively as radar data visualization aids. In them, the signal amplitude is used to modulate the brilliance of a spot the radial displacement direction of which depends on the antenna orientation (three-variable display). These screens have a very good resolution, but they are expensive and enable only a summary representation of the information (particularly the number of levels that can be represented is weak and the dynamics is only 10 to 15 dB, largely insufficient for the meteorological fields). They have been almost abandoned in favor of color computer screens. The PPI and RHI presentations, handy and familiar, often are preserved. The use of computer technique enables the composition of representations applying possibly to surfaces different from those scanned by the radar. They are obtained, after a volume scanning sequence, from tridimensional interpolation (for example, vertical planes or constant altitude planes). The quantities represented concern the reflectivity, polarization, or velocity fields. Other possibilities often are available: zoom; animation on several pictures; picture superimposition; addition of geographical marks, alpha numeric indication, or color scales; hard copy; and so forth.

### 1.2.4 Sensitivity of the Receiver

*Minimum Detectable Signal*

The sensitivity of the receiver is its capacity to detect weak signals; it is characterized by the minimum detectable signal, $s_{min}$, the power of which is on the order of $10^{-13}$ to $10^{-14}$ W; that is, in a ratio of $10^{-17}$ to $10^{-20}$ with the transmitted power. We could detect any signal, however weak, by amplifying it sufficiently, if the value of $s_{min}$ were not limited by the noise present at the input to the receiver: external noise of cosmic, atmospheric (emission by oxygen and water vapor), human origin, *et cetera* and the internal noise of the receiver.[5] In certain high-performance receivers using a "low-noise" amplifying stage before the mixer, the internal noise may be brought

[5]In fact, any signal can be detected in noise by sufficient integration provided that, unlike radar observing one "meteorological target," the spectra of the signal fluctuation and noise are different (cf. Section 1.5.6).

to a level at which the external noise becomes the main contribution and therefore constitutes a natural limit. At microwave frequencies, the external noise is relatively weak and, in most applications, the sensitivity is determined by the noise due to the receiver itself.

In an ideal receiver, the only cause of intrinsic noise is the thermal motion of the conduction electrons in the input stages. The power per Hertz of this thermal noise is equal to $kT$ (in W $Hz^{-1}$) where $k$ is the Boltzmann's constant ($k = 1.38 \times 10^{-23}$ J $K^{-1}$) and $T$ is the temperature (in K). At an ordinary temperature of 290 K, the factor $kT$ is equal to $4 \times 10^{-21}$ W $Hz^{-1}$. At the frequencies used for radar, the thermal noise is independent of frequency; it is white noise. Under these conditions, the average theoretical power of the noise $N_e$ at the input of a receiver closed on a matched resistance is proportional to its bandwidth $B_n$, and we have

$$N_e = k\,T\,B_n \tag{1.15}$$

$B_n$ is narrower than the band occupied by the thermal noise.

Certain ultrasensitive receivers, such as parametric amplifiers or masers (also often called *amplifiers with negative resistance*), can have an input noise that corresponds to a temperature lower than their physical temperature. Even in this case (rare in radar meteorology), extra components will degrade the signal in the reception chain. Finally, we can regroup all the causes of degradation and characterize the equipment by a noise factor $F_n$ or by an equivalent noise temperature at the input of this receiver $T_e$. The noise factor is defined as the ratio of the noise power at the output of the actual receiver $N_0$ to the noise power at the output of an equivalent ideal receiver whose temperature $T_0$ is conventionally fixed at the standard value of 290 K:

$$F_n = \frac{N_0}{k\,T_0\,B_n\,G} \tag{1.16}$$

where $G$ is the gain of the receiver; that is, the ratio of the output signal power $S_0$ to the input signal power $S_e$.

If the extra noise is $\Delta N$, the equivalent noise temperature $T_0$ can be written $\Delta N = k\,T_e\,B_n\,G$. Thus, from (1.16) we have

$$F_n = \frac{k\,T_0\,B_n\,G + \Delta N}{k\,T_0\,B_n\,G} = \frac{T_0 + T_e}{T_0} \tag{1.17}$$

The noise power at the output of the receiver is

$$N_0 = F_n\,k\,T_0\,B_n\,G = k(T_e + T_0)\,B_n\,G \tag{1.18}$$

We also can write $F_n = S_e/N_e/S_0/N_0$ and interpret the noise factor as a performance criterion expressing the diminution of the *signal-to-noise* (S/N) ratio on passing through the receiver; $F_n$ usually is expressed in decibels.

The power of the signal at the input to the receiver therefore can be expressed by $S_e = kT_0B_nF_nS_0/N_0$. If the signal-to-noise ratio necessary for detection is $(S_0/N_0)_{min}$; the minimum detectable signal $S_{min}$ can be written

$$S_{min} = k\, T_0\, B_n\, F_n\, (S_0/N_0)_{min} \tag{1.19}$$

$S_{min}$ is a radar characteristic that varies relatively little over time; it frequently is used as a reference for measuring the level of the received signal (see also Section 1.5.6 for the detection of weak signals).The preceding relationships easily can be generalized to a cascade of systems.

### *Matched Filter*

We obviously try to produce receivers the noise of which is as low as possible. We therefore are led to limit the width $B$ of the passband[6] (i.e., to be filtered) to the minimum value compatible with the band occupied by the signal and ensuring the best signal-to-noise ratio; in fact, the received pulse signal may be expressed in terms of its Fourier components, which form a spectrum the width of which increases as the pulses become shorter and their sides become steeper. If $B$ is large compared to the width of the band occupied by the signal, the S/N ratio decreases because $N$ increases to no avail. On the other hand, if $B$ is smaller than the bandwidth of the signal, the noise is reduced but so is the energy of the signal; the result is a distortion and loss of information. Analysis of this problem has led to the concept of matched filter receiver developed to improve the detectability of weak echoes immersed in the receiver noise; this is used in most meteorological radars. A receiver with matched filtering makes it possible to obtain the best signal-to-noise ratio at the output for a transmitted pulse with a given shape in the presence of a stationary Gaussian white noise.

In conventional radars, filtering is carried out most conveniently at intermediate frequencies; the amplifier at intermediate frequency is equivalent to a filter with gain. This filter is characterized by its frequency response function, the form of which is not affected by the value of the gain.

[6] $B$ denotes the passband at half-power (−3 dB). The frequency response characteristics of most radar receivers are such that the band $B$ differs little from the band $B_n$ used in the expression of noise.

The matched filter has a frequency response function $H(f)$ the amplitude of which is the Fourier transform of the transmitted pulse form $u_e(t)$ [Skolnik, 1962]; that is,

$$H(f) = G\,\mathrm{e}^{-\mathrm{i}2\pi f t_0}\,U^*(f) \tag{1.20}$$

with

$$U(f) = \int_{-\infty}^{+\infty} u_e(t)\,\mathrm{e}^{-\mathrm{i}2\pi f t}\,\mathrm{d}t$$

$U(f)$ is the Fourier transform of the signal at the input and $U^*(f)$ is the complex conjugate; $t_0$ denotes the time when the signal is observed at its maximum value; $G$ is the gain of the filter, usually taken as equal to 1.

The function $H(f)$ defined in this way, which provides the best signal-to-noise ratio at the output for the pulse form $u_e(t)$, is the complex conjugate of the received spectrum with a phase shift ($\mathrm{e}^{-\mathrm{i}2\pi f t_0}$) that varies uniformly with frequency. For example, if the transmitted signal is a square-wave pulse with duration $\tau$, the matched filter has a frequency response of the form $\sin x/x$ with a bandwidth at 3 dB given by $B = 0.88/\tau$.

Typically, the bandwidth may differ by 30–50% of the optimum value without the detectability being affected by a factor greater than 0.5 dB. Therefore, the maximum S/N as a function of $B\tau$ is not very sharp. For a radar transmitting square-wave pulses, matching is obtained approximately for a bandwidth (at 6 dB) of the receiver filter equal to the reciprocal of the temporal width of the pulse; that is, $B_6\tau \approx 1$. This is a reasonable approximation for conventional superheterodyne receivers. A slightly higher value often is adopted (for example, $B_6\tau = 1.5$–$2$ for a filter and pulses of rectangular form) to limit the losses due to any drifting of the radar frequency.

However, the matched filtering technique influences the shape of the pulse. At the input of the receiver, the echo returned by a point target has the same shape as the transmitted pulse. In square-wave pulses, usually used in meteorological radars, the amplitude has approximately a triangular shape at the output of the matched filter (Figure 1.6). This pulse shape represents the range-weighting function of the signal applied to the contribution of the scatterers. In radar meteorological applications, we consider the average power of the signal for which the weighting function is the square of the function relating to the amplitude; this weighting function therefore is more peaked than the function related to the amplitude.

The result of matched filtering therefore is a reduction in the effective radial depth of the scattering region contributing to the echo power from a given range with respect to what it would be for a filter with infinite bandwidth [Nathanson and Smith, 1972; see Section 1.2.6 on range resolution]. The corresponding effect on

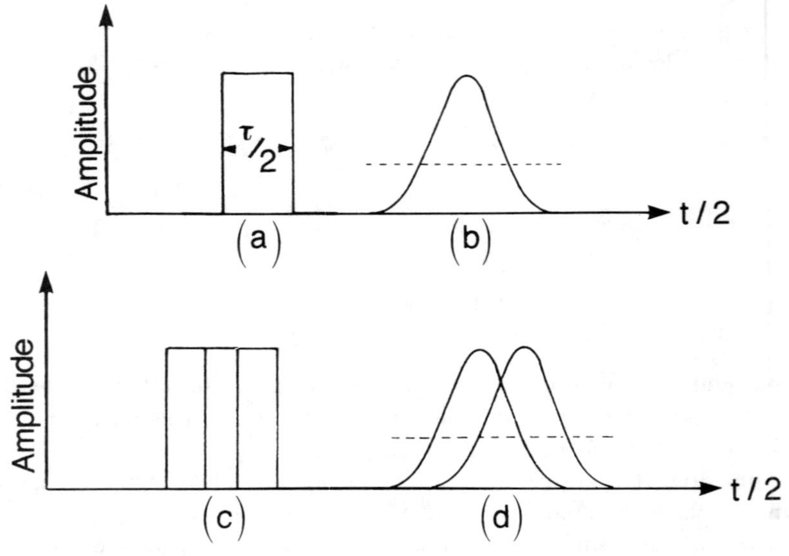


**Figure 1.6** Representation of the pulse shape (amplitude as a function of time): (a) at the input of the receiver and (b) at the output; (c) and (d) for two scatterers separated by less than $h/2$ (the dashed line represents the detection threshold).

the echo power at the receiver output, in square-wave pulses passing into a linear receiver with matched filter, is a reduction of approximately 1.8 dB. If the bandwidth is doubled, the reduction is only 0.9 dB. For a bandwidth greater than $4/\tau$, the reduction can be neglected in all cases. Actual receivers give only approximations to the theoretical form of matched filters. There are differences therefore in the response characteristics of different types of radar so we cannot determine a correction factor applying to all cases. However, a factor on the order of 1.0–1.5 dB is sufficient for most receivers responding approximately as matched filters, because $B$ is wider than the theoretical matched width in most meteorological radars and the pulses are not perfectly rectangular [Nathanson and Smith, 1972]. This loss factor due to the finite width of the receiver band will be denoted by $L_r$ in what follows.

### 1.2.5 Antenna

Volume sounding, which is the main application of atmospheric observation radar, uses narrow beams with simple shapes, usually circular around the maximum radiation axis. Such beams are obtained with the help of large parabolic reflectors of circular aperture, illuminated symmetrically by the primary radiation source, an electromagnetic horn placed at the focal point (Figure 1.7). There is an alternative design, the Cassegrain antenna, in which the primary source is placed at the apex of the paraboloid, illuminated by a small secondary hyperbolic reflector (see, among others, Skolnik, 1970, and Ulaby, Moore, and Fung, 1981).

These configurations have two disadvantages. First, the illumination system of the parabolic reflector intercepts some of the reflected radiation, which alters the antenna's aperture and the geometry of the beam. Second, the intercepted energy interacts with the primary source; stationary waves develop along the feed line creating an impedance mismatch, which degrades the emission performance. This mismatch may be partly corrected with the help of various auxiliary devices.

These two disadvantages are eliminated with offset antennas. The primary source always is at the focus of the paraboloid reflector, but it is inclined in relation to the axis of the paraboloid; only part of the paraboloid is preserved. This configuration modifies the polarization properties of the antenna, and this can cause a deterioration in its performances. At large wavelengths and when it is necessary to move the beam while keeping the antenna fixed, array antennas are used (see later).

The ratio of the focal length $f$ to the diameter $d$ of the aperture is an important parameter for reflector antennas. Its value governs the depth of the reflector and the aperture of the primary beam. The current values for meteorological radars lie between 0.4 and 0.5 for centered antennas and up to 0.8 for offset antennas.

The electromagnetic horn ensures the transition between the guided wave propagating along the transmission line (coaxial line or waveguide) and the electromagnetic wave propagating in free space. It is designed to adapt the illumination function

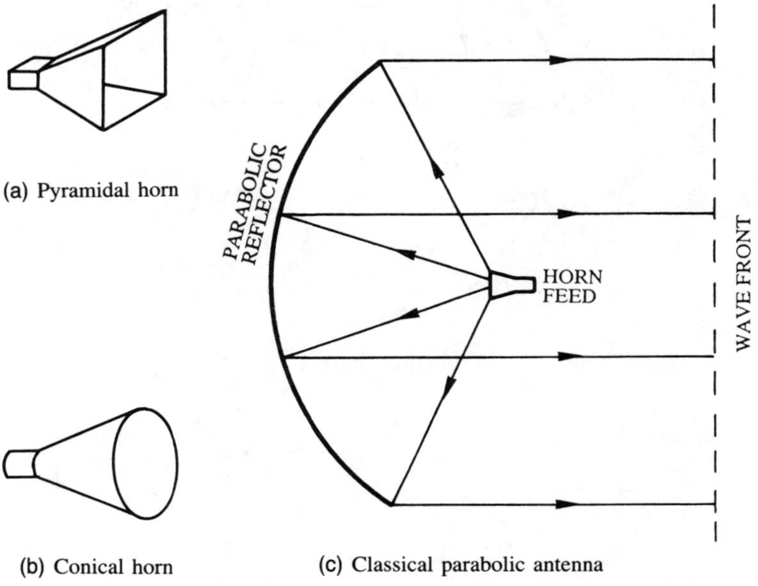


**Figure 1.7** Schematic diagram of commonly used radar antennas.

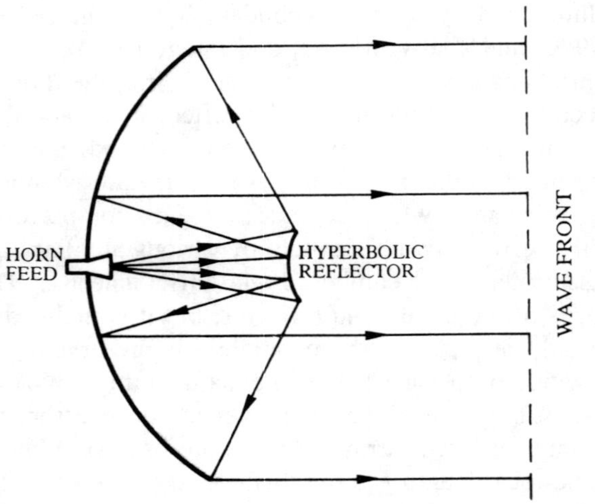


(d) Cassegrain antenna

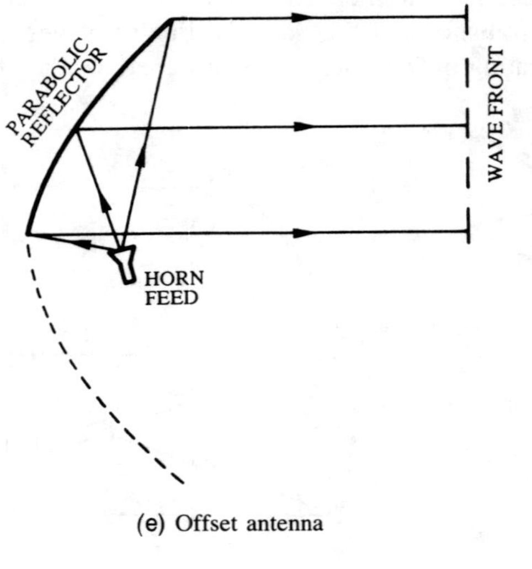


(e) Offset antenna

**Figure 1.7** (continued)

to the shape and the size of the radiative aperture (the parabolic reflector). This illumination function describes the distribution of electromagnetic energy across the aperture: it contains a phase term and an amplitude term.

The energy distribution in the beam is determined by the diffraction figure of the radiative aperture; therefore, it depends on the size and shape of the aperture as well as on the illumination function applied to it. When the aperture is large compared to the wavelength (as in current radar antennas), the radiated field may be calculated with the help of Kirchhoff's scalar diffraction theory. In the vicinity of the antenna, in the so-called near-field region, then in the Fresnel region that follows it, the rays between the radiative aperture and a point of observation do not run parallel, and the energy distribution in the beam is not constant with distance. Beyond these regions is the Fraunhofer region or far-field region in which the rays between the aperture and the target are almost parallel to each other; locally, the wave may be regarded as a planar wave that allows us to use various approximations, simplifying the calculation of the radiated field. Radar antennas are used in the Fraunhofer region. The Fraunhofer region starts at a distance $r$ from the antenna so that

$$r = \frac{2\,d^2}{\lambda} \tag{1.21}$$

where $d$ is the diameter of the aperture of the antenna and $\lambda$ is the wavelength. At this distance, the maximum directivity of the antenna reaches 99% of its value at infinity. For example, with a wavelength of 5 cm and for an antenna with circular aperture 4 m in diameter, the Fraunhofer region starts at 640 m from the antenna. The definitions and relationships that follow apply only to the far-field region.

The direction of propagation of the electromagnetic field emitted by the antenna is described by the complex Poynting vector that has an associated power density $\mathbf{S}_a$ given by

$$\mathbf{S}_a = \frac{1}{2}\,\mathrm{Re}\,(\mathbf{E} \times \mathbf{H}^*) \tag{1.22}$$

where $\mathbf{E}$ and $\mathbf{H}$ are the electric and magnetic field vectors, respectively; the asterisk denotes the complex conjugate; and Re is the real part. The term $\mathbf{S}_a$ is the time average of the Poynting vector. The directional dependence of $\mathbf{S}_a$ defines the antenna's directivity properties. Because of the spherical nature of the waves, they are described by using spherical coordinates $r$, $\theta$, and $\phi$ (range, elevation, and azimuth) with the origin as the (point) source of radiation.

In the far-field region, $\mathbf{S}_a$ has only one radial component $S_r$, which can be written as

$$S_r = \frac{1}{2\eta}\,|E(\theta, \phi)|^2 \tag{1.23}$$

where $\eta$ is the intrinsic impedance of the medium in which the wave is propagating. In space or the earth's atmosphere, $\eta$ is constant and equal to 377 $\Omega$ and $S_r$ is measured in W $m^{-2}$.

At a long distance from the antenna, the radiated field may be described by a spherical propagation function of the form $e^{-jkr/r}$, where $k = 2\pi/\lambda$, multiplied by a directional function $f(\theta, \phi)$, so that

$$E(\theta, \phi) = \frac{e^{-ikr}}{\eta} f(\theta, \phi) \tag{1.24}$$

The power density in the far-field region therefore is

$$S_r(r, \theta, \phi) = \frac{1}{2\eta r^2} |f(\theta, \phi)|^2 \tag{1.25}$$

However, instead of the power density $S_r(r, \theta, \phi)$, the directional properties of the antenna are usually described by a function independent of $r$, the normalized radiation intensity, also called the *radiation pattern* (or *power density pattern*), designated by $F_n(\theta, \phi)$ or by $|f_n(\theta, \phi)|^2$, so that

$$F_n(\theta, \phi) = |f_n(\theta, \phi)|^2 = \frac{S_r(r, \theta, \phi)}{S_r(r, \theta, \phi)_{max}} \tag{1.26}$$

This function applies in the same way whether the antenna is used for transmitting or receiving. Its variation makes possible the plotting of the normalized radiation pattern.

Obviously, we choose $\theta = 0$ and $\phi = 0$ in the direction of maximum $S_r$. This direction defines the beam axis for which $f_n^2(\theta, \phi) = 1$. In a revolving beam, we have $\theta = \phi = 0$ on the revolution axis.

A complete representation of $F_n(\theta, \phi)$ requires a three-dimensional pattern. However, in most practical cases concerning meteorological radar, a representation in two perpendicular planes passing through the beam axis is sufficient, or even a single plane in the case of a revolving beam. These planes are called the *principal planes* of the antenna.

Two examples of normalized radiation patterns are shown in Figure 1.8. The power is always represented by using a logarithmic scale. Figure 1.8(a) is a representation in polar coordinates that gives a good account of the spatial distribution of the energy. Another type of pattern frequently used for partial representation of narrow beams has rectangular coordinates (Figure 1.8(b)) and allows the useful part of the beam to be displayed. It is possible to distinguish several patterns in each plane for the different polarization modes of the radiated wave.

The radiation pattern reveals an energy distribution into several lobes. Most of the energy is concentrated around the beam axis in the main lobe. Outside this are

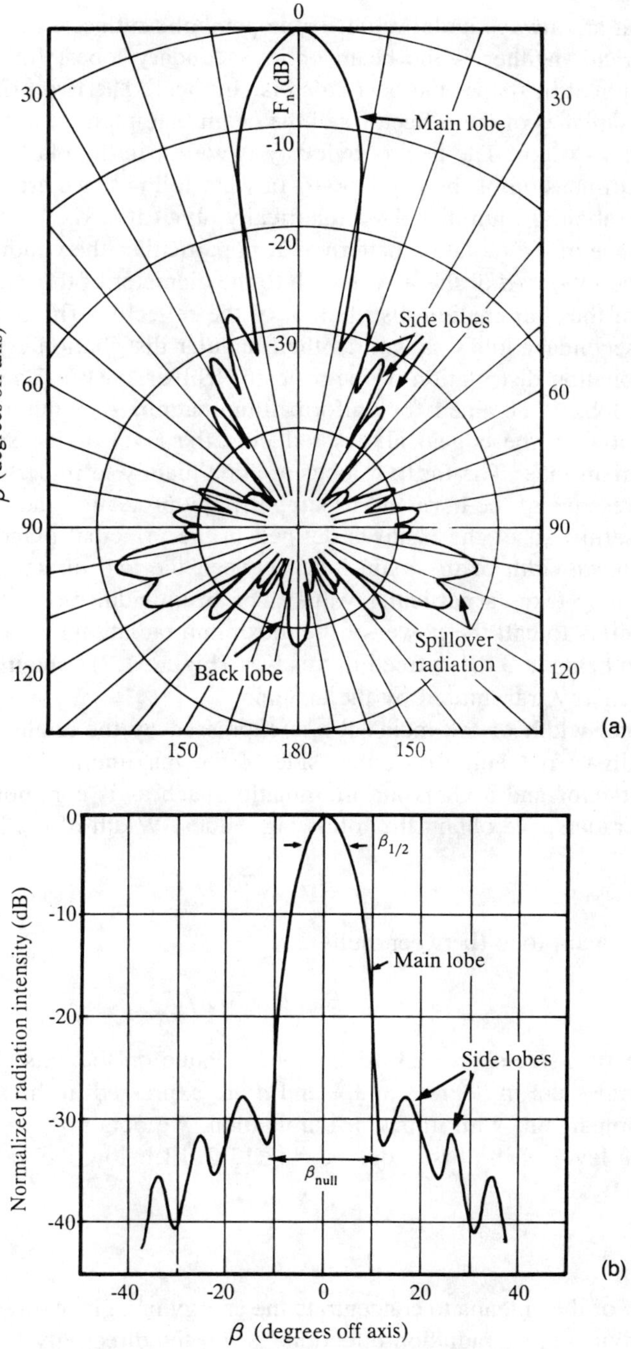


**Figure 1.8** Normalized radiation pattern of a radar antenna: (a) polar form; (b) rectangular form. Spillover is direct radiation from the feed that is not intercepted by the reflector.

sidelobes that are always undesirable. If the parabolic reflector or its illumination is not symmetrical, neither is the beam or its secondary lobes; for example, with a reflector elliptical in shape, the beam also is elliptical. The two principal planes are the symmetry planes of the reflector and the beam that intersect at right angles along the axis of the system. The beam obviously is widest in the plane corresponding to the smallest dimension of the paraboloid. In what follows, we are interested only in a circular paraboloid illuminated symmetrically about its axis.

The shape of the radiation pattern and, in particular, the width of the main lobe depend on the ratio $\lambda/d$ of the wavelength to the diameter of the paraboloid reflector as well as on the illumination distribution of the reflector. The relative importance of the first secondary lobes as well as their angular distribution essentially depends on the illumination distribution of the reflector [Silver, 1949]. The minimum width of the main lobe is obtained for uniform illumination. A nonuniform illumination (decreasing toward the edges) allows reducing the level of the sidelobes but also widens the main lobe. The farthest lobes are due mainly to imperfections in the antenna, the presence of the horn and its support, any protection radome, and so forth.

The useful part of the beam is defined in the principal planes by the angular aperture or beamwidth of the main lobe between the two directions for which the radiation density takes a particular value. The beamwidth most commonly used is that which refers to half the power of the maximum radiation on the beam axis (i.e., $-3$ dB). The beam at 3 dB defined in this way (Figure 1.8) contains more than 80% of the total energy transmitted by the antenna.

The total width of the main lobe is expressed by the angle between the first relative minima (first nulls) on either side of the maximum. For example, with a parabolic reflector and a Gaussian illumination such as is commonly used for meteorological radars, we obtain the following values. Width of the beam at 3 dB:

$$\beta_{1/2} \approx 70\ \lambda/d \tag{1.27}$$

Width of the main lobe (between nulls):

$$\beta_{\text{null}} \approx 170\ \lambda/d \tag{1.28}$$

Level of the first sidelobe: $-23$ dB below the value on the axis. In these relationships, the angles are in degrees and $\lambda$ and $d$ are expressed in the same units.

For comparison, with uniform illumination, we obtain $\beta_{1/2} = 51\ \lambda/d$, $\beta_{\text{null}} = 114\ \lambda/d$, the level of the first sidelobe $= -13.2$ dB below the value on the axis.

*Directivity*

The capacity of the antenna to concentrate the energy in a given direction is measured by its directivity. For a radiation direction $(\theta, \phi)$, the directivity $D(\theta, \phi)$ is the ratio of the power density radiated in the direction $\theta$, $\phi$ per unit of solid angle $S_r(\theta, \phi)$

to the average value corresponding to the power density radiated isotropically by a source of the same power $S_{r,\text{iso}}$; that is,

$$D(\theta, \phi) = \frac{S_r(\theta, \phi)}{S_{r,\text{iso}}} \tag{1.29}$$

with

$$S_{r,\text{iso}} = \frac{1}{4\pi} \iint_{4\pi} S_r(\theta, \phi)\, d\Omega \tag{1.30}$$

where $d\Omega$ is the elemental solid angle given by $d\Omega = \sin\theta \, d\theta \, d\phi$.

By combining (1.26) and (1.29), $D(\theta, \phi)$ may also be written, denoting the directivity on the axis ($\theta = \phi = 0$) by $D_0$:

$$D(\theta, \phi) = D_0 F_n(\theta, \phi) \tag{1.31}$$

*Gain*

Actual antennas always have losses, in particular due to a dissipation of thermal energy in the antenna structure. Thus, the power $P_0$ radiated in space is less than the available power $P_t$ available at the input to the antenna. The ratio of these two quantities defines the radiation efficiency of the antenna $\eta_l$, namely,

$$\eta_l = \frac{P_0}{P_t} \tag{1.32}$$

For a high-quality antenna, $\eta_l$ lies between 0.5 and 0.7.

To express the capacity of the antenna to concentrate the energy in a given direction $(\theta, \phi)$, taking into account the losses, the function $G(\theta, \phi)$ is used; it is called the *gain* of the antenna. The gain is the ratio of the power density radiated in the direction $\theta$, $\phi$ per unit solid angle to the average power density radiated isotropically by a source of the same power with no losses. In a similar manner to (1.26), we can write

$$G(\theta, \phi) = \frac{\eta_l S_r(\theta, \phi)}{\dfrac{1}{4\pi} \displaystyle\iint_{4\pi} S_r(\theta, \phi)\, d\Omega} \tag{1.33}$$

or

$$G(\theta, \phi) = \eta_l D(\theta, \phi) \tag{1.34}$$

If $G_0$ denotes the maximum gain on the beam axis, we can write

$$G(\theta, \phi) = G_0 F_n(\theta, \phi) \tag{1.35}$$

This is why $F_n(\theta, \phi)$ is also called the *normalized distribution* of the antenna gain.

The antenna gain function is involved in the energy transfer on transmission and again on reception; we say that $G(\theta, \phi)$ is the antenna gain for a path. When different antennas are used for transmission and reception, the total gain for the system for a given direction and for the two paths is the product of the gain of each antenna. When the same antenna is used for both operations, the total gain is

$$G^2(\theta, \phi) = G_0^2 F_n^2(\theta, \phi) = G_0^2 |f_n(\theta, \phi)|^4 \tag{1.36}$$

*Antenna Effective Area*

In the uniform illumination of a circular parabolic antenna, we can show [Silver, 1949] that the maximum directivity is related to the physical aperture of the antenna $A_p$ by the following relationship:

$$D_0 = \frac{4\pi A_p}{\lambda^2}$$

However, when the illumination is not uniform, as in meteorological radars, $D_0$ is less than the preceding theoretical value and an effective aperture $A_e$ is defined so that

$$D_0 = \frac{4\pi A_e}{\lambda^2} \tag{1.37a}$$

with $A_e = \eta_a A_p$, where $\eta_a$ is the aperture efficiency of the antenna. Also,

$$G_0 = \eta_l \frac{4\pi A_e}{\lambda^2} \tag{1.37b}$$

If we consider one radiation direction $(\theta, \phi)$, combining (1.36) and (1.37b), we can write

$$A_e(\theta, \phi) = \frac{G_0}{\eta_l} \frac{\lambda^2}{4\pi} |f_n(\theta, \phi)|^2 \tag{1.38}$$

These relationships apply in a manner similar to transmission (subscript $t$) and reception (subscript $r$) if a single antenna is used for the two functions:

$$G_t = G_r = G \qquad \text{and} \qquad A_t = A_r = A \tag{1.39}$$

*Array Antennas*

For atmospheric observation by radar, the classical antenna with a parabolic reflector is not suitable. This is the case, for example, for metric radars developed for clear air observation (ST radar; cf. Chapter 4); in fact, at metric wavelengths, the dimensions necessary for reflectors to obtain sufficiently directive beams become prohibitive (equation 1.27). This is the case, too, for certain on-board radars that lack the necessary thickness to use a parabolic antenna; array antennas are then used.

Array antennas consist of individual, suitably spaced radiating elements, such as dipoles, horns, or waveguide slots. The radiating elements may be arranged in a linear array or on a plane, cylindrical, or spherical surface. However, in meteorology, planar arrays are used most often. The concepts and definitions previously discussed in relation to antenna apertures apply to arrays. In the first case, the radiative properties of the antenna are determined by vector addition of the contributions of the elementary radiating subdomains distributed over the aperture; furthermore, the wave transmitted by the primary source is spherical—the reflector converts it into a planar wave. In the second case, the elementary subdomains are replaced by the array elements that taken together constitute the aperture; the concept of the illumination function is replaced by that of the phase and amplitude distribution for the signals used to excite the elements of the array. This distribution is controlled to produce the radiation pattern required. By operating on the amplitude distribution, the shape of the beam is determined; and by a suitable variation of the relative phases, we can steer the radiation electronically without moving the antenna (Figure 1.9). An array of $N$ elements equally spaced, excited in phase, produces a beam the principal lobe of which lies perpendicular (broadside) to the array. Such an array is sometimes called a *broadside array*. When the direction of maximum radiation is parallel to the array, we have an *endfire array*.

Let us consider a linear array of $N$ identical radiating elements, as shown in Figure 1.9. A feeding coefficient $A_n$ can be associated with each element, as the product of an amplitude term $a_n$ and a phase term $\psi_n$ so that

$$A_n = a_n \, e^{i\varphi_n} \tag{1.40}$$

For a direction of radiation (transmitted or received) $\theta$, the phase difference for two adjacent elements is $\varphi = 2\pi(s/\lambda)\sin\theta$. In the Fraunhofer region, the field and the power density radiated by the $n$th element are given by expressions analogous to (1.24) and (1.25), respectively. The radiation pattern for the array resulting from the vector addition of the contributions of the $N$ elements is the product of the radiation pattern of the base element $F_e(\theta)$ with an array factor $F_a(\theta)$; that is,

$$F(\theta) = F_a(\theta)\, F_e(\theta) \tag{1.41}$$

with

$$F_a(\theta) = F_a(\varphi) = \left| \sum_{n=0}^{N-1} a_n \, e^{i\varphi_n} \, e^{in\varphi} \right|^2 \tag{1.42}$$

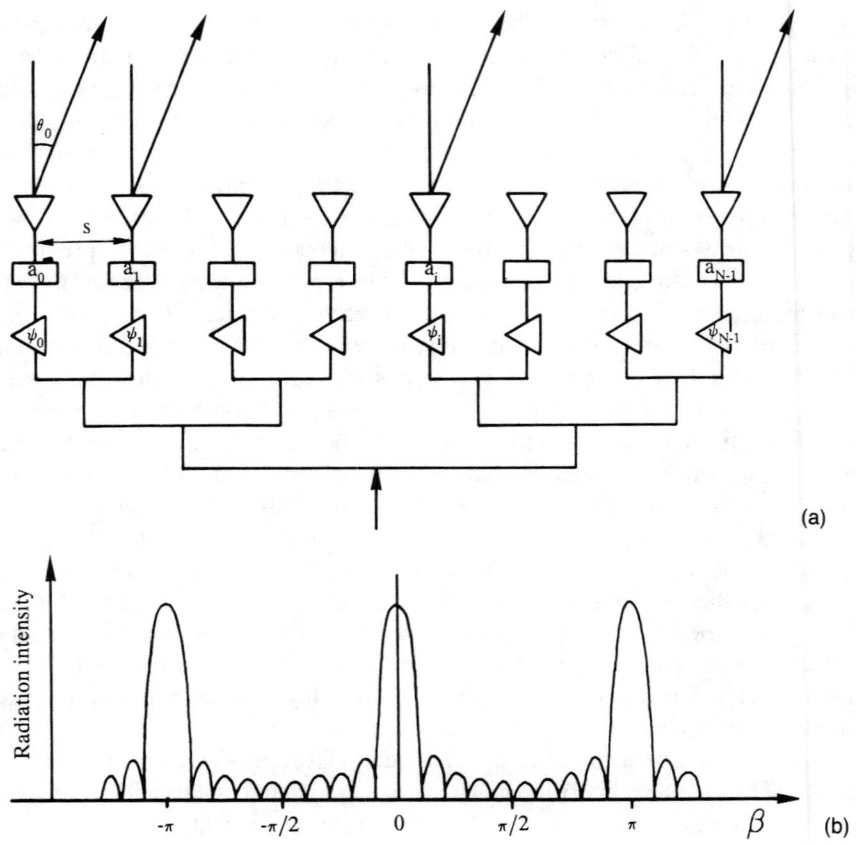


**Figure 1.9** (a) $N$-element linear array antenna with phase and amplitude control on each element: $\Psi$ = phase shifters, $a$ = amplifiers or attenuators. (b) Normalized radiation pattern of a 10-element linear array in rectangular form.

If we assume that the elements are isotropic radiators, the radiation pattern for a given range $r$ is equal to the array factor apart from a constant.

If the array is excited in a uniform manner in terms of phase ($\Psi_n = \Psi_0$) and in amplitude, with $a_n = 1$, equation (1.42) is a geometric series that can be written in the following form:

$$F_a(\varphi) = \frac{\sin^2(N\varphi/2)}{\sin^2(\varphi/2)} \tag{1.43}$$

The maximum value of $F_a(\varphi)$ corresponding to the principal lobe is obtained for $\varphi = \theta = 0$ and is equal to $N^2$. The normalized radiation pattern therefore becomes

$$F_{an}(\theta) = \frac{\sin^2[N\pi(s/\lambda)\sin\theta]}{N^2 \sin^2[\pi(s/\lambda)\sin\theta]} \tag{1.44}$$

When the numerator and denominator of this function are zero at the same time—that is, for $\sin\theta = 0$ or $\pm n\,\lambda/s$ with n an integer—the ratio of the sines is equal to 1 (the indeterminacy is removed by differentiating the numerator and denominator separately) and $F_{an}(\theta)$ passes through a maximum with amplitude equal to the principal lobe. When only the denominator is zero—that is, for $\sin\theta = \pm\lambda/Ns$ and $n \neq mN$ with $n$ and $m$ integers—the radiation pattern passes through zero.

Maxima that do not correspond to the principal lobe are called *grating lobes*; they usually are unwanted and must be avoided. For $s = \lambda$, the first grating lobes are situated at $\theta = \pm\pi/2$. For $s < \lambda$, there is no grating lobe in the half space $-\pi/2 < \theta < \pi/2$.

When $N$ is sufficiently large and $s = \lambda/2$, the radiation pattern of a linear array the elements of which are excited in a uniform manner is similar to that of an antenna with uniformly illuminated aperture; in particular, the first sidelobe lies 13.2 dB below the radiation maximum and the 3 dB aperture of the beam is approximately $\beta_{1/2} \approx 51\,\lambda/d = 102/N$, where $d$ is the length of the array, $N$ is the number of elements, and $\beta$ is in degrees. By reducing the value of $s$, we reduce the relative value of the sidelobes with respect to the principal lobe.

In a two-dimensional array, the radiation pattern can be described as the product of the radiation patterns in the two planes that pass through the principal axes of the array:

$$F_n(\theta_e, \theta_a) = F_{1n}(\theta_e)\,F_{2n}(\theta_a) \tag{1.45}$$

When a phase shift is produced between the signals, exciting the elements of the array, the principal lobe is no longer perpendicular to the array. If the phase difference between signals exciting the adjacent elements is $\Psi$, the radiation maximum of the main beam will lie along a direction $\theta_0$ so that $\Psi = 2\pi(s/\lambda)\sin\theta_0$. The expression for the radiation pattern is then

$$F_{an}(\theta) = \frac{\sin^2[N\pi(s/\lambda)(\sin\theta - \sin\theta_0)]}{N^2 \sin^2[\pi(s/\lambda)(\sin\theta - \sin\theta_0)]} \tag{1.46}$$

We can easily show that when the main beam deviates from the perpendicular (broadside) direction, its width increases. We also can show with the help of various approximations that, for such an array (excitation of uniform amplitude with phase shift $\Psi$ between adjacent elements), the 3 dB beam width is given by

$$\beta_{1/2} \sim \frac{0.886\,\lambda}{Ns\cos\theta_0} \tag{1.47}$$

where $\theta_0$ is the direction in which the beam is pointing. The validity of this formula deteriorates as we move away from the perpendicular.

### *Polarization*

Polarization is an important characteristic of meteorological radar. It is defined by considering the direction of the electric field vector transmitted and received by the antenna and can be linear, circular, or elliptic. We will see in Chapter 2 that most of the hydrometers are not spheric, and that their shape, orientation, and dielectric properties influence the radar signal polarization. The study of polarization therefore offers potentialities in the characterization of the atmospheric environment. This question will be discussed in Section 2.5.3.

Most of the conventional meteorological radars work with a fixed linear polarization identical at transmission and reception. The preferential orientation direction of the hydrometeors, largest dimension, because the action of the resisting forces exerted by the air are horizontal (as notably the case of the raindrops that flatten when falling), the polarization giving the highest retrodiffusion is the horizontal linear, the one generally used. It is obtained with a simple monomode horn.

In antennas with polarization diversity, polarizing devices under electrical or mechanical control generally are used, enabling the combination of two orthogonal linear components to produce the various forms of polarization wanted at the transmission or reception [see synthesis in Bringi and Hendry, 1990].

### *Radome*

The radar antenna is sometimes put inside a rigid sphere, called *radome,* to protect the antenna from mechanical constraints from the wind (large size paraboloid, regions exposed to strong winds) or corrosion and bad weather. The radome is made out of materials that are not very absorbent of the electromagnetic waves (for example, Teflon-coated fiberglass a few millimeters thick). The radome can alter the beam slightly (secondary lobes) and entail a loss of transmission. Its effects need to be measured.

During precipitation, the radome, although treated to decrease its wettability, is coverd with hydrometeors (raindrops, snow, frost) in a nonuniform way because of the wind. With rain, which is most frequent, a water coat forms on the radome, its depth depending on the rain rate. The attenuation by a dry radome must be under 1 dB for two paths. The additional attenuation due to the rain is certainly under 1 dB for a rate under 10 mm $h^{-1}$.

### 1.2.6 Resolution

The resolution of a radar, in the conventional meaning of the term, is its capacity to discriminate between two point targets close to each other. A distinction is made between the radial resolution for targets situated in the same direction of sight but at different ranges and the transverse resolution for targets situated at the same range of the radar but in different directions. They are related to the pulse length and the beamwidth, respectively. However, in the case of distributed targets, resolution has to be understood in a somewhat different sense. On the one hand, it is necessary to define independent scattering volumes and, on the other hand, to determine the distortions affecting the spatial distribution of the echoes as a result of the finite pulse length and the shape of the beam.

*Radial Resolution*

In a radar transmitting square-wave pulses, we have seen (Section 1.2.4) that, for a receiver free of noise and with infinite bandwidth, the echo from a point target would be a square-wave pulse with length $h = c\tau$. On a radar screen where the radar-target distance $r$ is presented directly and not the double distance traveled by the waves, the radial dimension of the "echo pulse" is $h/2$.

Let us consider targets consisting of multiple-point scatterers. When the radial distance separating two scatterers is less than $h/2$, the echo pulses from each partly overlap; in fact, the leading edge of the pulse scattered by the furthest scatterer from the radar will arrive at the receiver before the trailing edge of the pulse scattered by the nearest scatterer (Figure 1.6(c)). If we generalize to a large number of point scatterers, we can easily see that the echo corresponding to a given distance in reality will result from the combination of the contributions from all the scatterers situated in the beam over a radial depth of $h/2$ due to the overlapping echo pulses from these scatterers. In other words, the leading edge of the pulse scattered by elements situated at a distance $r + h/2$ will arrive at the radar at the same time as the trailing edge of the pulse scattered by elements situated at a distance $r$, and because of continuity, we also observe, at the same time, instantaneous contributions of echo pulses from all the scatterers situated between $r$ and $r + h/2$. The value $h/2 = c\tau/2$ defines the radial resolution independent of the distance from the target; but what is most important is that $h/2$ determines the radial extension of the region contributing to the received echo at a given time or the depth of the resolution volume of the pulse.

At the output of an actual receiver with matched filtering, the particular form of the pulse appears as a weighting function of the contributions to the echo from scatterers situated at different distances within the resolution volume associated with this echo (Figure 1.6(d)). In the common case of pulses with a central peak and sloping edges, the result is a reduction in the effective depth of the resolution volume with respect to $h/2$.

### *Transverse Resolution*

Let us consider a radar beam encountering a stationary point target or a slowly moving point target during a scanning movement; for example, at increasing azimuth ($\beta$) and constant angle of elevation. The power $P$ of the signal received as a function of the azimuth of the antenna qualitatively reproduces the angular radiation pattern because the power is proportional to the square of the gain; that is, $G^2(\theta, \phi)$. Displayed on a PPI screen, the echo has the shape of an arc of a circle with thickness $h/2$ in which extent increases with the width of the beam and the power of the signal; the exact direction of the target is that of the middle of the arc, which is also the direction corresponding to the maximum signal (for a symmetrical beam).

For an echo of medium power, the extent of the arc corresponds approximately to that of the main lobe. For a signal with power (above the detectable minimum) greater than the relative level of the sidelobes with respect to the maximum radiation, the target, after an interruption, may be detected again by the sidelobes. The transverse resolution therefore is determined by the width of the beam and the radiation pattern; it decreases with distance.

In the observation of a scattering volume of great extent, the finite width of the beam results in a smoothing of the spatial fluctuations of the transverse distribution of the signal and a lateral extension of the echoes outside the actual scattering zone. The magnitude of the error for a given antenna depends on the shape of the target and the power of the echoes. This error, which increases with distance, is difficult to correct if the wave scattering properties of the target are ignored. For example, the measurement errors made as a result of the distortions produced by the main lobe and the sidelobes in the determination of the dimensions of thunderstorm clouds, the properties of which are assumed to be known, do not exceed $\pm 1500$ m at distances less than 50 km for beamwidths ranging from 1 to 2°. In the extreme case of very intense thunderstorms at distances greater than 100 km, the altitude of the top of the echoes can be observed up to 6000 m above the actual top [Donaldson, 1964]. This error therefore can be considerable, and the interpretation of the results obtained at great distances for intense precipitation cells with the help of wide-beam antennas must be made with great care.

In principle, we can correct for distorted profiles if the antenna pattern is known. However, this correction (or deconvolution) presents problems in most real cases. Approximation methods are available, with the help of nomograms, to reduce the estimation error of the altitude at the top of echoes by the main lobe, whereas the error possibly due to the sidelobes is indicated by a reduction in the sensitivity of the receiver that makes the echoes collected by the sidelobes disappear.

*Pulse Resolution Volume*

The resolution volume of the pulse ($V$) is what produces signals received by the radar at the same time, the sum of which determines the instantaneous intensity of the signal. This volume has the same cross section as the beam at distance $r$ and a depth approximately equal to half the length of the pulse. If we take into account the beam at 3 dB, we obtain the approximate formula:

$$V = \frac{\pi r^2 \, \theta_0 \, \phi_0}{4} \frac{c\tau}{2} \tag{1.48}$$

where $\theta_0$ and $\phi_0$ represent the width at 3 dB (in radians) of a beam with elliptical cross section in the two perpendicular planes of symmetry (for a circular beam $\theta_0 = \phi_0$) and $c \approx 3 \times 10^8$ m s$^{-1}$ is the velocity of propagation.

The shape of the pulse volume varies with distance. For example, with an antenna creating a circular beam with 1° width at 3 dB transmitting pulses with lengths equal to 1 $\mu$s, the resolution volume at a distance of 100 km is approximately equal to $3.6 \times 10^8$ m$^3$ with a diameter of 1745 m and a depth eleven times smaller (150 m). At 10 km, the diameter is almost equal to the depth. The distribution of the energy is not uniform in the pulse volume. It varies radially (cf. matched filter Section 1.2.4) and transversally (as $G^2(\theta, \phi)$).

### 1.2.7 Refraction

In a homogeneous medium, the phase velocity of an electromagnetic wave is given by $v = (\varepsilon\mu)^{-1/2}$ where $\varepsilon$ is the dielectric permittivity and $\mu$ is the magnetic permeability of the medium. In free space, $v = c = (\varepsilon_0 \, \mu_0)^{-1/2} = 3 \times 10^8$ m s$^{-1}$ and the waves are propagated rectilinearly. In the atmosphere, the relative magnetic permeability $\mu_r = \mu/\mu_0$ is approximately equal to 1 but the relative dielectric permittivity $\varepsilon_r = \varepsilon/\varepsilon_0$ varies with the temperature, pressure, and the specific humidity. The variations in $\varepsilon_r$, although small and negligible for measurements in the range and velocity of targets, are sufficient to affect the direction of propagation of the waves and to curve the rays. This is the phenomenon of atmospheric refraction.

To analyze it, we consider the phase refractive index $n$ (or absolute refractive index) of the medium, defined by the ratio of the velocity of electromagnetic waves in vacuum to their phase velocity in the medium under consideration; that is,

$$n = \frac{c}{v} = \varepsilon_r^{1/2} \tag{1.49}$$

For microwaves, $n$ does not vary significantly with the wavelength.

A more general form for the refractive index is used for media in which the waves are attenuated as they are propagated. In this case, the refractive index is written in the form of a complex quantity:

$$m = n - ik \tag{1.50}$$

The imaginary part is associated with the absorption. It is equal to zero in a perfect dielectric. We will see that it makes an important contribution in clouds and precipitation (cf. Sections 2–3). In this discussion of refraction, we consider only the real refractive index given by equation (1.49).

In the troposphere, at radioelectric frequencies, $n$ differs very little from the refractive index in a vacuum; therefore, it is very close to 1 (on the order of 1.0003 close to the surface of the earth). This is why it is often replaced by the term $N$, called the *radioelectric refractivity* or the *refractivity* $N$, defined by

$$N = (n - 1) \times 10^6 \tag{1.51}$$

The unit $N$ is sometimes associated with this definition.

At radioelectric wavelengths, for most applications, the relationship between the refractivity, the atmospheric presure $p$, the absolute temperature $T$, and the partial pressure of water vapor $e$, can be written approximately with an accuracy of $0.1N$, in the following simplified form [Bean and Dutton, 1968]:

$$N = (77.6/T)\,(p + 4810\,e/T), \tag{1.52}$$

where $p$ and $e$ are in hPa and $T$ is in K.

In the atmosphere, the pressure decreases exponentially with height, and the temperature profiles are more complex. However, if we consider an average state of atmosphere, the temperature also decreases with the height up to the tropopause ($\approx$10 km at middle latitudes); the pressure term overrides the temperature; and normally, in the troposphere, $N$ decreases with altitude.

Knowing the distribution of $p$, $T$, and $e$ in space makes it possible to calculate the distribution of $N$ by using equation (1.52). These three variables vary in the atmosphere for all scales of time and space. We shall see in Chapter 4 that the small-scale fluctuations of refractivity scatter the electromagnetic waves and are used for radar detection in clear air.

However, for most applications concerning the direction of propagation of the microwave beam, we restrict ourselves to considering average effects so that small-scale fluctuations are neglected. We can then assume that the atmosphere is stratified horizontally (i.e., spherically around the planet) and therefore completely described once the vertical profiles are known. In this case, $N$ depends only on the altitude $z$ and, for example, radio sounding is sufficient to determine its profile.

The relationship between the curvature[7] of a ray and the vertical gradient of refractivity may be established simply [Bean and Dutton, 1968]. Let us consider (Figure 1.10) a wave surface moving from $AB$ to $A'B'$ in a stratified atmosphere represented by surfaces of constant refractive index $n$ and $n + \mathrm{d}n$ situated at altitudes $z$ and $z + \mathrm{d}z$. Let us call $v$ and $v + \mathrm{d}v$ the phase velocity along the rays $AA'$ and $BB'$ whose radii of curvature are $\rho$ and $\rho + \mathrm{d}\rho$, respectively. The angular velocity is $v/\rho = (v + \mathrm{d}v)/(\rho + \mathrm{d}\rho)$. By combining this equation with (1.49) we can write

$$\frac{\mathrm{d}v}{v} = \frac{\mathrm{d}\rho}{\rho} = -\frac{\mathrm{d}n}{n} \tag{1.53}$$

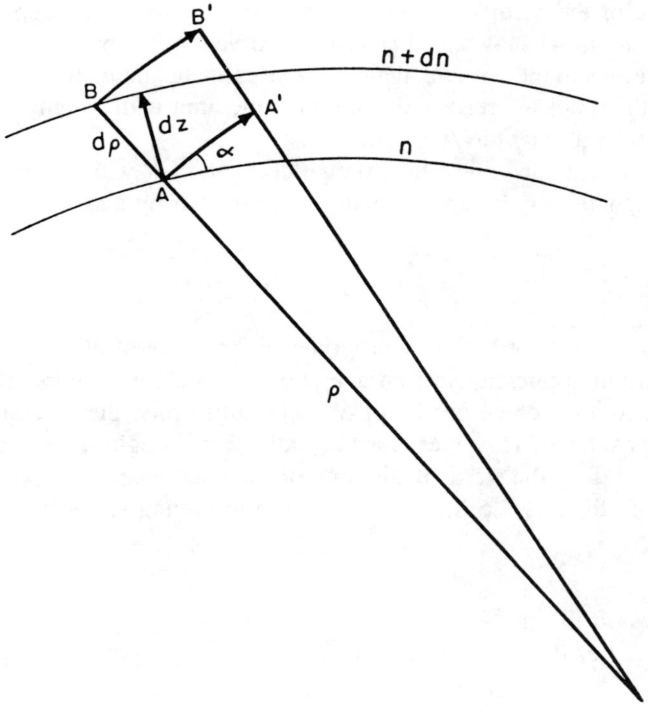


**Figure 1.10** Differential geometry used in the derivation of ray curvature [from Bean and Dutton, 1968].

[7]Consider an arc d$s$ along a locally planar curve and denote d$\alpha$ the corresponding variation of the angle of the tangent to this curve. The curvature $C$ is equal to $\mathrm{d}\alpha/\mathrm{d}s$ and the radius of curvature $\rho = 1/C$.

If $\alpha(z)$ is the angle between the ray and the surface of constant $n$, we have $dz = d\rho \cos\alpha$, and (1.53) can be written with $dn/dz = dN/dz \times 10^{-6}$:

$$\frac{1}{\rho} = -\frac{1}{n}\frac{dN}{dz}\cos\alpha \times 10^{-6} \tag{1.54}$$

with $\rho$ in km.

This relationship makes possible calculating the curvature $1/\rho$ of the rays and plotting their trajectory step by step, using an iterative method from the profile of $dN/dz$.

Relationship (1.54) also shows that, when the gradient of the refractive index is equal to $1/a$ where $a$ is the earth's radius (average value 6371 km)—$dN/dz = -157\ N\ \text{km}^{-1}$—the curvature of horizontally transmitted rays ($\alpha = 0$) is equal to the curvature of the earth's surface. We say that the rays are trapped.

Equation (1.54) shows further that the curvature is a maximum for beams with small inclinations to the horizontal and decreases as the inclination increases. Finally, it shows that the rays are refracted toward the medium with a higher refractive index; that is, downward when $dN/dz < 0$.

Under normal conditions, the average decrease of $N$ with altitude over the height of the troposphere may be approximately represented by an exponential function of the form

$$N(z) = N_0\, e^{-z/H} \tag{1.55}$$

where $H$ is a "scale" of height appropriate to the value of $N$ for $z = 0$; that is, $N_0$.

For certain applications of an analytic or speculative nature, it is useful to be able to refer to an average profile of $N$; to establish this, the Standard Atmosphere[8] usually is considered. However, for the radioelectric region, we must make an assumption regarding the vertical distribution of water vapor. If we take $e = 10.2\,(1 - 0.0640z)$ for $z \leq 7.5$ km, we obtain the following simplified expression:

$$N(z) = 315\, e^{-z/7.36} \tag{1.56}$$

with $z$ expressed in km.

In this case, the vertical gradient is

$$\frac{dN}{dz} = -43\, e^{-z/7.36} \tag{1.57}$$

For the first kilometer above sea level, the average value of the decrease of $N$ for

[8]The Standard Atmosphere is a hypothetical average distribution of the atmospheric temperature, pressure, and density considered by international agreement to be representative of the atmosphere for various applications (for example, calibrating altimeters in aeronautics).

the standard atmosphere is approximately $-40\ N\ \mathrm{km}^{-1}$, or $dN/dz \approx [-1/(4a)] \times 10^6$. This value is the standard gradient defining the standard radioelectric atmosphere; it serves as a basis for comparing for phase and refraction studies.

When $dN/dz > -40\ N\ \mathrm{km}^{-1}$, the rays are curved downward less than under standard conditions; this is subrefraction. When $dN/dz < -40\ N\ \mathrm{km}^{-1}$, the curvature is stronger than normal, and we have superrefraction. When $dN/dz \leq -157\ N\ \mathrm{km}^{-1}$, we have ducting conditions that may lead to the beam being confined to a thin layer and propagated over a long distance with low loss [Hall, 1979].

A current practice involves replacing the earth's radius with a fictive radius $a_e$ associated with an atmosphere in which the waves would be propagated in straight lines; $a_e$ is called the *effective radius* of the earth. In other words, the curvature of the earth $C_a$ is decreased by the value of the curvature of the rays $C_\rho$; the fictive curvature $C_e$ therefore is such that $C_e = C_a - C_\rho$. We can write

$$a_e = ka \tag{1.58}$$

$$\frac{1}{a_e} = \frac{1}{a} - \frac{1}{\rho} \tag{1.59}$$

That is, by combining (1.58), (1.59), and (1.54) and for $\alpha = 0$:

$$k = \frac{a_e}{a} = \left[1 - \frac{a}{n}\frac{dN}{dz} \times 10^{-6}\right]^{-1} \tag{1.60}$$

For the standard value of the gradient $dN/dz = -40\ N\ \mathrm{km}^{-1}$, we find that $k = 4/3$, or $a_e \approx 8500$ km.

When the rays curve downward, the radar horizon is pushed beyond the geometric horizon. Under the conditions of standard refraction, we can show on the basis of simple geometrical considerations that for a radar at height $h$ above the ground with $h \ll a$, the radar horizon is approximately at the distance

$$d = (2\ kah)^{1/2} \tag{1.61}$$

with

$$k = 4/3 \quad \text{and} \quad d = 130\ h^{1/2} \tag{1.62}$$

where $d$ and $h$ are in km.

Under current applications of meteorological radar, the observations are aimed at scanning scattering volumes at moderate ranges and not at precisely locating specific points. Quantitative applications (for example, measurement of rain; cf. Chapter 2) are usually made at ranges below approximately 150 km (for current radars) so that the scattering medium remains homogeneous in the pulse volume (cf. Sections 1–3 on the radar equation). Moreover, the precise state of the local distribution of

$N$ is generally unknown. The effects of refraction therefore most often are neglected. Figure 1.11 shows the trajectory of the rays for the exponential decrease of $N$ described by equation (1.56). However, it is important to "recognize" situations in which the refractivity gradients depart significantly from normal and where, because of this, we must either correct the trajectories or reject spurious data.

The high values for the refractivity gradients are associated with temperature inversions (layers where the temperature increases with height), because in such cases the variations in pressure and temperature act in the same direction (equation (1.52)). Temperature inversions often result from radiation effects (night radiation from the ground in clear air or from the top of clouds), advection (such as warm air above a sheet of cold water), subsidence, or a combination of these processes. In the presence of inversions and for aiming angles close to the horizontal, not only the average direction but also the shape of the beam may be modified considerably. The alterations of the mean direction may lead to spectacular effects due to detection of ground echoes and, so, easily identified but also to effects that are more difficult to detect in the presence of many diffuse echoes. The alterations in shape are because not all the rays making up the beam are deviated in the same manner. This may lead to a significant modification of the angular aperture and pulse volume, resulting in distorted observations.

Significant alterations of the beam also may be produced in and around convective storms, because of the intense vertical exchanges creating strong refractivity

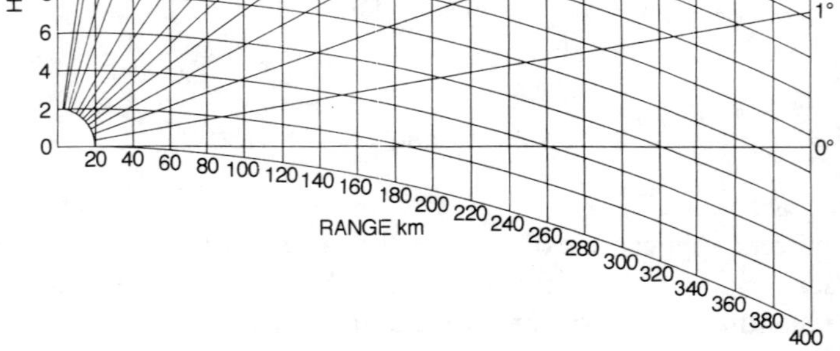


**Figure 1.11** Radar range-height-angle chart with refracted rays represented as straight lines calculated from (1.55) with $N_0 = 313$ and $H = 7$ [from Blake, 1968].

gradients at the boundaries between the warm air and the cold air. A quasi-mirror reflection of the beam may result, so that the storm is detected in a wrong direction.

### 1.2.8 Attenuation

Atmospheric attenuation can distort observations by radar affecting the shortest wavelengths. The attenuation of radiation during its progression through the atmosphere deforms the contours and reduces the intensity of the signals. When the attenuation is not too significant, an iterative calculation can correct for its effect in an approximate manner (Section 2.5.7).

By definition, the loss in energy $dP$ as a function of the traversed distance $dr$ in the attenuating medium is given, for a double path, by an expression of the form

$$dP = -2\, a'P\, dr \tag{1.63}$$

or, with $P = P_0$ at $r = r_0$,

$$\ln \frac{P}{P_0} = -2 \int_0^r a'\, dr \tag{1.64}$$

In this relation, $a'$ is expressed in Neper[9] (Np) per unit of length for one path. If we call $L$ the attenuation for one path between 0 and $r$, we have

$$L^2 = \frac{P}{P_0} = \exp\left(-2 \int_0^r a'\, dr\right) \tag{1.65}$$

The attenuation rate usually is expressed in dB per unit of length for one path. We therefore have

$$10 \log \frac{P}{P_0} = \frac{10}{\ln 10} \ln \frac{P}{P_0} = -4.343 \times 2 \int_0^r a'\, dr = -2 \int_0^r a\, dr \tag{1.66}$$

where $a(\text{dB}) = 4.343\, a'(\text{Np})$ and

$$L^2 = 10^{-0.2\int_0^r a\, dr} \tag{1.67}$$

## 1.3 THE RADAR EQUATION: GENERAL FORMS

### 1.3.1 Single Scatterer

The radar equation expresses the relationship between the average power of the received signal and the properties of the scattering volume or point target, situated at

[9] The Neper is the expression by its neperian logarithm of the ratio of two powers, with 1 Np = 4.343 dB.

range $r$, as a function of the technical characteristics of the radar and the conditions of propagation between the radar and the target. The power flux density per unit area perpendicular to the radiation direction or intensity of radiation, in the direction $(\theta, \phi)$ at distance $r$ from a radar transmitting a power $P_t$ in an attenuating medium with the help of a transmission antenna with gain $G(\theta, \phi)$ (1.36), is equal to

$$S_i(\theta, \phi) = \frac{P_t}{4\pi r^2} G_0 |f_n(\theta, \phi)|^2 L \tag{1.68}$$

In this equation, the term $1/(4\pi r^2)$ expresses the intensity reduction associated with the power spreading on a sphere with a radius $r$ (isotropic source), sometimes called the *spreading loss*. $P_t$ is the available power before the antenna. $L$ is the attenuation on the radar-target path.

It is convenient to assume that the target, the properties of which are unknown, radiates isotropically the intercepted power and to characterize it by its effective radar backscattering cross section $\sigma$, which is defined as the area intercepting an amount of incident power equal to that which radiated isotropically corresponds to the power effectively backscattered to the radar; that is,

$$S_i \sigma = 4\pi r^2 S_r \tag{1.69}$$

where $S_r$ is the power flux density backscattered at the reception antenna. Therefore we assume that the scattering processes are linear. The value of $\sigma$ is different from the geometric cross section of the target and depends on many factors: shape and dimension of the target, orientation with respect to the direction of radiation, dielectric factor, and conductivity of the material.

If the target $\sigma$ is in the direction $(\theta, \phi)$, at distance $r$, the power flux density returned by the target and arriving at the reception antenna is

$$S_r = \frac{P_t G_0 |f_n(\theta, \phi)|^2 L^2 \sigma}{(4\pi r^2)^2} \tag{1.70}$$

These relations make it possible to calculate the order of magnitude of the incident and backscattered powers. For example, if $G_0 = 10^5$ and $P_t = 100$ kW, the power flux density incident on a target situated at 100 km on the beam axis, in a nonattenuating medium is approximately $8 \times 10^{-2}$ W m$^{-2}$. If the target has a radar backscattering cross section of 0.01 m$^2$, the backscattered power density is only $6.3 \times 10^{-15}$ W m$^{-2}$ at the antenna. For an antenna whose effective aperture $A_e$ is 20 m$^2$, the collected power is $1.26 \times 10^{-13}$ W or $-99$ dBm.

The power of the echo collected by the antenna with effective aperture $A_e$ is $P_r = A_e S_r$. Therefore, assuming that a single antenna is used for transmission and reception, the received power is

$$P_r = \left(\frac{P_t G_0^2 \,|f_n(\theta, \phi)|^2 \lambda^2 L_r}{(4\pi)^3}\right) L^2 \frac{\sigma}{r^4} \tag{1.71a}$$

where $L_r$ is the attenuation due to the finite width of the pass band of the receiver. All the radar factors are regrouped between the parentheses. The expression (1.71) works for a point target situated in the $(\theta, \phi)$ direction.

For a point target situated on the axis of the beam, $|f_n(\theta, \phi)|^2 = 1$ and

$$P_r = C_1 L^2 \frac{\sigma}{r^4} \tag{1.71b}$$

where $C_1$ is a constant characteristic of the radar used. Equations (1.71) are general and apply to any type of point target, such as a bird, airplane, or isolated sphere.

They show that the received power is proportional to the transmitted power and the square of the antenna gain. They also suggest that $P_r$ is proportional to the square of the wavelength, which is deceptive because, in fact, the gain of an antenna with given aperture, the radar backscattering cross section, and the maximum power of the transmitter are parameters that depend on wavelength. This is why it is preferable, to make explicit the relationship between $P_r$ and $\lambda$, to represent the antenna by its effective aperture and write (1.71) in the form

$$P_r = \left(\frac{P_t \, \eta_l^2 A_e^2 \, L_r}{4 \, \pi \, \lambda^2}\right) L^2 \frac{\sigma}{r^4} \tag{1.71c}$$

which shows that $P_r$ increases when $\lambda$ decreases, the other quantities remaining unchanged. We also can see that, for a point target, the received power is proportional to the radar backscattering cross section and inversely proportional to the fourth power of the distance.

### 1.3.2 Distributed Target

For a "distributed target" (i.e., consisting of a large number of scatterers simultaneously present in the same resolution volume $V$ with randomly distributed phases and a distribution assumed homogeneous, for example, raindrops), the radar backscattering cross section is equal to $\Sigma_V \, \sigma_i$, the summation made over the entire resolution volume. This term may be written in the form

$$\sum_V \sigma_i = V\left(\sum_V \sigma_i / V\right) = V\eta \tag{1.72}$$

The quantity in parentheses $\eta$ is the average cross section of the target per unit volume. The term $\eta$ is called the *radar reflectivity*. It is in $m^2\ m^{-3}$ but is commonly expressed in $cm^{-1}$.

Returning to equation (1.71a), we therefore have for a distributed and homogeneous target totally filling the pulse volume

$$\bar{P}_r = \frac{P_t\, G_0^2\, \lambda^2\, L_r\, L^2\, \eta}{(4\pi)^3} \int_V \frac{|f_n\,(\theta,\, \phi)|^4}{r^4}\, \mathrm{d}V \tag{1.73a}$$

where $V$, the volume of integration is that of the pulse with $\mathrm{d}V = r^2\, \mathrm{d}r\, \mathrm{d}\Omega$, and $\mathrm{d}\Omega$ is the element of solid angle. $P_r$ is replaced by the average power $\bar{P}_r$ (this point will be justified in Section 1.5).

When $r$ is large compared to the depth of the resolution volume $c\tau/2$, with $\mathrm{d}r \approx c\tau/2$, we have

$$\bar{P}_r = \frac{P_t\, G_0^2\, \lambda^2\, L_r\, L^2}{(4\pi)^3} \frac{c\tau}{2} \frac{\eta}{r^2} \int_\Omega |f_n\,(\theta,\, \phi)|^4\, \mathrm{d}\Omega \tag{1.73b}$$

In most cases, we can assume that the power distribution in the main lobe can be conveniently represented by a Gaussian function of the form [Nathanson and Reilly, 1968]

$$|f_\eta\,(\theta,\, \phi)|^2 = \exp\left[ -\left( \frac{\theta^2}{2\,\sigma_\theta^2} + \frac{\phi^2}{2\,\sigma_\phi^2} \right) \right] \tag{1.74}$$

where $\sigma_\theta$ and $\sigma_\phi$ are standard deviations of the distribution. Under these conditions, neglecting sidelobes, we have [Probert-Jones, 1962]

$$\int_\Omega |f^4\,(\theta,\, \phi)|^4\, \mathrm{d}\Omega = \frac{\pi\theta_0\, \phi_0}{8\, \ln 2} \tag{1.75}$$

where $\theta_0$ and $\phi_0$ are the 3 dB beamwidths and ln2 is the natural logarithm of 2. The error made in neglecting the sidelobes is less than 0.2 dB, which therefore is negligible.

We finally obtain

$$\bar{P}_r = \left( \frac{P_t\, \lambda^2\, G_0^2\, \theta_0\, \phi_0\, c\tau\, L_r}{1024\, \pi^2\, \ln 2} \right) L^2 \frac{\eta}{r^2} \tag{1.76a}$$

that is,

$$\bar{P}_r = C_2\, L^2 \frac{\eta}{r^2} \tag{1.76b}$$

where $C_2$ is a radar constant. The particular forms of this equation applied to meteorological targets and the choice of units are discussed in Section 2.5.7.

Before 1962, the radar equation for the distributed target was derived without involving the distribution function of the antenna gain, by limiting the pulse volumes to the fraction situated within the beam at 3 dB and by assuming that the value of the gain in the solid angle defined in this way is constant. Although the error made by taking the gain as constant is compensated partly by the error introduced by neglecting the energy coming from targets situated outside the beam at 3 dB, a difference from the exact expression still remains. The approximate equation overestimated $P_r$ by 3.5 dB. This difference, estimated by calibration, was compensated for by an empirical factor introduced into the equation. Probert-Jones [1962], who established the exact equation analyzes in detail the terms of the error.

Quite obviously, these equations apply only if the hypotheses used to establish them are respected. It is sometimes necessary to modify their terms to take into account the peculiar conditions, such as temporary or permanent alterations to the beam or attenuation. When the antenna is not protected, the hydrometeors can affect its functioning; for example, because water or frost accumulates and distorts or disadapts the reflector or the primary source. When a protection radome is used, the permanent attenuation term must be added. The term is around 1–2 dB when the radome is dry; it increases when the radome is wet, the value of the increase depending on the rain intensity. The same occurs with snow or frost.

Equation (1.76a) is used when meteorological targets of great extent are observed that satisfy the assumption of a scattering medium occupying the entire pulse volume and presenting a homogeneous distribution of the radar reflectivity. Note that the decrease in the received power as a function of the distance from the target, which varies as $r^{-4}$ for a point target (equation (1.71)) is $r^{-2}$ for a target filling the beam (the volume of the target increases proportionally to $r^{-2}$, thus cancelling the spreading loss on the incident energy).

According to (1.76), the received power is related to the properties of the beam in an inverse manner by the terms $G_0^2$ and $(\theta_0, \phi_0)$. In reality, because $G_0$ is proportional to $1/(\theta_0, \phi_0)$, the product $G_0^2\ (\theta_0, \phi_0)$ is proportional to $G_0$. Therefore, for a distributed target filling the beam, $\bar{P}_r$ is proportional to the antenna gain whereas it is proportional to its square for a point target (1.71). In other words, for a distributed target, the increase in $G_0$ increases the contribution to the signal of each particle contained in the target, proportionally to the square of the gain, but at the same time, reduces the volume of the beam and thus the number of elements contributing to the signal. According to (1.37), under the same conditions, $\bar{P}_r$ is proportional to the aperture of the antenna.

Equation (1.76) also indicates that $\bar{P}_r$ increases proportionally to the product of the duration of the pulse $\tau$ by the transmitted power $P_t$. If we denote the average transmitted power by $P_{av}$, we have

$$P_t\,\tau = P_{av}/f_r$$

which shows that the received power is proportional to the ratio of the transmitted average power to the pulse repetition frequency $f_r$.

## 1.4 CALIBRATION AND SETTING UP

To use the radar as a measuring instrument, we must calibrate it with care to know the precise values of the constants in the radar equation. The main parameters to be checked are the transmitter power, the microwave losses on transmission and reception, the noise level and gain of the receiver as well as its dynamic range (i.e., the range of the power transmitted without distortion). The same problems arise for non-meteorological radars; and they are dealt with in general works [Skolnik, 1970]. This subject therefore will not be discussed in detail.

No single method is universally recommended for calibration. A large number of procedures have been proposed, most of which fall into two categories: (1) those that have recourse to separate measurements on the "modules" of the radar with the aim of then calculating the overall performance; (2) those that lead directly to an overall evaluation of the performance by measuring the echo of a known target without any reference to the equipment.

Each method has advantages and disadvantages. The measurement of the properties of the radar modules makes it possible to pinpoint faults and keep track of developments (for example, the power of the transmitter that depends on the aging of the oscillator) but this is possible only if suitable test equipment is available. Overall calibration with a known target sometimes is difficult even under optimal conditions.

From a practical point of view, all the elements of the radar do not require the same attention. Therefore, the performance of the antenna does not change appreciably over time; the transmitter power varies little from one day to another, and intermittent checks are sufficient in most cases. The reception quality, which depends in particular on frequency tuning, and the noise level must be checked frequently. For this purpose, a stable microwave generator is used, producing standard pulses of the same frequency as the transmitter. These pulses arrive at the receiver in pulse repetition frequency. The measurement of the pulse level makes it possible to test the receiver and the data acquisition system. This procedure is used in coherent and noncoherent radars. In coherent radars, we sometimes introduce a frequency-shifted signal at the level of the average frequency to test the procedures for velocity data processing [Nutten *et al.*, 1979].

Calibration on a known target is the most common and oldest method. It is a direct application of equation (1.71b), which makes it possible to calculate $C_1$. The target is often a metal sphere of diameter $D$ so that $\pi D/\lambda > 10$; that is, it satisfies the laws of geometrical optics for the wavelength considered and, as a consequence, its scattering cross section is equal to its geometric cross section.

However, any other type of target or reflector with known properties can be used to carry out the measurement. To avoid perturbations by ground echoes, the target is either suspended underneath a balloon, although it is sometimes difficult to maintain it in the beam axis, or placed in the vicinity of the ground at such a point that the radar does not see the ground at the distance where the target is situated. Thus, to pick out the contribution of the known targets from the surrounding echoes we can use a device that eclipses it regularly; the fluctuating part of the echo is the useful signal. Another possibility involves firing balls of known dimension into the beam [Stratman *et al.*, 1971]. More sophisticated is the device using a target with coded transponder, which lets us carry out a separate calibration of the transmission and reception of doppler and nondoppler radars by avoiding most of the disadvantages of other methods [Glover *et al.*, 1976]: the transmission of the radar is measured by the transponder placed at ground. The measurement is returned to the radar with a delay, which prevents the signal from being simultaneous to ground echos, in the form of calibrated pulses and possibly with a shift in frequency and variation in amplitude. Finally, a rapid check of the receiver and antenna performance can be made easily with the transmitter switch off, by an aiming onto a radio source such as the sun, which is a source of noise with well-known properties. At short wavelengths, we may need to take into account atmospheric attenuation [Whitten, Smith, and Harbuck, 1976].

With the meteorological radars, it is difficult to obtain a more precise calibration than about 1 dB. In measurements of precipitation rate by radar, calibration is often made starting from an almost simultaneous measurement of the drop-size distributions (which determines the backscattering cross section of the target) or the precipitation rate, or both, with the help of ground sensors, the precipitation echo being measured just above the sensors (see Section 2.6).

In a general manner, calibration and the problems of radar measurement lead us to consider the conditions for setting up ground equipment. A good implantation of meteorological radar is incompatible with a quantitative use; that is, it permits us to correctly apply the radar equation. The two main conditions are the absence of any obstacle liable to modify the beam shape by total or partial occultation, and the absence of ground echoes at the distance where the measurements must be made. We can correct the effects of a partial occultation by modifying the radar equation accordingly; for example, by using a correction factor expressing the occulted beam fraction (when it is known). However, the obstacle can also reflect part of the energy in a direction different from the one aimed at by the antenna, in which case "parasitical" targets can be detected. Ground echoes can be due to the main lobe, the sidelobes, or the direct radiation between the primary source and the ground (spillover). Very short distance echoes (a few kilometers) are not used for meteorological observations for various reasons, among which are

- Blinding due to the commutation time of the duplexor;
- The Fresnel region;

- Ground echoes;
- The smallness of the pulse volume.

In prospecting for a site for a meteorological radar, the problem of the obstacles and ground echoes can be treated summarily with topographical maps and more sharply by computer, using a ground numerical model (file of data of the altitude field of a given area). It is then easier to vary the position, test the consequence of the choice of the height of the work that sustains the antenna, the contribution of the sidelobes, the effects of the refractivity, and so forth.

Although the generally accepted idea that radars must be installed at elevated points to clear obscuration created by nearby obstacles is correct, it is important to take care in relying on this. In fact, let us consider a radar situated on a high point well clear of surrounding objects. For an almost horizontal elevation, the beam and the sidelobes encounter the ground at several kilometers or tens of kilometers from the radar, distances at which useful echoes are observed and the measurements perturbed. It is therefore preferable to install the radar antennas on the ground, which may be high in altitude but locally flat or even have a small depression, so that the antenna regardless of its orientation does not see any distant ground-based obstacles. Here, the perceived ground echoes are very close and do not perturb measurements at a distance.

## 1.5 METEOROLOGICAL SIGNALS

### 1.5.1 Meteorological Targets

With meteorological targets, the resolution volume is occupied by a high number of individual scatterers, for example, raindrops. The voltage corresponding to the scatterer $i$, if we disregard the carrying frequency filtered in the receptor, is the real part of

$$v_i(t) = Ai\, e^{i\varphi_i t} \tag{1.77}$$

$A_i$ is the amplitude that depends on the backscattering cross section of the scatterer and its location in the pulse volume. In this section, $A_i$ is considered constant; that is, the fluctuations that could result for example from the whirling of nonspherical particles (snowflakes) or nonpermanent distortion of large drops as they fall do not intervene significantly. The term $\varphi_i(t)$ represents the total phase, cf. (1.4):

$$\varphi_i(t) = -\frac{4\pi r_i(t)}{\lambda} + \Psi_i + \Psi \tag{1.78}$$

where $\Psi_i$ is the phase associated with the diffusion process and $\Psi$ is a constant initial

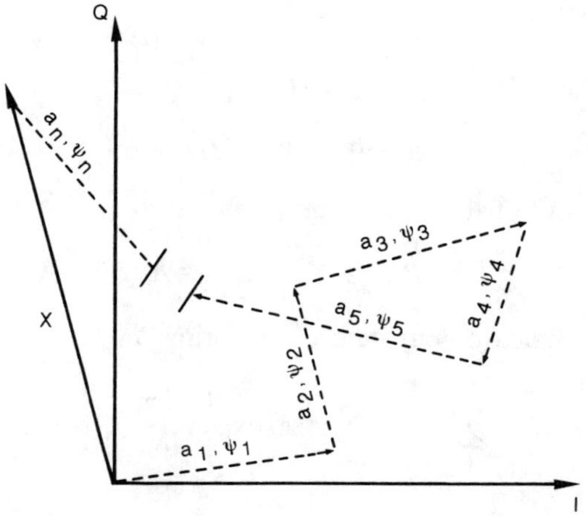


**Figure 1.12** Representation of the received signal in the complex plane.

phase. The echo received at the instant $t$ is the vectorial sum of the waves returned by each of $N$ scatterers present in the pulse volume between the distances $r$ and $r + h/2$. The resulting voltage is

$$X(t) = \sum_i v_i(t) = \sum_i A_i \, e^{i\varphi_i(t)} = A \, e^{i\phi(t)} \qquad (1.79)$$

with $i = 1$ to $N$. $X(t)$ is called the complex amplitude. $A$ is the real amplitude or envelope voltage of the echo modulation. A graphic representation of $X(t)$ is given in Figure 1.12.

The scatterers move with respect to each other principally due to the turbulence and the dispersion of the terminal fall velocity, so that the value of $r_i$ and thus of $\varphi_i$ have a random distribution. As the resolution volume is high with respect to the wavelength, the values of $\varphi_i$ spread to the $(0.2\ \pi)$ interval. The components that result are

$$I(t) = \sum_i A_i \cos\varphi_i(t) = A \cos\phi(t) \qquad (1.80a)$$

$$Q(t) = \sum_i A_i \sin\varphi_i(t) = A \sin\phi(t) \qquad (1.80b)$$

with

$$A = (I^2 + Q^2)^{1/2} \tag{1.81}$$

$$\phi = \tan^{-1}(Q/I) \tag{1.82}$$

The power (or intensity) of the echo is proportional to

$$P(t) = X(t)X^*(t) = A^2(t)$$

where $X^*(t)$ is the complex conjugated; or, ignoring $\Psi_i$,

$$P(t) = \sum_i A_i^2 + \sum_{i \neq j} A_i A_j \exp\left[ i \frac{4\pi}{\lambda} (r_i(t) - r_j(t)) \right] \tag{1.83}$$

with $j = 1, N$.

Equation (1.83) shows that $P(t)$ is the sum of a continuous component $P_{dc}$ and of a fluctuating component $P_{ac}$ equal to the first and second sommation of (1.78) respectively. $P_{dc}$ is an incoherent diffusion term, which depends on the number and backscattering cross section of scatterers and is independent of their respective positions. $P_{ac}$ is a coherent diffusion term to which the echo variance is associated.

From one pulse to the next, for a scatterer couple $(i, j)$, the argument of the exponential of $P_{ac}$ varies, cf. (1.4–1.7):

$$\frac{4\pi}{\lambda}\left(\frac{\Delta r_i - \Delta r_j}{T_s}\right) = \frac{4\pi}{\lambda}(v_i - v_j) = \frac{4\pi}{\lambda}\Delta v_{ij} = -2\pi(f_i - f_j) = -2\pi F_{ij} \tag{1.84}$$

$F_{ij}$ is the doppler frequency difference of all the scatterers, the radial velocities of which present a difference equal to $\Delta v_{ij}$. The scatterers, which have the same velocity compared to the radar ($v_i = v_j$), do not create any fluctuation between them. The value of $P_{ac}$ thus does not depend on the mean movement but only on the displacement of one scatterer in relation to another. The average on a period $F_{ij}^{-1}$ of a component of $F_{ij}$ frequency is equal to 0. Therefore, when successive samples of the instantaneous signal are averaged, $P_{ac}$ tends to 0 and the average power tends to $P_{dc}$. The decrease of $P_{ac}$ is more important when the averaged series is long. Figure 1.13 shows an example of temporal series of values of $P(t)$. The decorrelation between two consecutive samples increases with the length of the sampling interval $T_s$ and the difference between the radial speeds; that is, the importance of the $\Delta r_{ij}$. If the $\Delta r_{ij}$ are small compared to $\lambda$, the samples are correlated and their average does not reduce $P_{ac}$. When the $\Delta r_{ij}$ correspond to a significant fraction of the wavelength, the samples are decorrelated; they are said to be independent, and their average reduces the variance. If partially correlated samples are averaged, the only point that counts for the reduction of the variance is the equivalent number of independent samples corresponding to the series duration. Contrariwise, to measure $\Delta\phi$ with the

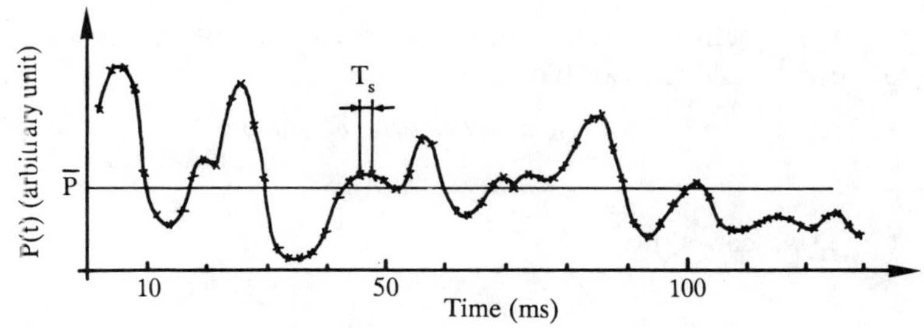


**Figure 1.13** Temporal series of the received power $P(t)$ in rain for $\lambda = 5$ cm and $T_s = 2$ m s.

doppler radar, the samples must be correlated; if not, $\Delta\phi$ is ambiguous (cf. Section 1.2.2 and Figure 3.2).

It appears that, without the disordered displacement of the scatterers relative to each other inside the atmospheric targets—that is to say, for a distributed target strictly stationary—the measurement of the average power by temporal integration would not be possible, the second summation of (1.83) being constant. Moreover, if the diffusers were distributed uniformly, there would be no echo, as the result would equal 0. In what follows, the discussion will bear first on the statistical properties of the echo of a meteorological target (i.e., distributed), then on the independence of the samples, and finally on the reduction of the variance for the estimation of the average. Doppler measurement is dealt with in Chapter 3.

### 1.5.2 Signal Statistics

The term $X(t)$ and the various functions deduced from it, $I(t)$, $Q(t)$, $P(t)$, are random functions of the time that can be deduced only statistically. The statistical properties of the fluctuating echoes created by distributed targets such as clouds and precipitation and the estimation of their average intensity (or power) were discussed in detail by Marshall and Hitschfeld [1953], Wallace [1953], and Smith [1964], among others.

The distribution of $\varphi_i$ being random and $N$ large, we can apply the central limit theorem to $X(t)$; according to it the sum of a high number of independent random variables tends to a Gaussian distribution if none of the variables is dominant compared to the others. Thus we can admit that $I$ and $Q$ are normally distributed and have the same variance $\sigma^2$. $A$ and $\phi$ are independent random variables, and $\phi$ is uniformly distributed between 0 and $2\pi$. The joint normal probability density function of $I$ and $Q$ is the product of the individual probability densities and can be written

$$p(I, Q) = \frac{1}{2\pi\sigma^2} e^{-(I^2+Q^2)/2\sigma^2} \tag{1.85}$$

The probability density function of $A$ and $\phi$ can be obtained from (1.85) by changing the variable. Indeed, we have

$$p(I, Q)\, dIdQ = p(A, \phi)dAd\phi$$

$$dIdQ = AdAd\phi$$

and, after (1.85), we can write

$$p(A, \phi) = \frac{A}{2\pi\sigma^2} e^{-A^2/2\sigma^2} \quad \text{for} \quad \begin{cases} 0 \le A \le \infty \\ 0 \le \phi \le 2\pi \end{cases} \tag{1.86}$$

The individual probability densities for $A$ and $\phi$ are deduced from it:

$$p(A) = \int_0^{2\pi} p(A, \phi)d\phi = \frac{A}{\sigma^2} e^{-A^2/2\sigma^2} \quad \text{with} \quad A \ge 0 \tag{1.87}$$

$$p(\phi) = \int_0^{\infty} p(A, \phi)dA = \frac{1}{2\pi} \tag{1.88}$$

where $p(A)$ is a Rayleigh probability density. Its first two moments, see (2.3), are

$$\bar{A} = \sqrt{\frac{\pi}{2}}\,\sigma \tag{1.89}$$

$$\overline{A^2} = 2\sigma^2 \tag{1.90}$$

The variance is

$$\sigma_A^2 = \overline{A^2} - \bar{A}^2 = \left(2 - \frac{\pi}{2}\right)\sigma^2 = 0.429\,\sigma^2 \tag{1.91}$$

The probability density of the power $P = A^2$ is obtained by changing the variable from (1.87), noting that $dP = 2AdA$ and $p(P)dP = p(A)dA$. We have

$$p(P) = \frac{1}{2\sigma^2} e^{-P/2\sigma^2} \tag{1.92a}$$

which is an exponential distribution. The first two moments are

$$\bar{P} = \overline{A^2} = 2\sigma^2 \tag{1.93}$$

$$\overline{P^2} = 2\bar{P}^2 \tag{1.94}$$

The variance is

$$\sigma_P^2 = \overline{P^2} - \bar{P}^2 = \bar{P}^2 \tag{1.95a}$$

$$\sigma_P = \bar{P} \tag{1.95b}$$

which shows that the standard deviation of the power is equal to the average.

Also note that the ratio of the mean power to the amplitude variance is constant:

$$\frac{\bar{P}}{\sigma_A^2} = \frac{2\sigma^2}{0.429\,\sigma^2} = 4.66 \tag{1.96}$$

With (1.93), equation (1.92) can be written

$$p(P) = \bar{P}^{-1}\,e^{-P/\bar{P}}, \qquad P \geq 0 \tag{1.92b}$$

If the range of the values of the power are very large, to compress the signal dynamic, it is often preferable to use its logarithm compared with a reference power $P_1$:

$$L = 10\,\log(P/P_1) \tag{1.97}$$

Often the value of $P_1$ taken is equal to 1 mW and the power level $L$ is expressed in dBm. The mean power is also expressed by its level in relation to $P_1$; that is,

$$L_0 = 10\,\log(\bar{P}/P_1)$$

The probability density of $L$ is obtained from (1.92b) by the following change of variable $L - L_0 = 10\,\log(P/\bar{P})$:

$$p(L) = m\,\exp[m(L - L_0) - e^{m(L-L_0)}] \tag{1.98}$$

with $m = (\ln 10)/10$.

The most probable value is $L_0$. The mean value is $L_0 - 2.51$, which means that, to obtain $L_0$, we must add 2.51 dB to the value found by averaging a series of observations of $L$. The properties of these various distributions are summed up in Table 1.1. The given numerical values can be found easily from the previous relations. A few independent scatterers simultaneously present in the pulse volume are sufficient to enable the observation of such distributions.

### 1.5.3 Decorrelation Time: Independent Samples

The time $\tau$ necessary for reshuffling the scatterers to ensure the independence of the samples depends on the characteristics of the target (variance of the terminal fall velocities of the hydrometeors, wind shear, turbulence in the resolution volume) and the characteristics of the radar, principally its wavelength. The doppler spectrum of the meteorological signals, for uniform reflectivity regions and when the widening of the spectrum is due to independent causes, is observed to be approximately Gaussian; and the power spectral density (cf. Chapter 3) is

$$S(f) = S_0\,e^{-f^2/2\sigma_f^2} \tag{1.99}$$

**Table 1.1.**
Properties of the Probability Distributions Used for Meteorological Radar Signal Description (from Marshall and Hitschfeld, 1953)

| *Signal function* | | | *Probability density function P(j)* | *Most probable value* $j_p$ | *Mean value* $\bar{j}$ | *Standard deviation* $\sigma_j = [\overline{(j - \bar{j})^2}]^{1/2}$ |
|---|---|---|---|---|---|---|
| Gaussian | 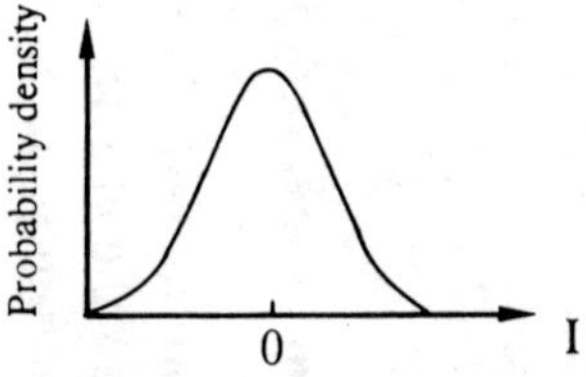  | I or Q | $\dfrac{e^{-I^2/\bar{P}}}{\sqrt{\pi \bar{P}}}$ | 0 | 0 | $0.707/\bar{P}$ |
| Rayleigh | 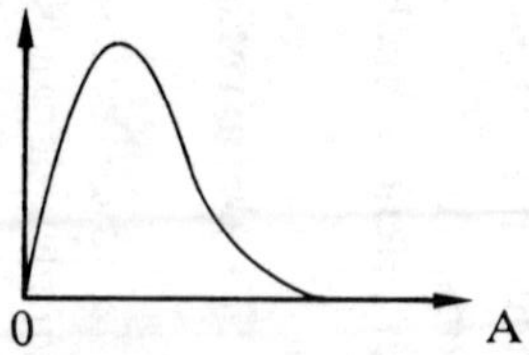  | A | $\dfrac{2A}{\bar{P}} e^{-A^2/\bar{P}}$ | $0.707\sqrt{\bar{P}}$ | $0.886\sqrt{\bar{P}}$ | $0.463\sqrt{\bar{P}}$ |

| | | | | | | |
|---|---|---|---|---|---|---|
| Exponential | 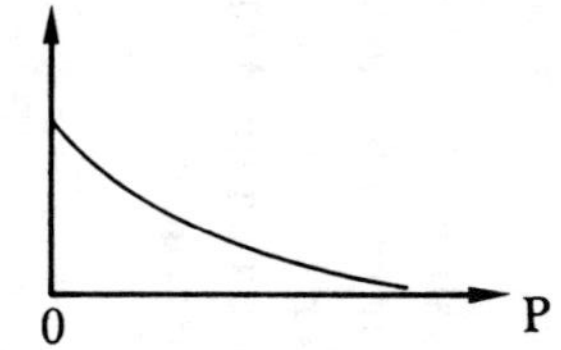  | P | $\dfrac{e^{-P/\bar{P}}}{\bar{P}}$ | 0 | $\bar{P}$ | $P$ |
| Logarithmic | 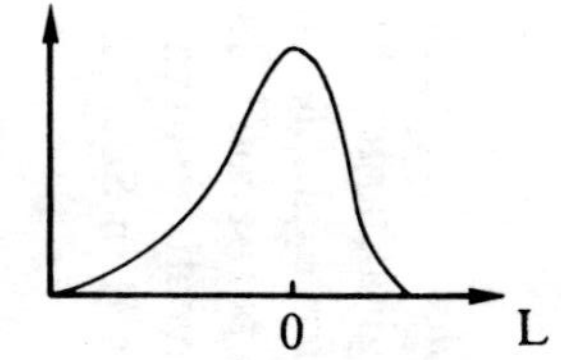  | L | $m \exp[m(L - L_0) - e^{m(L-L0)}]$ | $L_0$ | $L_0 - 2.51$ | 0.557 |
| | 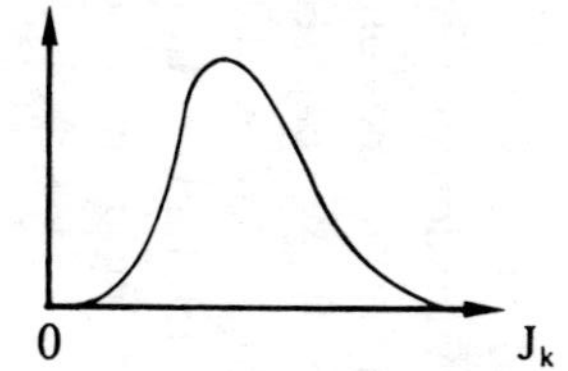  | $J_k$ | $\dfrac{k^k}{\bar{P}^k(k-1)!} J_k^{k-1} \exp(-kJ_k/\bar{P})$ | $\bar{P}(1 - 1/k)$ | $\bar{P}$ | $\bar{P}/\sqrt{k}$ |

where $f$ is the doppler frequency and $\sigma_f^2$ is the variance of the spectrum.

From the Fourier transform of $S(f)$, the normalized autocorrelation function $R(\tau)$ is obtained; it too, is Gaussian and can be written

$$R(\tau) = e^{-\tau^2/2\sigma_\tau^2} \tag{1.100}$$

where $\sigma_\tau^2$ is the variance of the autocorrelation function, with

$$\sigma_\tau = (2\pi\sigma_f)^{-1} \tag{1.101}$$

We will see (3.15) that $\sigma_v$ and $\sigma_f$ are linked by the relation

$$\sigma_v = \frac{\lambda}{2}\sigma_f \tag{1.102}$$

Thus, we can rewrite (1.100) as [Nathanson, 1969]

$$R(\tau) = \exp\left[-8\left(\frac{\pi\sigma_v\tau}{\lambda}\right)^2\right] \tag{1.103}$$

The value of $R(\tau)$ decreases more quickly with $\tau$, from 1, when $\sigma_v$ is high. Nathanson shows that the samples are independent when $R(\tau) < 0.02$. The time $\tau_i$ necessary to reach this value is called the *independence* or *decorrelation time*. From (1.103) we have

$$\tau_i = 2\lambda/\sigma_v \tag{1.104}$$

with $\lambda$ in cm, $\sigma_v$ in m s$^{-1}$ and $\tau$ in m s.

Figure 1.14 shows the variation of $\tau_i$ *versus* $\sigma_v$ for various values of $\lambda$. The values of $\sigma_v$ to be considered for the meteorological targets range from 0.5 (snow) to 1 m s$^{-1}$ (rain) (cf. Chapter 3). We can see, for example, that for $\lambda = 10$ cm in rain, only 50 independent samples are available per second. In some perturbed fast flows, the values of $\sigma_v$ are much higher. In convective storms $\sigma_v$ reaching 5 m s$^{-1}$ are found. The factors influencing $\sigma_v$ are discussed in Section 3.14.

### 1.5.4 Sample Time Averaging: Reducing the Variance of the Mean

The output voltage of the receiver, $W$, is related to the power of the signal at the input by a functional relationship, the transfer function of the receiver, which in meteorological radars can take the following forms.

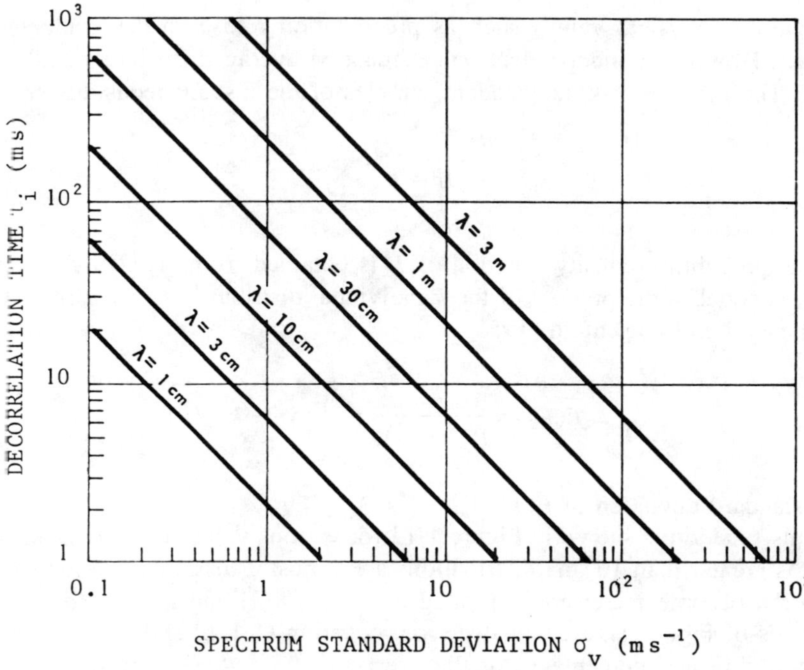


**Figure 1.14** Equivalent independent sampling time $\tau_i$ calculated from eq. (1.104) *versus* $\sigma_v$.

1. Linear form: The voltage at the output of the receiver is proportional to the amplitude of the input signal; that is, to the square root of its power:

$$W \propto P_r^{1/2} \propto A$$

2. Quadratic form: The output is proportional to the power of the signal at the input:

$$W \propto P_r \propto A^2$$

3. Logarithmic form:

$$W \propto \log P_r \propto \log A$$

The proportionality constants depend on the gains and losses in the various stages of the receptor. These must be determined by calibration. The value of $W$ for these various types of receptors is described by one of the probability distributions given in Section 1.5.3.

For meteorological applications, the useful quantity is the mean power. We will see indeed in Chapter 2 that $\bar{P}_r$ is proportional to the radar reflectivity from which various physical values such as precipitation rate or water content can be measured. How many independent pulses must be averaged (or integrated) to evaluate $P_r$? The average of $k$ independent samples of the instantaneous power is

$$J_k = k^{-1} \sum_{i=1}^{k} P_i \qquad (1.105)$$

The probability density function of $J_k$ is obtained from (1.92) by writing that $P(J_k)dJ_k$ is equal to the product of the $k$ individual distributions $P(A_i^2)d(A_i^2)$. We obtain [Marshall and Hitschfeld, 1953]

$$P(J_k) = \frac{k^k}{\bar{P}^k(k-1)!} J_k^{k-1} \exp[-kJ_k/\bar{P}] \qquad (1.106)$$

with a standard deviation of $\bar{P}/k^{1/2}$.

This function is shown in Figure 1.15 for various values of $k$. We can see that when $k$ is greater than 10, the distributions are almost Gaussian. As $k$ increases, the distribution become increasingly peaked at $\bar{P}(1 - 1/k)$ and narrow ($\sigma = \bar{P}/k^{1/2}$). The curves of Figure 1.16 calculated from equation (1.106) show the limits below which the indicated percentages of the averages of $k$ independent samples are expected to fall. For instance for $k = 30$, 10% of the averages are below 0.75 $\bar{P}$ and 90% below 1.25 $\bar{P}$; that is, the 80% confidence limits are $\bar{P} \pm 0.25\,\bar{P}$. These limits correspond to +1 dB to −1.3 dB and are close to the instrumental accuracy in the measurement of $\bar{P}$. These results let us determine the duration of the observations necessary to arrive at an estimation with a given accuracy.

Similar calculations can be made based on the statistics corresponding to linear and logarithmic receivers. The comparative Table 1.2 [Zrnic', 1975] shows the mathematical expectations and the standard deviations of the estimation of $\bar{P}$ obtained with the three types of receivers. Note that the standard deviation of the estimation depends on the type of receiver and is proportional to $\bar{P}/k^{1/2}$.

### 1.5.5 Reducing the Integration Time

The time necessary to acquire the number of independent samples needed for estimating the average power imposes a limitation on the beam scanning speed. Sometimes, we wish to reduce this time, or in the case of a convective storm, to ensure that the target was stationary during the volume acquisition, or for airborne radar, as the platform moves at a high and imposed speed. Various techniques can be used to increase the independence of the samples obtained with $T_s < \tau_i$.

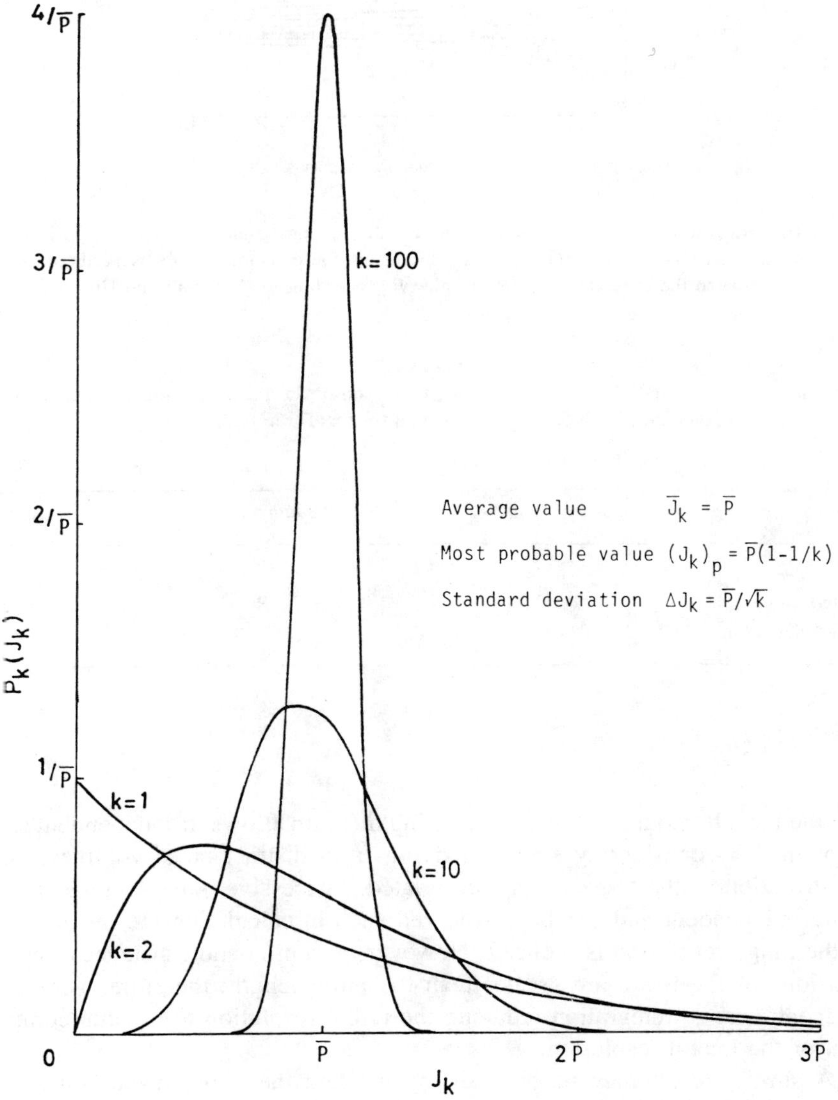


**Figure 1.15** Probability distribution of the mean of $k$ independent values of the signal power when the average power is $\bar{P}$ (According to [Marshall and Hitschfeld, 1953]).

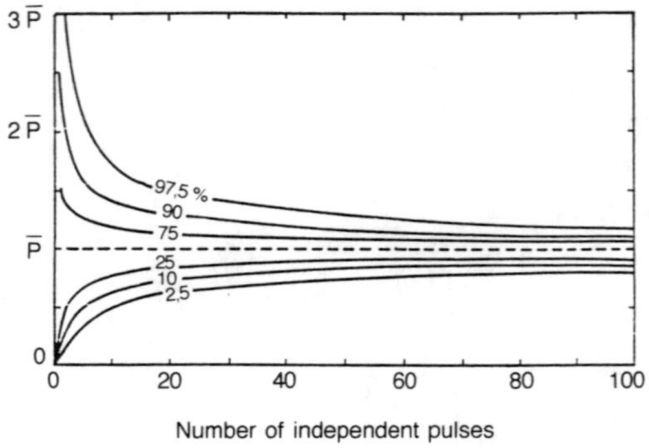


**Figure 1.16** Estimation of $\bar{P}$ as a function of the number of independent pulses used to calculate the average. The figures shown on the curves are the percentages of observed values situated between the curve and the exact value $\bar{P}$ (according to [Marshall and Hitschfeld, 1953]).

**Table 1.2.**
Comparison of Expected Values and Standard Deviation of the Power Estimates Obtained by an Average of $k$ Independent Samples for the Three Receiver Types

| *Transfer Function* | *Linear* | *Square Law* | *Logarithmic* |
|---|---|---|---|
| Expected value | $0.79\bar{P}$ | $\bar{P}$ | $0.56\bar{P}$ |
| Standard deviation | $1.05\bar{P}/\sqrt{k}$ | $\bar{P}/\sqrt{k}$ | $1.28\bar{P}/\sqrt{k}$ |

## *Space Averaging*

When the local homogeneity of the scattering medium allows it, independent samples are obtained by considering signals arriving from distinct pulse volumes, because the distribution of the targets is not correlated. Successive pulse volumes of length $h/2$ are independent and can be considered such in calculating the mean. In doing this, the range resolution is reduced. However, except at short distances, the lateral dimensions of the beam are greater than the pulse length; therefore, generally we can accept a range integration reducing the radial resolution to a value comparable to that of the lateral resolution.

A similar result may be obtained by limiting the radial resolution loss by a reduction in the pulse depth, but then the radar sensitivity also is reduced, eq. (1.107). Independent samples can be obtained as well by taking into account contiguous volumes but at different elevations [Walker *et al.*, 1980].

*Frequency Diversity*

If the frequency transmitted is shifted from one pulse to the next, the spectrum of the corresponding samples is also shifted and their decorrelation increases. If $\Delta f > 1/\tau$, where $\Delta f$ is the frequency shift and $\tau$ is the pulse duration, the samples are shown to be decorrelated [Nathanson, 1969] in the sense explained in Section 1.5.3. With pulses of 1 $\mu$s, for example, a value of $\Delta f$ wider than 1 MHz is enough for the samples to be independent. When doppler measurements have to be made, pairs of pulses of the same frequency are transmitted.

*Broadband Noise Technique*

Instead of a single frequency for each pulse, the radar transmits a noise occupying a frequency band $B$ around $f$. This signal is generated by a noise source whose band is limited, and that is amplified. With such a signal, the structure of the second term of (1.83) is modified and fluctuates with $r$ (i.e., with time) so that the echo is decorrelated on a distance smaller than $c\tau/2$. If the noise is assimilated to a frequency continuum, the echoes sampled at radial distances of $B^{-1}c/2$ are shown to be decorrelated. The result is that in a pulse $c\tau/2$ long, there are $B\tau$ independent samples that can be used to calculate the average. For example with $B$ = 40 MHz and $\tau$ = 1 $\mu$s, 40 independent samples are obtained. This technique was proposed by Krehbiel and Brook [1979] for a fast scanning radar.

### 1.5.6 Detecting Weak Signals

In most applications of meteorological radar concerning precipitations, the problem of detecting weak signals does not arise. In fact, at ranges for which the beam resolution remains acceptable, the echoes are usually strong. When observing clouds or targets in clear air, for example, the situation is different. It may be necessary to improve the signal-to-noise ratio. Examination of the radar equation and the discussion of Sections 1.2.4 and 1.3 show that the S/N ratio increases when the transmitted power is increased and the pulse is widened because the noise is lessened. Increasing the transmitted power is costly and restricted for a large number of technical reasons. Widening the pulse to reduce the band occupied by the signal leads to *frequency-modulated continuous-wave* (FM-CW) radar and pulse-compression radar designs. Samples averaging or integration also improves the signal-to-noise ratio.

*Integration*

The integration may be made at intermediate frequencies or video frequencies. In the first case, the integration is said to be a predetection or coherent integration (the information relating to the phase of the signal is still present); and in the second case it is said to be a postdetection or noncoherent integration. Postdetection integration is less efficient than an ideal predetection integration (in particular, due to the losses resulting from the nonlinear effects of the video detector that convert part of the signal energy to noise during the detection process); but it is easier to carry out.

The ideal coherent integration is such that the signal components of the successive samples (i.e., the echo) combine in phase, whereas the noise components combine with random phases. In this case, the increase of the S/N ratio would be proportionate to the number of averaged samples. Because the meteorological echoes have random phases for distributed targets, the conditions of a coherent integration are not met. A quasi-coherent integration can be performed when the signal phase changes very slowly over the integration time, as in slowly moving targets observed with long waves (ST radar, Chapter 4). In the microwave meteorological radars, even the doppler radars, $\bar{P}$ often is estimated on a noncoherent canal and not from the doppler spectrum.

Signal power $P$ and noise power $N$ obey Rayleigh distributions, and their average value for $k$ independent samples has an almost Gaussian distribution for $k \geqslant 10$. Thus their sum also has a Gaussian distribution of the mean $(\bar{P} + \bar{N})$. On the basis of this statement, we can study the number of pulses necessary to form the average to distinguish $(\bar{P} + \bar{N})$ from $\bar{N}$ alone. Marshall and Hitschfeld [1953] found that the sum of the power of the signal and noise has a 97.3% probability of lying above the power of the noise when

$$\bar{P}/\bar{N} = 2/(k^{1/2} - 1) \tag{1.107}$$

For $k = 9$, we have $\bar{P} = \bar{N}$, which means that a small number of pulses is sufficient to distinguish a meteorological signal in noise with the same power. When $k$ increases, we can detect a signal with power less than that of the noise. For example, for $k = 100$, a signal whose power is $-6.54$ dB below the power of the noise can be detected. For $k \geqslant 100$, the ratio $\bar{P}/\bar{N}$ at the detection limit decreases as the reciprocal of $k^{1/2}$, which is the characteristic gain factor for the signal-to-noise ratio of the postdetection integration. For $k = 441$, a meteorological signal whose intensity is one-tenth that of the noise is detected with a probability of 97.3%.

### *FM-CW Radar*

The principle of FM-CW radar has been known for a long time (see [Skolnik, 1962]): the radar transmits a continuous wave whose frequency varies linearly between $f$ and $f + F$ with a constant repetition period $T_M$ (Figure 1.17). For meterological observations, typical values of $F$ and $T_M$ are of the order of 10 MHz and ms, respectively. The wave returned by a target situated at range $r$ is received after time $\Delta t = 2r/c$ and instantaneously mixed with the wave being transmitted (homodyne montage).

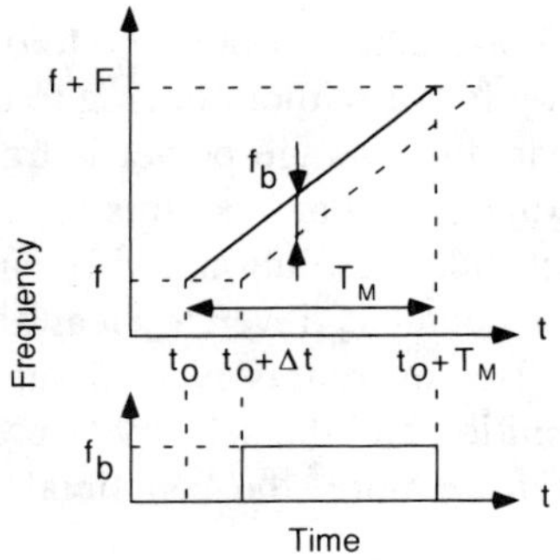


**Figure 1.17** Frequency-time diagram showing the principle of the FM-CW radar in the case of linear modulation. In the top diagram, the solid line represents the transmitted wave and the dashed line is the wave received from a point target. The diagram at the bottom represents the oscillation frequency.

Thus, beats are made to appear with frequency $f_b = F\ \Delta t/T_M$ that determines the range of the target; the signal corresponding to range $r$ is a sinusoidal wave of frequency $f_b$ with

$$r = \frac{T_M c}{2F} f_b \tag{1.108}$$

In multiple targets, a spectral analysis of the frequency of the beats allows discriminating between the received signals in terms of range.

One difficulty *a priori* of using the FM-CW radar for measurements is because both the range and the Doppler effect influence the frequency of the echo. However, the Doppler effect on atmospheric targets introduces frequency variations ($\Delta f_b$) small enough to neglect in measuring the range. For example, for $\lambda$ = 10 cm, a radial velocity of 10 m $s^{-1}$ produces a variation $\Delta f_b$ of 200 Hz. If $F$ = 200 MHz and $T_M$ = 50 m s, this shift corresponds to only $\Delta r$ = 7.5 m. The doppler frequencies can be extracted by measuring the phase shift of the signal from one frequency scan to the next. This measurement is possible because the FM-CW radar is coherent throughout the duration of a repetition. The doppler spectrum is obtained by Fourier transform from a series of measurements obtained from several consecutive repetitions [Barrick, 1973; Chadwick *et al.*, 1976].

An important advantage of the FM-CW radar is that the range resolution, which depends on the resolution of the measurement of $f_b$, is inversely proportional to the frequency modulation $F$ and so can be regulated. To increase it, it is sufficient to increase the value of $F$. Therefore, we can obtain a very high resolution compared to that of pulse radars, which is not very good.

Furthermore, FM-CW radar makes it possible to obtain a significant signal-to-noise ratio through a high duty factor without having to use strong peak power transmitters, such as in pulse radar (the average power is transmitted). Because of these various advantages, most altimetric radars use this type of modulation.

Observations with spectacular resolution and sensitivity have been carried out on structures of the planetary boundary layer in clear air [Richter, 1969; Richter *et al.* 1973] (see Section 5.2). However, this type of radar is still not used much in meteorology because its resolution and the sensitivity are not of decisive importance in observing precipitation. Furthermore, the beneficial effect on the S/N ratio disappears in doppler mode.

### *Pulse Compression Radar*

Another technique making it possible to improve the S/N ratio and the range resolution is pulse compression. This involves coding the transmitted pulse to reduce (or compress) its length and increase its amplitude at reception. This result can be obtained by modulating the frequency of the wave train forming the transmitted pulse (for example, a linear decrease of $f$ from the front to the rear of the pulse, as in Figure 1.18). If a filter is used at reception (acting as a decoder), transmitting the pulse at a speed depending on the frequency of the wave (speed $\propto f^{-1}$ in our example), at the output of the filter, we have a shorter pulse of the form $\sin x/x$, with power that increases with the compression ratio (i.e., the ratio of the length $\tau_p$ of

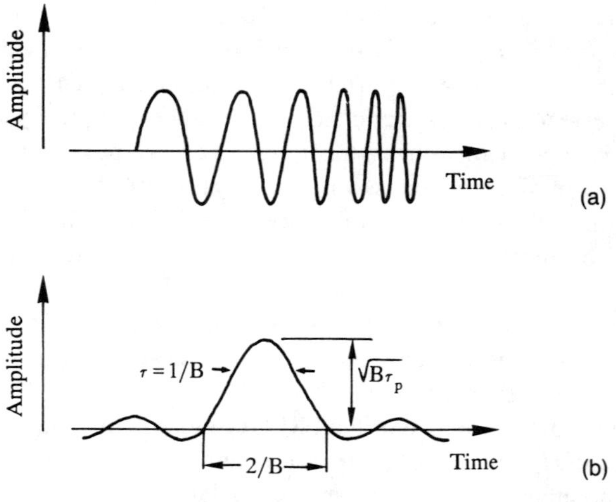


**Figure 1.18** Principle of linear frequency modulated pulse compression. a) Transmitted waveform. b) Output of the pulse compression filter.

the pulse at the input of the decoder filter to that at the output). The amplitude increases as the square root of this ratio.

If $B$ is the bandwidth of the pulse at the input of the compression filter, the width between zero points of the pulse at the output is $2/B$, but the effective width is approximately equal to $\tau = 1/B$ so that the compression ratio is equal to

$$\frac{\tau_p}{\tau} = \tau_p B \tag{1.109}$$

$\tau_p B$ is called the *time-bandwidth* product. Its value expresses the ratio of the peak power required by a radar with a short nonmodulated pulse to that of a radar with pulse compression to obtain the same resolution or gain in resolution due to compression as the pulse relative to the nonmodulated pulse.

This technique was originally proposed to carry out compression. In sonars where this procedure is used, the frequency modulation produces a chirping noise in the audio frequency region. This is why pulse compression radars are often called *chirp radars,* and this term is also used when the compression is not achieved by frequency coding.

The modulation also can be obtained by binary coding of the pulse. The transmitted pulse of duration $\tau_p$ is segmented into $N$ subpulses of equal duration $\tau_p/N$. Binary modulation is most often obtained by modifying the phase of the subpulses by 180° (+ when the phase is unchanged, − when the phase is modified). $N$ is the number of code bits. The filtering matched to such a pulse is obtained by a cross correlation between the sample of the transmitted code stored and the received signal. For this, a tapped delay line is used, segmented into $N$ cells of width $\tau_p/N$ with a total delay $\tau$. The output of the $N$ taps are added after having undergone phase shifts arranged to reproduce the coding sequence of the transmitted pulse (unchanged phase for the +; with a change of 180° for the −). When, during their progression along the delay line, the coded segments of the received pulse are in perfect agreement with the filter cells, all the segments arrive in-phase at the adder the output of which presents a peak equal to the sum of $N$ signals. This coincidence is produced for a duration of $\tau_p/N$. The code is chosen so that, for all the other positions of the pulse passing along the delay line, the signals in-phase and at 180° compensate each other to give the weakest possible signal at the output. The compression ratio therefore is equal to $N = B\tau_p$, where $B = 1/\tau$ is the bandwidth. The power of the signal increases from 1 to $N^2$ over the length $\tau_p/N$ of the compressed pulse. However, the compression also increases the power of the noise by a factor $N$ by summing the noncorrelated noise samples, so that the compression increases the signal-to-noise ratio by only a factor $N$. Figure 1.19 shows an example of a 13-bit code giving an increase of signal power of 22.3 dB. Pulse compression is compatible with the doppler technique. Other coding techniques are available too.

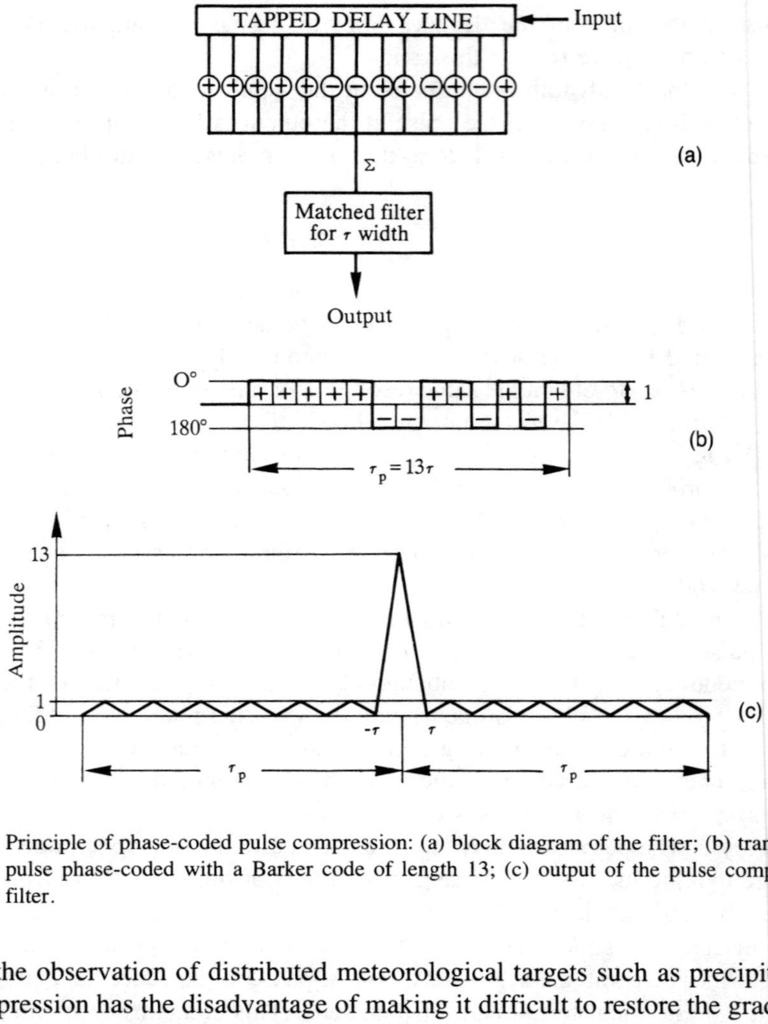


**Figure 1.19** Principle of phase-coded pulse compression: (a) block diagram of the filter; (b) transmitted pulse phase-coded with a Barker code of length 13; (c) output of the pulse compression filter.

For the observation of distributed meteorological targets such as precipitation, pulse compression has the disadvantage of making it difficult to restore the gradients. Furthermore, as mentioned for FM-CW radar, the problem of the signal-to-noise ratio does not arise in the observation of precipitation. Therefore, this procedure is not in ground-based microwave meteorological radars. The possibilities they offer, however, may find use in future orbital radars and for the observation of targets of low reflectivity, such as targets in clean air.

### 1.5.7 Sampling and Demultiplexing

To extract the information from the radar signal (i.e., to calculate $\bar{P}$, the doppler spectrum, *et cetera*), the echo fraction relative to each distance $r$ must be isolated

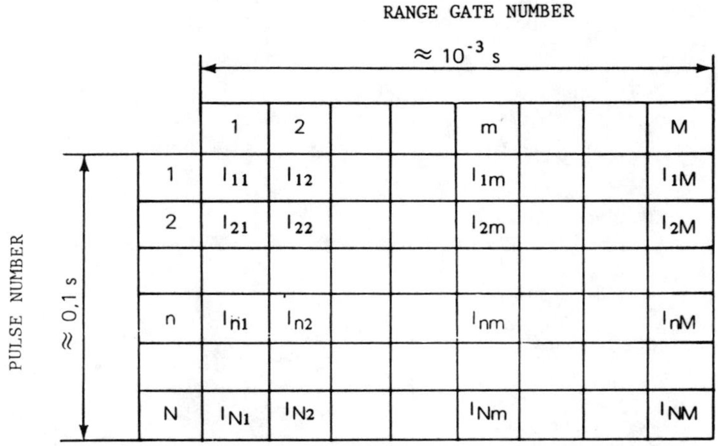


**Figure 1.20** Demultiplexing of signals for calculating mean power at $M$ points (from [Serafin, 1975]).

for a certain number of successive pulses, then each temporal series at a constant distance is treated independently. Therefore, the signal has to be demultiplexed. To do this, the received signal is sampled synchronously with the transmission of the pulses so that, for a pulse of order $n$, a sequence of discrete values of the measured quantity ($P$, $\varphi$, *et cetera*) is obtained; that is $I_{n,m}$ (Figure 1.20) where $m$ is the rank of the sample that determines the distance from the region in question. If we consider a succession of $N$ pulses, each sampled at $M$ distances, we obtain a matrix of $N$ rows and $M$ columns in which the rows consist of the sequence of sampled values between two successive pulses and where the columns are the samples arriving at a frequency $f_r$ from a region situated at distance $r$. The matrix is written following the rows (during the time $N/f_r$) and processed following the columns. $N$ is the number of values in a column. The hardware for this operation is a memory that can store the $N \times M$ data.

# Chapter 2
# Hydrological Measurements

## 2.1 INTRODUCTION

Radar applied to hydrometeor observation provides two types of information. On the one hand, the signal arriving from a small region, defined by the pulse resolution volume, gives quantitative information on the properties of the scattering medium; essentially, reflectivity and speed. On the other hand, all the signals arriving from different points in the space make it possible to define the contours of the targets and, inside them, the distribution of the measured quantities; that is, their small-scale and medium-scale structure, their evolution, and their movement, as well as other information whose interpretation calls on the experience of the analyst. The second type of information is often considered qualitative because of the difficulty of incorporating it in quantitative models at the present time.

Radar is a powerful research instrument in meteorology and also for telecommunications at frequencies higher than 1 GHz. In fact, radar enables us to gather considerable quantities of data on the three-dimensional structure of the atmosphere in a flexible, efficient, and rapid manner. These data can be used to describe atmospheric phenomena and study radio propagation, such as the provision of various statistics (for example, statistics on precipitation and attenuation). In this chapter, we are concerned with the basic mechanisms influencing the propagation of radar waves in clouds and precipitation and with the "measurement physics" of reflectivity.

In radar observations at short and medium range, only the following atmospheric elements will be taken into consideration:

- A gas phase composed of oxygen and water vapor that interacts with the radiation (the other constituents of air are not involved);
- Cloud and precipitation particles in the liquid and solid form that absorb and scatter the waves in all directions, in particular in that of the radar;
- Variations in the refractive index of air that are not involved in the microwave frequencies under current observation (these will be examined in Chapter 4).

The characteristics of the received signal therefore depend on the backscattering by the target and the attenuation due to absorption and scattering that the radiation undergoes throughout its journey between the radar and the target.

## 2.2 CLOUDS AND PRECIPITATION

### 2.2.1 Physical Processes of Formation

Cloud particles (droplets or crystals) are formed mainly by condensation or deposition of water vapor on hygroscopic or ice nucleii acting as seeding centers. Other particles, called secondary particles, are formed from existing elements by simple mechanical break-up or in clouds at minus Celsius temperatures by various physical processes accompanying the heterogeneous freezing of supercooled droplets[1] (the burst effect).

During the first growth phase, the terminal fall velocities of the cloud particles are negligible and the increase in mass solely due to the condensation or deposition of water vapor; it is therefore very slow. Gradually, the fall velocities increase until their value is sufficient for the growth to proceed by large particles capturing small particles because of the differences in the terminal fall velocity. The growth is produced by combining either liquid elements (coalescence) or solid elements (aggregation), or by ice particles capturing of supercooled droplets in an environment at minus Celsius temperatures (accretion, icing) [see Rogers, 1976; Pruppacher and Klett, 1978]. Figure 2.1 shows, in a simplified manner, the main routes of growth of the precipitation elements.

### 2.2.2 Hydrometeor Size Distributions: General Forms

The term hydrometeor designates all kinds of liquid, solid, or mixed atmospheric particles constituting the clouds and precipitation [Huschke, 1959]. Natural hydrometeors occupy a very wide range of dimensions. Clouds are composed of little hydrometeors with a negligible limit fall speed. The boundary between cloud and precipitation elements is conventionally fixed at a value of the diameter 200 $\mu$m. This value corresponds to a velocity of fall of 0.75 m s$^{-1}$ for a water droplet (at a pressure of 700 millibars, or approximately 3000 m altitude) that lies in the vicinity of the threshold starting from which growth by capture ensures rapid development of precipitation.

As they grow, the raindrops lose their spherical shape and oscillate around a mean shape which is that of a flat ellipsoid (cf. Section 2.2.6). The drops break up when the superficial tension forces are not sufficient to maintain the cohesion any

[1]Liquids, although at minus Celsius temperatures.

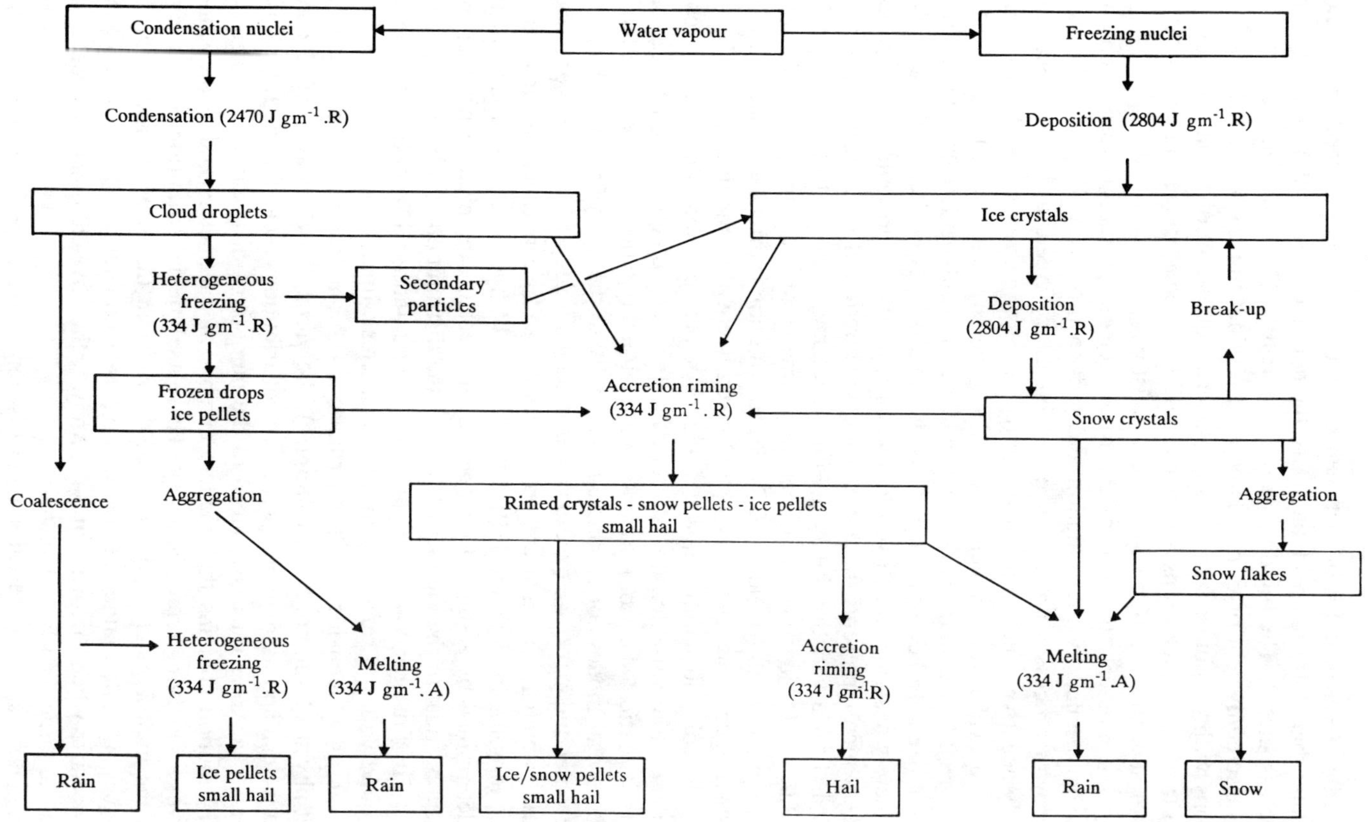


**Figure 2.1** Simplified representation of the main processes for the formation of precipitation. In parentheses is the latent heat absorbed (*A*) or released (*R*).

longer or as the effect of collision between drops (in this case, several drops and droplets smaller than 200 $\mu$m are created). To the notion of size, that of life duration must thus be associated. The big drops have a short life. The limit size seems to be around a spheric equivalent of 6 mm. However, it is not impossible that, for a few instants, drops with a volume larger than a spheric equivalent of 6 mm may exist; it is difficult to observe such drops, of course. Solid hydrometeors can reach physical dimensions of several cm (large snowflakes, hailstones).

At a given moment, the size of the hydrometeors presents a dispersion, or distribution, determined by the initiation and growth processes. For rain, the two important growth processes (coalescence and break-up) create a decrease with an approximately negative exponential shape for the increasing sizes above about 1 mm. They could also explain the existence of peaks at preferential sizes (0.7, 1.0, 1.9, 3.2 mm) observed by some investigators [Steiner and Waldvogel, 1987]; the reality of these peaks needs to be confirmed. This distribution, also called *size spectrum,* almost completely describes the "hydrological targets." It is represented by the function $N(D)\mathrm{d}D$, which represents the numerical concentration of the hydrometeors per volume unit and for a variation interval with a $D$ diameter, going from $D - 1/2\mathrm{d}D$ to $D + 1/2\mathrm{d}D - D$; in general, this is the diameter of the equivalent spherical particle. The dimension of $N(D)\mathrm{d}D$ is $L^{-4}$. The distribution is often designated simply by $N(D)$, which means that the value of $\mathrm{d}D$ taken is equal to the unit.

From one point to another inside a cloud or precipitation, the hydrometeor distributions are quite variable, almost random (cf. among others Section 5.2.2, on the effect of the vertical shear of horizontal wind on precipitation). We can also have spectra with several modes corresponding essentially to the hydrometeors under the two phases: water and ice. If we exclude measurements made with a vertical pointing doppler radar, which let us measure precipitation size distributions (under certain restrictive conditions, cf. Section 3.2), a distribution of hydrometeors of any shape cannot be determined by a single frequency radar, because the number of independent variables is equal to the number of lines of the distribution. Such a distribution is obtained by in situ observation. Figure 2.2 shows raindrop distribution measured from the ground with a disdrometer in a tropical squall line.

However, if we consider mean distributions (i.e., averaged in time or space on large but physically homogeneous samples), the variability decreases (cf. Figure 2.2(c)), and we can approximately represent the distributions by analytical functions. The number of variables thus is reduced to the number of distribution parameters. The best distribution contains the smallest number of parameters and at the same time preserves a satisfying representativity. These analytical distributions play a pivotal role in the hydrological applications of the radar in the measurement methodologies, as they are often aimed at determining its parameters, and also in the data processing because when the distribution is known, the physical quantities useful to characterize the target can be calculated.

The general shape of the hydrometeor distribution is given in Figures 2.3 and 2.4. For increasing dimensions, the concentration first increases very quickly until

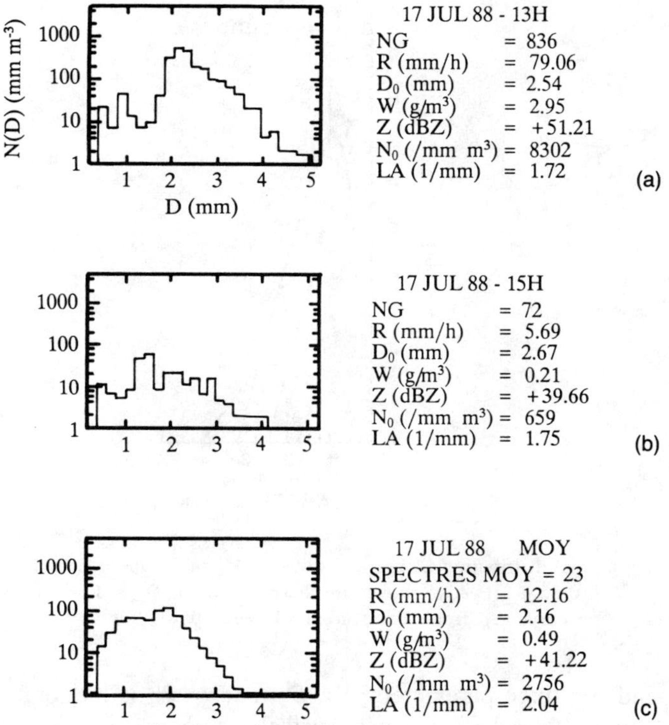


**Figure 2.2** Examples of drop-size distribution $N(D)$ observed at ground level in a tropical squall line at Niamey (Niger) on 17 July 1987 by using a disdrometer. Drop concentration $N(D)$ is in $mm^{-1}$ $m^{-3}$. For each distribution integral parameters are listed on the right. $NG$ = number of drops observed by the sensor, $R$ = rainfall rate, $D_0$ = mean volume diameter, $W$ = water content, $Z$ = radar reflectivity factor, $N_0$ and $LA$ (for lambda) are the parameters of the exponential distribution (eq. (2.15)). Panels (a) and (b) are instantaneous distributions (i.e., for sampling time 1 mn) for convective ($R$ = 79 mm $h^{-1}$) and stratiform ($R$ = 5.69 mm $h^{-1}$) parts of the squall line; (c) is the averaged distribution for 233 samples of 1 mn each.

it reaches the mode situated on the side of the small elements, then it decreases more slowly, with a slope that diminishes as it approaches large dimensions. Most often, the proportion of the large elements increases with the water content. Among the functions proposed to represent them, the most frequently used ones are the log normal distribution and modified gamma distribution, which in simplified form is the exponential distribution.

The general form of the log normal distribution is

$$N(D) = \frac{N_T}{\sqrt{2\pi}\,(\ln\sigma)D} \exp[-\ln^2(D/D_g)/(2\ln^2\sigma)] \tag{2.1}$$

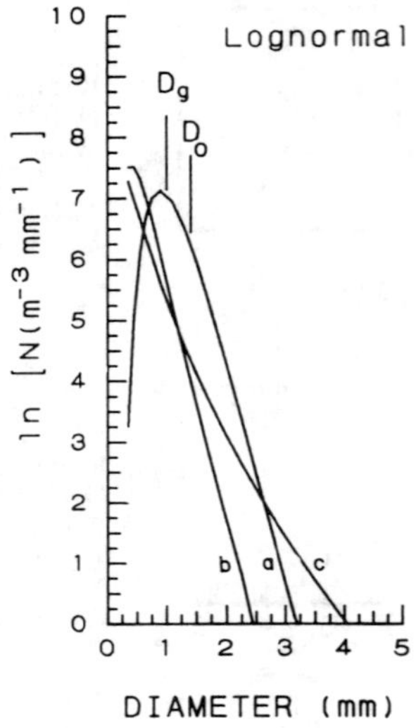


**Figure 2.3** Log normal distribution (2.1) for various values of $D_g$ and $\sigma$: (a) (1.0; 1.4); (b) (0.5; 1.6); and (c) (0.35; 2.2). Also shown are the positions of $D_G$ and $D_0$ (median volume diameter (2.8)) for curve (a) (from [Feingold and Levin, 1986]).

which depends on three parameters: $N_T$, $D_g$, and $\sigma$. $N_T$ is the total number of drops per volume unit; $D_g$ is the mean geometrical diameter, such as $\ln D_g = \overline{\ln D}$; and $\sigma$ is the standard deviation such as $\ln^2\sigma = \overline{(\ln D - \ln D_g)^2}$.

The modified gamma distribution can be written

$$N(D) = N_0 D^{\mu} \exp(-\Lambda D) \tag{2.2}$$

which depends on the three parameters: $N_0$, $\mu$, and $\Lambda$. Parameter $\mu$ can be either positive or negative. The exponential distribution can be obtained from (2.2) with $\mu = 0$ (cf. Section 2.2.5). Figures 2.3 and 2.4 show the shape of these distributions for various values of the parameters.

### 2.2.3 Integral Parameters

Knowing the size distribution of a hydrometeor population lets us compute, from the distribution moments, most of the terms useful to characterize the target.

The moment on the order $n$ of a distribution $N(D)$ is defined by

$$m_n = \int_0^{\infty} D^n N(D)\,\mathrm{d}D \tag{2.3}$$

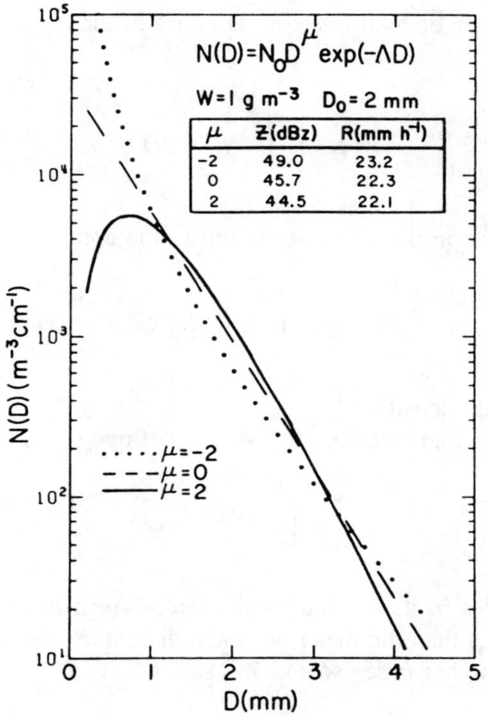


**Figure 2.4** Gamma distribution (2.2) for $\mu = -2$, 0, and 2 with liquid water content $W = 1$ g m$^{-3}$ and median volume diameter $D_0 = 2$ mm. The inset table shows the corresponding values of radar reflectivity factor $Z$ (Rayleigh approximation) and rainfall rate $R$ (from [Ulbrich, 1983]).

The integration limits of equation (2.3) are $D_{min}$ and $D_{max}$. The choice of $D_{min} = 0$ and $D_{max} = \infty$ simplifies the analytical integration of the exponential and introduces a negligible error [Sekhon and Srivastava, 1970].

By using distributions (2.1) and (2.2) in (2.3), we obtain

$$m_n = N_T D_g^n \exp[(n^2/2) \ln^2\sigma] \quad \text{(log normal)} \tag{2.4}$$

$$m_n = N_0 \frac{\Gamma(n + \mu + 1)}{\Lambda^{n+\mu+1}} \quad \text{(modified gamma)} \tag{2.5}$$

where $\Gamma$ is the complete gamma function with $\Gamma(n + 1) = n\Gamma(n)$ and in the particular case when $n$ is integer $\Gamma(n + 1) = n!$

The total Number of hydrometeors is equal to the moment of the zero order; that is,

$$N_T = \int_0^\infty N(D)\, dD = m_0 \tag{2.6}$$

Water content is proportional to the third moment:

$$M = \frac{\pi}{6}\rho \int_0^\infty D^3 N(D)\, dD = \frac{\pi}{6}\rho m_3 \tag{2.7}$$

where $\rho$ is the particle density.

Median volume diameter, $D_0$, for M, we define

$$\frac{\pi}{6}\rho \int_0^{D_0} D^3 N(D)\, dD = \frac{M}{2} \tag{2.8}$$

The interest of $D_0$ lies in its greater independence from the extremes of the distribution ($D_{min}$ and $D_{max}$) than the mean or modal diameters and its greater dependence on the moments of higher order $M$ and $Z$.

*Precipitation Rate*

It is convenient to characterize precipitation by intensity or rate $R$, defined as the amount of water crossing a horizontal cross section of unit area over a certain interval of time. Rate is expressed by a height (that of the equivalent volume of liquid water with unit cross section) divided by time, given by the expression

$$R = \frac{\pi}{6}\rho \int_{D_{min}}^{D_{max}} N(D) D^3 [v(D) - w]\, dD \tag{2.9}$$

where $v(D)$ is the terminal fall velocity in still air of the hydrometeors and $w$ is the vertical component of the air speed counted positively upward (for ground measurements, $w = 0$). This relation is general and applies to any type of hydrometeor on the condition that $D$ is equal to the diameter of the equivalent spheric and liquid particle. In this case, $\rho$ is the water density. $R$ is usually expressed in mm $h^{-1}$ of water.

If $w = 0$ and if $v(D)$ can be written in the form

$$v(D) = aD^b \tag{2.10}$$

where $a$ and $b$ are coefficients the value of which is discussed in Section 2.2.5 (we will see that $b$ ranges from 0.3 to 0.8), we have

$$R = \frac{\pi}{6} a \int_{D_{\min}}^{D_{\max}} D^{3+b} N(D) \, \mathrm{d}D = \frac{\pi}{6} a m_{3+b} \tag{2.11}$$

*Radar Reflectivity Factor*

Its use will be explained and justified in Section 2.5.1. This factor is equal to the momentum of the sixth order of the distribution; that is,

$$Z = \int_0^\infty D^6 N(D) \, \mathrm{d}D \tag{2.12}$$

*Other Terms*

Other terms can be calculated in the same way. Attenuation (cf. eq. (2.63)):

$$a = 0.4343 \int_0^\infty D^n N(D) \, \mathrm{d}D \tag{2.13}$$

Optical extinction has the following form [Van de Hulst, 1957]:

$$E = \frac{\pi}{2} \int_0^\infty D^2 N(D) \, \mathrm{d}D \tag{2.14}$$

### 2.2.4 Clouds

To analytically represent the size distribution of cloud particles (i.e., nonprecipitating), relations of the log normal or modified gamma type are used, (2.1) and (2.2). The values of the parameters to use in these relations depend on the type of air mass considered, the dynamic of the ambient air, and the temperature. For example, due to differences of size and concentration of the condensation nuclei, the cumuli formed with maritime air have typical concentrations of a few tens of relatively large droplets per $cm^3$, whereas the continental clouds have typical concentrations of a few hundred comparatively smaller droplets per $cm^3$. The growth of the cloudy hydrometeors depends on the humidity available for the condensation and deposition. For clouds under a single phase, the humidity depends on the vertical movement of the ambient air, the ascent increases the relative humidity, whereas the descent decreases it. In mixed (water and ice) clouds, the ice particles rapidly grow at the expense of the population of supercooled droplets.[2]

Whether in homogeneous or mixed phase, the development of a precipitation in a cloud creates a bimodal distribution because the precipitating hydrometeors, distribution in a larger size mode is superimposed on the initial distribution of a cloud in the very small hydrometeor range.

### 2.2.5 Precipitation

The *drop size distribution* (DSD) can be represented by log normal or modified gamma distributions. However, we saw that the precipitation has a very wide distribution, and that the rate, eq. (2.9), is proportional to $D^{\alpha}$, with $\alpha$ ranging from 3.3 to 3.8. This means that small hydrometeors, because their volume and terminal fall speed are weak, contribute little to precipitation compared to large hydrometeors, which represent the main part. We will also see in Section 2.3.1 that the backscattering cross sections of hydrometeors are proportional to $D^6$. Therefore the contribution to the echo of the small hydrometeors is negligible compared to that of the big ones. For all these reasons, for application relative to the precipitations we largely use a simplified form of the general distribution, which is obtained when we take the term $D^{\mu} = 1$; that is, $\mu = 0$ in eq. (2.2). The resulting distribution is exponential.

The exponential distribution was proposed by Marshall and Palmer (MP) as early as 1948, long before eqs. (2.1) and (2.2):

$$N(D) = N_0 \, e^{-\Lambda D} \tag{2.15}$$

with

$$\Lambda = aR^b \tag{2.16}$$

$N_0$, $a$, and $b$ are parameters depending on the considered precipitation.

Exponential distribution is interesting because it introduces the quantity $R$, which is very useful in hydrology and measured the most easily and the most frequently at ground level. However, when, for the considered application, it is not legitimate to neglect the small drops, we must come back to expressions like (2.1) or (2.2), for low intensity rains (such as drizzle).

The moments on the $n$ order of (2.15) are given by formula (2.5) with $\mu = 0$. The expression of the integral parameters is deduced from it.

[2]The saturation vapor pressure of ice is lower than that of water at the same temperature. Therefore, the two phases cannot coexist; crystals grow by deposition, whereas droplets evaporate.

*Rain*

For the rain, which is the case treated most frequently, the use of peculiar values of the parameters was proposed [Marshall and Palmer, 1948]:

$$N_0 = 0.08$$
$$\Lambda = 41R^{-0.21} \tag{2.17}$$

where $N_0$ is in $cm^{-4}$, $\Lambda$ in $cm^{-1}$, and $R$ in mm $h^{-1}$.

Relations (2.15) and (2.17) define the MP distribution, considerd a good representation of the mean distributions of stratiform rain at medium latitudes. Distributions of this type are used widely because they depend only on parameter $R$, which greatly simplifies the application. Figure 2.5 gives the aspect of $N(D)$ for various values of $R$. In the figure, the distribution of Laws and Parson [1943] (LP) are shown, too, because they are often mentioned in the literature.

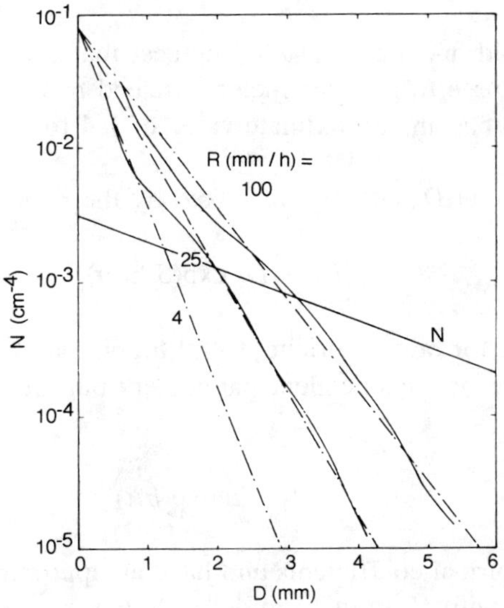


**Figure 2.5** Distribution of raindrops proposed by Laws and Parson (continuous line) and Marshall and Palmer (dot and dash line) as a function of the diameter for different values of $R$. For snow at 10 mm $h^{-1}$, the straight line $N$ represents the distribution of the diameters of the equivalent melting particles given by (2.18).

Somewhat different coefficients have been obtained by other authors. For example, based on spectral observations by a radar with vertical incidence in precipitation with a stratiform character, Sekhon and Srivastava [1971] found

$$N_0 = 0.07R^{0.37}$$
$$\Lambda = 38R^{-0.14} \tag{2.18}$$

with the same units as (2.17).

For rain of a peculiar nature (for example, orographic or coastal), of course, we must adjust specific coefficients (or refer to the literature). In the problems of diffusion by the rain, we sometimes use the median volume diameter $D_0$ rather than $\Lambda$.

For the (2.15) distribution, it has been shown [Atlas 1964; Sekhon and Srivastava, 1970] that, for $D_{max}/D_0 \geqslant 2.5$,

$$\Lambda = 3.67/D_0 \tag{2.19}$$

and (2.15) can be rewritten as follows:

$$N(D) = N_0\, e^{-3.67D/D_0} \tag{2.20}$$

For rain, Feingold and Levin [1986] suggest that the log normal distribution can be reduced from three to two parameters. Indeed $\sigma$ on average, is independent of the rainfall rate and has an approximate value of 1.4 for a large spread of rainfall rate (5–150 mm $h^{-1}$).

They also show that $D_0$ and $D_g$ are linked by the relation

$$D_0 = D_g \exp(3 \ln^2\sigma) \tag{2.21}$$

In the same way, for rain, Ulbricht [1983] found that $N_0$ and $\mu$ of the modified gamma distribution are not independent parameters but are linked by a relation of the form

$$N_0 = a \exp(b\mu) \tag{2.22}$$

where $a$ and $b$ are empirical coefficients and have an approximate value of $6 \times 10^{-4}$ and 3.2, respectively (with $N_0$ in $m^{-3}\ cm^{-1-\mu}$). A two-parameter distribution is thus obtained. He also found the $\Lambda$ of equation (2.2) are $D^0$ are linked by the approximate relation

$$\Lambda D_0 = 3.67 + \mu \tag{2.23}$$

*Snow*

Snow has very different dynamic properties from rain. The fall speed spectrum is narrow, around 1 m $s^{-1}$. For the size distribution, an exponential distribution is usually used, with [Gunn and Marshall, 1958]

$$\begin{aligned} N_0 &= 0.038R^{-0.87} \\ \Lambda &= 25.5R^{-0.48} \end{aligned} \tag{2.24}$$

or [Sekhon and Srivastava, 1970]

$$\begin{aligned} N_0 &= 0.025R^{-0.94} \\ \Lambda &= 22.9R^{-0.45} \end{aligned} \tag{2.25}$$

with the same units as (2.17).

*Hail*

Due to the great natural variability in the size of hail, it is difficult to provide an expression with universal application. Chen and English [1983] propose a relation of the (2.15) form with

$$N_0 = 115\Lambda^{3.63} \tag{2.26a}$$

where $N(D)$ and $N_0$ are in $m^{-3}$ $mm^{-1}$, $\Lambda$ in $mm^{-1}$, and $D$ in mm with [Torlaschi, Humphries, and Barge, 1984]

$$\Lambda = \left(\ln \frac{88}{R_h}\right) \Big/ 3.45 \tag{2.26b}$$

where $R_h$ is the hail-fall rate in mm $h^{-1}$ of liquid water.

## 2.2.6 Terminal Fall Velocity of Hydrometeors

To calculate the intensity of the precipitation from (2.9), we must know the terminal fall velocity as a function of size. Air is a viscous fluid in which the falling movement of cloud and precipitation particles under the action of gravity reaches a terminal velocity. This velocity $v$ corresponds to the equilibrium between the inertial forces and the dragging forces acting on the particles; therefore, it is influenced by the properties of the air at the fall level and by the density, size, shape, and rigidity of the particle. There is an abundant literature on these questions, which are important in the physics of clouds and precipitation and in hydrological measurement by radar. The problem is in fact complex, and the explicit formulation of the solutions suggested is beyond the scope of this book [see Pruppacher and Klett, 1978].

The inertial forces emanate from the difference between the weight of the particle and the weight of the displaced air (buoyancy force). For a spherical element this force is equal to $(4/3)\ \pi a^3 g(\rho_w - \rho_a)$, where $a$ is the radius of the particle, $g$ is the acceleration due to gravity, and $\rho_w$ and $\rho_a$ are the densities of the element and air, respectively. For a velocity of fall with respect to the air, $v$, the resistant force on the sphere is equal to $6\pi\eta avN$, where $\eta$ is the dynamic viscosity of air. $N$ is the dimensionless quantity that is a function of two terms, the drag coefficient $C_d$ and the Reynolds number:[3]

$$N = C_d R_e / 24$$

For droplets with radius less than approximately 40 $\mu$m, N is close to 1 so that in this range of sizes, we can write that at equilibrium:

$$\frac{4}{3}\pi a^3 g(\rho_w - \rho_a) = 6\pi\eta a v$$

that is,

$$v = v_s = \frac{2a^2 g(\rho_w - \rho_a)}{9\eta}$$

where $v_s$ is the Stokes velocity, and its range of validity is the "Stokes region" in which the terminal velocity of fall is proportional to the square of the radius of the sphere. Complications arise when we attempt to describe the drag forces precisely for various types of nonspherical particles and for all the dimensions that exist in nature.

For water drops, three regimes of drag, with limits not defined with precision, have to be considered [Pruppacher and Klett, 1978] (Figures 2.6 and 2.7). At very low values of the Reynolds number ($10^{-8}$–$10^{-2}$) and for all small droplets (0.5

[3]The Reynolds number $R_e$ expresses the dimensionless ratio between the inertial forces and the viscosity forces. For the air flow around the sphere, we write

$$R_e = \frac{uD}{\nu}$$

where $D$ is the diameter of the sphere, $u$ the velocity, and $\nu$ the kinematic viscosity of air.

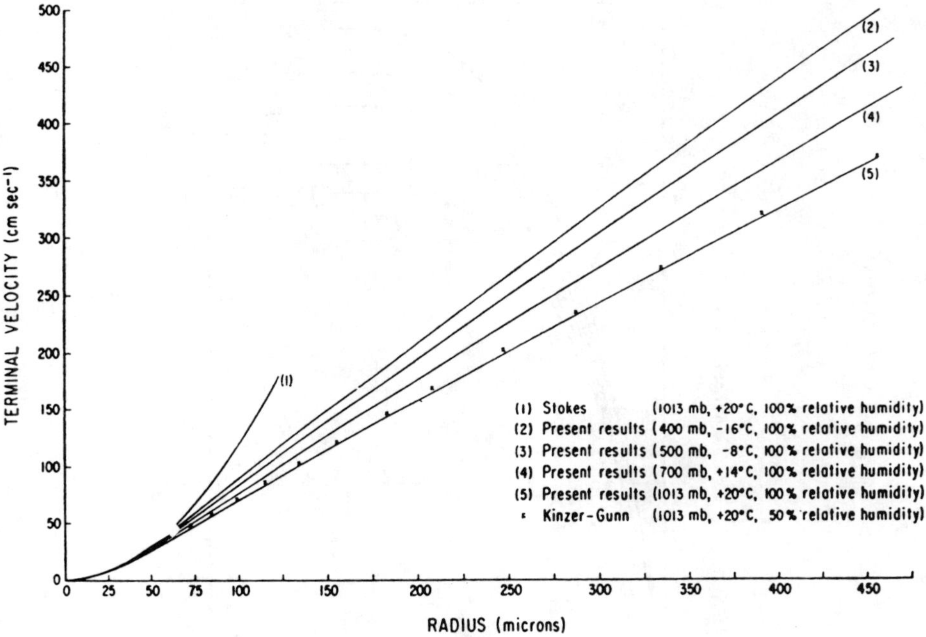


**Figure 2.6** Terminal velocity of water drops of various sizes under various environmental conditions. Curve 1 corresponds to Stokes's law. The points represent the values obtained experimentally by Gunn and Kinzer [1949] (from [Beard and Pruppacher, 1969]).

$\mu m \lesssim a \lesssim 10\ \mu m$), the description of the Stokes drag in which air is considered as a continuum is no longer satisfactory, and we must obtain $v$ to correct $v_s$ by a term proportional to the free molecular path and the reciprocal of the radius. The correction is approximately 1% for $a = 10\ \mu m$ and 17% for $a = 1\ \mu m$ (under normal pressure and at 20°C). The correction increases with altitude.

For intermediate values of $R_e$ (between $10^{-2}$ and $3 \times 10^2$) and average sizes ($10\ \mu m \lesssim a \lesssim 535\ \mu m$), the drops fall as rigid spheres in a medium in which the inertial forces and the viscosity are not negligible. In Figure 2.6, we can see that $v$ increases almost linearly with the radius between 100 and 500 $\mu m$. We can also see that, as the radius increases, the Stokes velocity $v_s$ overestimates $v$ more and more. For $a = 30\ \mu m$, the deviation is only 6%, whereas for $a = 42\ \mu m$, it is 13%.

For large sizes ($535\ \mu m \lesssim a \lesssim 3.5$ mm) and high values of $R_e$ (between $3 \times 10^2$ and $4 \times 10^3$), the problem of shape arises; the drops no longer can be considered spherical, and the calculation of the drag force has to take into account their deformation. In this region, $v$ is approximately proportional to the square root of the radius. For large drops, the velocity increases appreciably with the height from approximately 9 m s$^{-1}$ at sea level up to more than 12 m s$^{-1}$ at the level of 500 hPa

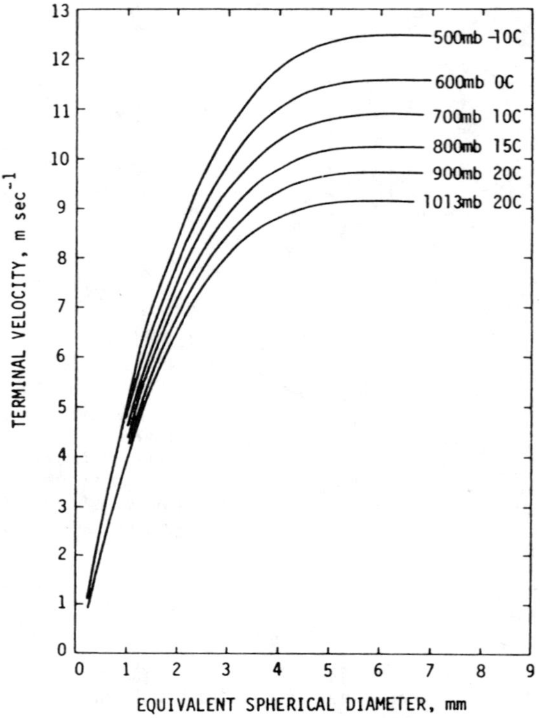


**Figure 2.7** Terminal velocity of raindrops at six pressure levels as a function of equivalent spherical diameter (from [Beard, 1976]).

pressure. We can see from Figure 2.8 that $v$ becomes independent of the size for $a \gtrsim 5$ mm. This is because the larger the drop, the more it is flattened and the greater is the master couple presented to the flow; the resulting increase in drag force compensates for the increase in the force due to gravity.

Exact but relatively complex analytical expressions have been established semi-empirically for various regimes [Foote and du Toit, 1969; Beard and Pruppacher, 1969; Wobus, Murray, and Keonig, 1971; Beard, 1976]. For example, Atlas *et al.* [1973] propose

$$v = 965 - 1030 \exp(-6D) \tag{2.27}$$

where $v$ is in cm s$^{-1}$ and $D$ is in cm.

Approximate expressions of the form

$$v = aD^b \tag{2.28}$$

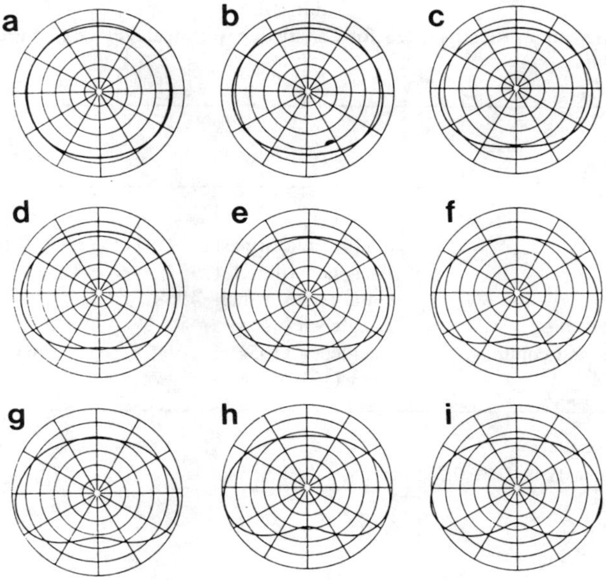


**Figure 2.8** Variation in the shape of water drops in air by size: (a) $a_0$ = 0.11 cm, (b) $a_0$ = 0.14 cm, (c) $a_0$ = 0.18 cm, (d) $a_0$ = 0.20 cm, (e) $a_0$ = 0.25 cm, (f) $a_0$ = 0.29 cm, (g) $a_0$ = 0.30 cm, (h) $a_0$ = 0.35 cm, (i) $a_0$ = 0.40 cm (from [Pruppacher and Pitter, 1971]).

where $a$ and $b$ are constants and $D$ is the diameter of the water sphere (or the equivalent diameter of the melted particle) often are used to facilitate the analytical treatment of particular problems. The main values used are given in Table 2.1. For ice crystals, the terminal velocity depends on the shape and the size of the crystal. Relationships of the same form are also established between $v$ and other integral parameters such as $Z$ (see Section 3.2).

The values given in Table 2.1 apply to around sea level. To use them at higher levels, the right-hand side of equation (2.28) must be multiplied by the correction factor $(\rho_0/\rho)^{0.4}$, where $\rho$ and $\rho_0$ are the densities of air at the level of the observation site and at sea level, respectively [Foote and du Toit, 1969]. The correction is about +10% at 2.5 km and +20% at 4.5 km for standard atmosphere.

### 2.2.7 The Shape of Hydrometeors

In clouds and precipitation, only the small water drops are spherical. Other species of hydrometeors can be spherical or move away from sphericity, sometimes very importantly. For a nonspheric individual hydrometeor, or population of hydrometeors, the three parameters influencing the propagation and polarization of the scattered wave are

**Table 2.1**
Value of the Coefficients in Relation 2.28 for Calculating Terminal Velocity of Precipitation Particles

| | *Source* | *a* $(cm^{1-b}\ s^{-1})$ | *b* |
|---|---|---|---|
| Rain[1] | Sekhon and Srivastava (1971) | 1690 | 0.6 |
| Rain[1] | Atlas and Ulbrich (1977) | 1767 | 0.67 |
| Snowflakes[2] | Gunn and Marshall (1958) | 200 | 0.3 |
| Ice crystals[2] (dendrites) | Langleben (1954) | 160 | 0.3 |
| Ice crystals[2] (plates and columns) | Langleben (1954) | 234 | 0.3 |
| Hail[3] | Auer (1974) | 9 | 0.8 |

*Notes:*
1. At 1013 hPa for $0.05 \lesssim D \lesssim 0.5$ cm.
2. Melted diameters.
3. At 800 hPa and 0°C.

- the shape of the hydrometeor, usually measured by a size ratio, the axial ratio;
- the orientation in relation with the incident electrical field;
- the dielectric constant of the material.

*Water*

The shape of the raindrops is determined by the superficial tension forces (which, if there was no other term, would act isotropically and maintain the sphericity) and aerodynamical constraints due to the fall, in which the main component is vertical. With the mass increase, the fall speed increases and, the balance of forces being modified, the drop flattens. Figure 2.8 shows the evolution of a drop shape in the vertical symmetry plane, according to the size of the equivalent sphere.

In perfectly still air, the shapes described in Figure 2.8 would be stable and the symmetry axes would remain vertical. In nature, things are different. First, the symmetry axes can move away from the vertical because of a dissymmetry in the forces applied and cant in various directions. The canting angles are variable and present a distribution whose average is almost vertical. Then, the shape of the drops can oscillate around the stable medium state. The amplitude of the periodic oscillations of the axial ratio increase but their frequency decreases for an increasing drop size.

The shape of the raindrops can be approximated by an oblate ellipsoid characterized by two axes: the symmetry axis, or minor axis, vertical, $2a$ long; and the perpendiculary axis, $2b$ long. The distortion remains weak until 1 mm diameter. For

$0.1 \text{ cm} \leq D_e \leq 0.6$ cm, the axes ratio is approximately given by the formula [Pruppacher and Pitter, 1971]:

$$\frac{a}{b} = 1.03 - 0.62D_e \tag{2.29}$$

where $D_e$ is the equivalent spherical diameter in cm. However, it seems that the small drops are slightly less oblate and the large drops slightly more oblate than expected by Pruppacher and Pitter [Beard and Chang, 1987].

For the frequency of the periodic oscillations around the average shape, Nelson and Gokhale [1972] proposed the empirical relation

$$f = 11.7D^{-1.47} \tag{2.30}$$

with $f$ in $s^{-1}$ and $D$ as the spherical equivalent diameter in cm.

The case of the raindrops thus is relatively simple because the orientation of the deformation is common (horizontal or quasi-horizontal major axis), the amplitude of the deformation depends on the size. Also the dielectric constant is known and homogeneous for the whole population.

*Ice*

Ice particles present a more complex case. The basic shape of the ice crystal depends on the temperature and relative humidity of the air in which it develops. In the (0, −3°C) and (−8, −25°C) approximate intervals, plates or dendritic crystals form if the humidity is high (growth in the plane defined by the three symmetry axes of the hexagonal structure or $a$ axes). In the (−3, −8°C) and (< −25°C) intervals, columns or needles form if the humidity is high (growth along the perpendicular to the hexagone plane or $c$ axis). The orientation of fall is stable for plates or dendrites with $a$ axes horizontal and, for the columns, the $c$ axis horizontal.

However, in nature, the homogeneous populations of plates or columns are exceptional. Most often, the population is composed of complexly shaped crystals resulting from a growth at various temperatures and humidities and from the aggregation of a simpler crystal, so that, in the general case, the shapes, orientations, and densities (thus, the dielectric factor) present a high variability (it is said that in nature no two crystals are identical). It is the same for snowflakes the density of which can be smaller than one-tenth that of ice. Such particles with complex and dissymmetrical shapes can fall either in whirls (nonstable conditions) or under any stable incidence, depending on shape and the distribution of the density in the particle. When they fall in stable conditions, they can oscillate around their mean orientation.

Melting crystals or snowflakes tend to adopt the shape of oblate ellipsoids with a vertical symmetry axis. Indeed the melted fraction spreads in the horizontal plane around the solid fraction. The eccentricities can be important.

For hail, which results from a growth by accretion (captation of supercooled water droplets) from a frozen drop or a dense agglomerate of crystals, forming an embryo, where the spherical shape is prevailing (hailstones whirl while falling). The hailstone presents concentric layers of variable density according to the type of growth, which can be dry (freezing of droplets on contact with the hailstone, which gives an opaque ice with air inclusions) or humid (the latent heat of freezing is not evacuated quickly enough because the accretion rate is too high, the temperature at the surface of the hailstone becomes positive, and a water layer forms that progressively freezes in transparent and dense ice). In this last case, the dielectric factor that must be considered depends on the depth of the water layer. The same occurs with a hailstone during the melting process.

## 2.3 SCATTERING AND ATTENUATION CROSS SECTIONS

### 2.3.1 Homogeneous Spherical Particles

*Mie Formulas*

The theory of the scattering of electromagnetic waves by particles has been the subject of a large number of studies since the end of the last century, when the optical properties of colloidal suspensions were discovered. A general solution has been given by Mie [1908] from previous results [historical survey, Kerker, 1969] and then reformulated by several authors, in particular Stratton [1941], Van de Hulst [1957], and Deirmendjian [1969]. It is based on the application of Maxwell's equations to the scattering of a planar wave by a homogeneous sphere in a nonabsorbing medium.

The basic parameters used for formulating the problem are

- The dimensionless product of the radius $a$ of the sphere and the propagation constant in vacuum of the incident wave $2\pi/\lambda$, namely $\alpha = 2\pi a/\lambda$ ($\alpha$ is sometimes called the *radioelectric size* of the sphere);
- The complex refractive index $m$ with respect to vacuum of the material of the sphere with $m = n - ik$, where $n$ is the ordinary refractive index ($n = \varepsilon_r^{1/2}$, where $\varepsilon_r$ is the permittivity dielectric, also called the *constant dielectric*) and $k$ the absorption coefficient of the material of the sphere;
- A scattering angle $\theta$ formed by the vectors of the incident and scattered waves.

When an electromagnetic wave meets a stationary and inert body (i.e., not charged, not excited, and with its own emission negligible) with dielectric properties different from those of the surrounding medium, an interaction develops so that some of the energy of the incident field is absorbed by the body and appears in the form of heat whereas another part is scattered in all directions in the form of an electromagnetic field with the same wavelength as the exciting energy. The scattering results from the combination of the oscillating electric and magnetic multipoles induced by the electromagnetic wave in the body.

Under these conditions, the scattered electric field at a great distance from the sphere can be described by a vector **A** whose two scalar components $A_1$ and $A_2$ are perpendicular and parallel, respectively, to the scattering plane defined by the vectors of the incident and scattered waves. The Mie solution gives the complex amplitude of these components in the form of convergent series that can be written

$$A_1 = \frac{\lambda}{2\pi} \sum_{n=1}^{\infty} \frac{2n+1}{n(n+1)} (a_n \pi_n + b_n \tau_n)$$
$$A_2 = \frac{\lambda}{2\pi} \sum_{n=1}^{\infty} \frac{2n+1}{n(n+1)} (b_n \pi_n + a_n \tau_n) \quad (2.31)$$

where $n$ is a positive integer, $a_n$ and $b_n$ are the "Mie coefficients" obtained from Bessel and Hankel functions with arguments $\alpha$ and $m$. These coefficients correspond to scattering by the induced magnetic and electric modes of order $n$, respectively. The quantities $\pi_n$ and $\tau_n$ only depend on $\theta$ and can be defined with the help of the Legendre polynomials. The amplitudes $A_1$ and $A_2$ make it possible to express the differential scattering cross section $d\sigma_d$ as a function of $m$, $\alpha$, and $\theta$. The term $\sigma_d$ is defined by the ratio of the incident flux to the scattered flux. We can show that for unit solid angle and a unit incident flux

$$d\sigma_d (\theta, \alpha, m) = \frac{1}{2} \mathbf{A}\mathbf{A}^* \, d\Omega \quad (2.32)$$

where $\mathbf{A}^*$ is the conjugate vector to **A** and $d\Omega$ is the solid angle element.

From this expression, we can define the following specific cross sections. The total scattering cross section, $Q_s$, is the surface intercepting the amount $P_s$ of incident power effectively scattered by the sphere. It is obtained by integrating (2.32) for the solid angle $\Omega = 4\pi$. If $S_i$ is the incident power density,

$$P_s = Q_s S_i$$

The radar backscattering cross section (already defined in Section 1.3), $\sigma$, is the equivalent surface of an isotropic scatterer, that is, scattering in all directions with an intensity equal to that scattered to the radar. It is obtained by multiplying the value of $\sigma_d$ taken from 2.32 for $\theta = \pi$, by $4\pi$ (in the particular case of isotropic scattering, $\sigma = Q_s$). The term $\sigma$ is specific to radar backscattering and different from cross sections usually used in physics because it is applied to a privileged scattering direction.

The total attenuation cross section, $Q_t$, is obtained from (2.32) when $\theta = 0$ and defined as the surface intercepting an amount $P_a$ of power equal to that taken in total by the sphere from the incident wave by absorption and scattering:

$$P_a = Q_t S_i \tag{2.33}$$

The absorption cross section $Q_a$, is the surface intercepting an amount of incident power equal to that absorbed in the form of heat by the sphere. It is equal to the difference $Q_t - Q_s$.

The Mie formulas are

$$Q_s = \frac{\lambda^2}{2\pi} \sum_{n=1}^{\infty} (2n + 1)(|a_n|^2 + |b_n|^2) \tag{2.34}$$

$$\sigma = \frac{\lambda^2}{4\pi} \left| \sum_{n=1}^{\infty} (-1)^n (2n + 1)(a_n - b_n) \right|^2 \tag{2.35}$$

$$Q_t = \frac{\lambda^2}{2\pi} (-\mathrm{Re}) \left[ \sum_{n=1}^{\infty} (2n + 1)(a_n + b_n) \right] \tag{2.36}$$

where $n$ is a positive integer.

Deirmendjian [1969] adapted the Mie expressions to machine computation. The proposed computational form involves no approximation and is simpler to use for machine processing than the expressions involving Bessel function of the complex argument. Using the recurring function

$$W_n = \left( \frac{2n - 1}{\alpha} \right) W_{n-1} - W_{n-2} \tag{2.37}$$

with

$$\begin{aligned} W_0 &= \sin\alpha + i \cos\alpha \\ W_{-1} &= \cos\alpha - i \sin\alpha \end{aligned} \tag{2.38}$$

the Mie coefficients are written in the form

$$a_n = \frac{\left(\frac{A_n}{m} + \frac{n}{\alpha}\right) R_e(W_n) - R_e(W_{n-1})}{\left(\frac{A_n}{m} + \frac{n}{\alpha}\right) W_n - W_{n-1}} \qquad (2.39)$$

$$b_n = \frac{\left(mA_n + \frac{n}{\alpha}\right) R_e(W_n) - R_e(W_{n-1})}{\left(mA_n + \frac{n}{\alpha}\right) W_n - W_{n-1}} \qquad (2.40)$$

where

$$A_n = -\frac{n}{y} + \left(\frac{n}{y} - A_{n-1}\right)^{-1} \qquad (2.41)$$

$$A_0 = \cot y \qquad (2.42)$$

$$y = m\alpha \qquad (2.43)$$

For $y = p - iq$, where $p = n\alpha$ and $q = k\alpha$, we have

$$A_0 = \frac{\sin p \cos p + i \sinh q \cosh q}{\sin^2 p + \sinh^2 q} \qquad (2.44)$$

*Rayleigh Approximation*

In general, the radioelectric size of the spheres considered $\alpha = \pi D/\lambda$ is small because the meteorological scatterers are small compared to the wavelengths used. Under these conditions, the multipolar moments can be neglected and the effective reflectivity of a spherical particle is obtained approximately by replacing the sphere with an equivalent dipole as long as the diameter of the sphere. The time average of the intensity of radiation scattered by an oscillating dipole with moment $\mu$ in the direction of the source is

$$S_r = \frac{2\pi^3 \mu^2 c}{\lambda^4 r^2} \qquad (2.45)$$

where $c$ is the velocity of propagation in vacuum, and $r$ is the distance of the radar from the target. The dipole moment of a particle of diameter $D$ and dielectric constant $\varepsilon$ is

$$\mu = \frac{\varepsilon - 1}{\varepsilon + 2} \frac{D^3}{8} E \tag{2.46}$$

where $E$ is the electric field acting on the dipole.

Substituting (2.46) in (2.45),

$$S_r = \frac{\pi^3}{32} \left( \frac{\varepsilon - 1}{\varepsilon + 2} \right)^2 \frac{cE^2 D^6}{\lambda^4 r^2} = \frac{\pi^4}{4} \left( \frac{\varepsilon - 1}{\varepsilon + 2} \right)^2 \frac{S_i D^6}{\lambda^4 r^2} \tag{2.47}$$

where $S_i = cE^2/8\pi$ is the intensity of the incident radiation. The backscattering cross section $\sigma$ is obtained by writing (cf. definition and Section 1.3) that $\sigma = 4\pi r^2 S_r/S_i$.

The same result is obtained by starting from the "exact" Mie expression, assuming that for $\alpha \ll \lambda$, the terms of an order higher than 5 in the expansion of $a_n$ and $b_n$ can be neglected as a power series in $\alpha$. Substituting the coefficients by their expansion in eqs. (2.34) to (2.36) then leads to the following simple expressions:

$$Q_s = \frac{2}{3} \frac{\pi^5}{\lambda^4} |K|^2 D^6 \tag{2.48}$$

$$\sigma = \frac{\pi^5}{\lambda^4} |K|^2 D^6 \tag{2.49}$$

$$Q_a = \frac{\pi^2}{\lambda} D^3 \, \mathrm{Im}(-K) \tag{2.50}$$

$$Q_t = Q_a + Q_s \tag{2.51}$$

where $K = m^2 - 1/m^2 + 2$; Im means that the imaginary part is taken.

Expression (2.49) is the Rayleigh law, according to which the scattered power is proportional to the reciprocal of the fourth power of the wavelength and to the sixth power of the diameter. It is analogous to the expression giving the effective scattering cross section of atmospheric molecules at optical wavelengths, which Rayleigh used to explain the color of the sky by increasing scattering for decreasing wavelengths.

In the Rayleigh region, the energy scattered by the particle is radiated almost isotropically. In the Mie region, the diffusion diagram is more complex and the maximum radiation direction is not backward.

Relations (2.48) to (2.51) are very important. They are at the basis of any discussion on the use of microwaves to observe clouds and precipitation.

*Influence of the Dielectric Properties*

Scattering and attenuation by homogeneous spheres are strongly influenced by $m$. This dependence is expressed in equations for spheres of small dimensions by the terms $|K|^2$ and $\mathrm{Im}(-K)$. The scattering and absorption cross sections are proportional to these terms, respectively. For a given substance, $m$ is a function of the wavelength and the temperature of the sphere. The values of $n$, $K$, $|K|^2$, and $\mathrm{Im}(-K)$ calculated by Gunn and East [1954] are given in Appendix 4.

For a dielectric without loss, $Q_a = 0$ and $Q_t = Q_s$. For a dielectric with high loss such as water and for spheres of small size, $Q_s \ll Q_a$ and so $Q_t \approx Q_a$. In the intermediate case of ice, $Q_a$ and $Q_s$ are comparable, and the two terms must be taken into account to obtain $Q_t$.

For water, $|K|^2$ is practically constant at the temperatures found in the atmosphere and wavelengths in the centimetric range, currently used in radar; it is approximately 0.93 for liquid water. For ice, $n$ and $k$ are independent of wavelength and $k$ is small. For ice with normal density ($\rho = 0.92$ gcm$^{-3}$), $|K|^2 \approx 0.176$ regardless of the temperature. For $\rho = 1$ gm$^{-3}$ we have $|K|^2 \approx 0.208$ (cf. Section 2.3.2).

Therefore, due to the value difference of $K^2$, the backscattering cross section $\sigma$ for spherical and homogeneous hydrometeors of the same diameter is about 5.3 times weaker for ice than for water; that is, 7.3 dB. The diameter of the ice hydrometeors has to be superior to that of the water hydrometeors by only 32% to compensate for this difference.

*Cross-Section Tables*

Computers allow easy calculation of these various cross sections. A large number of results have been published on their values as a function of $\alpha$ and $m$ [Gunn and East, 1954; Herman and Battan, 1961a and 1961b; Herman, Browning and Battan, 1961; Stephens, 1961; Battan, Browning and Herman, 1970a and 1970b; Battan, 1971].

*Backscattering*

Figure 2.9(a) shows the variation of the normalized[4] backscattering cross section $\sigma/\pi a^2$ ($a$ is the radius of the sphere) as a function of $\alpha$ for spheres of water and ice, calculated from (2.35). The curve for ice applies to all wavelengths. The curve for water is calculated for a temperature of 0°C and a wavelength of 3.2 cm. For low values of $\alpha$, within the range of validity of Rayleigh's approximation ($\alpha \leq 0.3$), the

[4]Usually, to facilitate graphic presentation of the results, the cross section is "normalized" by dividing $\sigma$ by the geometrical cross section (or surface of the large circle) of the scattering sphere.

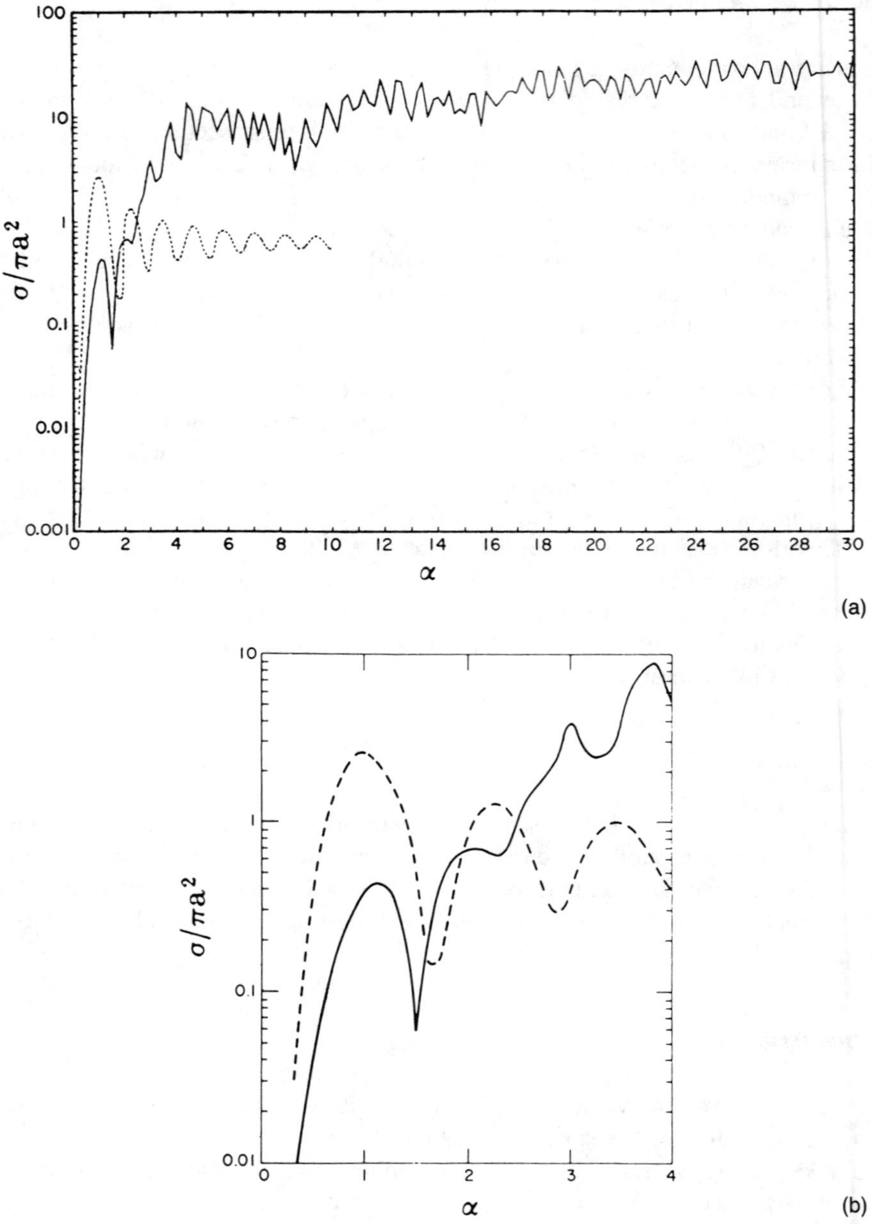


**Figure 2.9** (a) Normalized backscatter cross section $\sigma/\pi a^2$ as a function of $\alpha = \pi D/\lambda$ for water at $\lambda = 3.2$ cm and 0°C (dashed line) and for ice at $\lambda = 1$ to 10 cm and 0°C; (b) is an expansion of (a).

normalized backscattering cross section of the water spheres is appreciably greater than that of ice; the cross section of the water spheres exhibits oscillations around a mean value with decreasing amplitude for increasing $\alpha$; this is the resonance region. Then, the curve stabilizes as we enter the optical region ($D \gtrsim 3\lambda$): the cross sections depend only on the geometrical cross sections and the dielectric properties of the material. Below $D \approx \lambda$, the curve for ice reaches values higher than those of water, then oscillates around a mean value that increases slowly with $\alpha$ and reaches its limit only for a large $\alpha$. The anomalies of $\sigma$ for ice and not totally understood.

*Attenuation*

Figure 2.10 shows the normalized attenuation cross sections $Q_t/\pi a^2$ for water and ice. For water, the cross section increases rapidly until $\alpha \approx 1$ and then varies little. The wavelength has little effect on the curve. For ice, the attenuation has a maximum between $\alpha = 1.5$ and $\alpha = 4$, for which the values are almost double those of water. For $\alpha > 4$, the values are close to those of water.

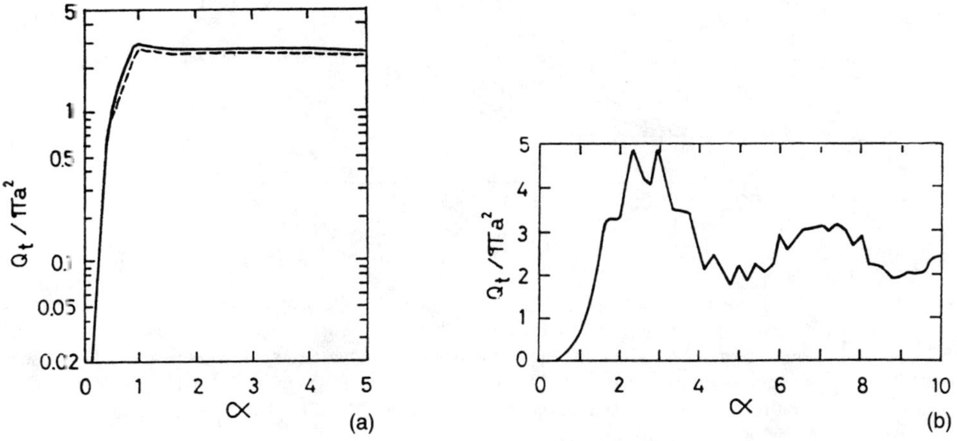


**Figure 2.10** Normalized attenuation cross sections of (a) water sphere at 0°C and $\lambda$ = 3.2 cm (solid line) and 5.5 cm (dashed line) (from [Herman *et al.* 1961]); and (b) of ice spheres (from [Herman and Battan, 1961]).

*Validity Range for the Rayleigh Approximation*

To determine the validity range for the Rayleigh approximation, the variations of the ratios $\sigma_{\text{Mie}}/\sigma_{\text{Rayleigh}}$ and $Q_{t\text{Mie}}/Q_{t\text{Rayleigh}}$ are examined as a function of $\alpha$. Figure 2.11a shows the values for backscattering by water and ice at wavelengths between 1 and 10 cm: the deviations increase with the wavelength and the Rayleigh region

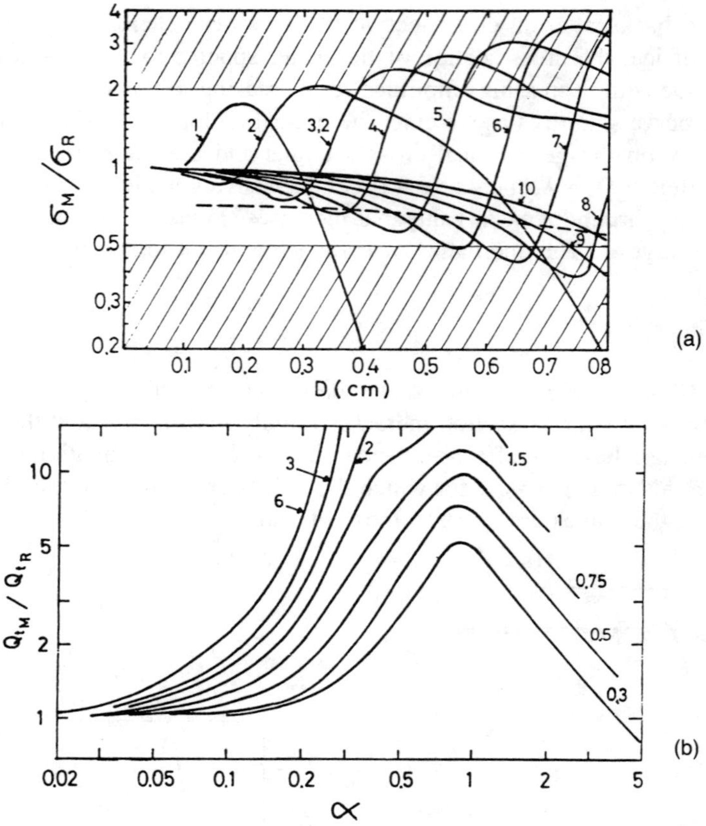


**Figure 2.11** (a) Variation of the ratio $\sigma_{\text{Mie}}/\sigma_{\text{Rayleigh}}$ as a function of the drop diameter. The figures indicate the wavelengths in cm. The part not hatch marked corresponds to values of the ratio between 0.5 and 2 to which the approximation range can be extended (see text). The dashed curve corresponds to ice at 0°C for $\alpha$ = 3.2 cm (from [Stephens, 1961]). (b) Variation of the ratio $Q_{t\text{Mie}}/Q_{t\text{Rayleigh}}$ as a function of $\alpha$ for water at 18°C. The figures indicate the wavelengths in cm (from [Gunn and East, 1954]).

incorporates particles of ever-increasing size. The curves in 2.11(b) relate to attenuation cross sections and merit the same comments.

Gunn and East [1954] suggest defining an approximation range for backscattering as that for which the ratio $\sigma_M/\sigma_R$ lies between 1/2 and 2. For $\lambda < 2$ cm, this region extends up to $\alpha \approx 1$, namely $D < 3.5$ mm for $\lambda = 1$ cm and $D < 6.5$ mm for $\lambda = 2$ cm. For $\lambda > 3$ cm, the Rayleigh region corresponds to $\alpha \leq 0.3$.

The Rayleigh region for the attenuation is less wide than that for retrodiffusion. For example, for wavelengths of 0.86 cm and 10 cm, the ratio $Q_{tM}/Q_{tR}$ reaches 1.5 for drop diameters of 0.4 mm and 2.5 mm, respectively. The clouds thus are in the

Rayleigh region for the attenuation, but the precipitation is excluded from it for the current radar wavelengths.

For ice, the application range for the Rayleigh approximation is appreciably broader than for water, due to the effect of the dielectric constant on the wavelength inside the material of the sphere. In fact, although the preceding formulas refer to the wavelength in a vacuum, the wavelength in the sphere that governs the width of the approximation range is $\lambda_i = \lambda\varepsilon_i^{-1/2}$. It is shorter in media with high dielectric constant, such as water, than in those with a low dielectric constant, such as ice or mixtures of water and ice (see also Section 2.5.1, the application of the Rayleigh approximation to natural media with a broad granulometric spectrum; that is, clouds and precipitation). In fact, most precipitation is formed from ice so that, at altitude, the conditions encountered are more complex and nonspherical and nonhomogeneous solid particles have to be considered.

### 2.3.2 Nonhomogeneous Particles

*Small, Dry Particles*

These are essentially ice crystals or snowflakes and small dry hail lying within the Rayleigh approximation region. Their shapes are generally far from spheric. However, such particles are composed of a mixture of air and ice. We will see later that, because of the weak value of their dielectrical factor, they do not depolarize much. When their size is small compared to the wavelength (Rayleigh region), they can be assimilated to spherical scatterers. The backscattering cross section is given by (2.49):

$$\sigma = \frac{\pi^5}{\lambda^4}|K|_m^2 D_m^6$$

where the subscript $m$ refers to the mixture. If $M_m$ is the particle mass equal to $(\pi/6)\rho_m D_m^3$, we have

$$\sigma = \frac{36\pi^3}{\lambda^4}\frac{|K|_m^2}{\rho_m^2}M_m^2 \tag{2.52}$$

To calculate $|K|_m^2$, Marshall and Gunn (1952) used Debye's theory (1929) according to which the $K/\rho$ ratio of the mixture is equal to the sum of the ratios corresponding to each substance pondered by their relative mass:

$$\frac{K_m}{\rho_m} = \frac{K_i}{\rho_i}\frac{M_i}{M_m} + \frac{K_a}{\rho_a}\frac{M_a}{M_m} \tag{2.53}$$

where the subscripts $i$ and $a$ refer to the ice and air and $M$ is the mass. Although the applicability of Debye's theory to the ice hydrometeors is questionable, no other treatment of this question was proposed. Because the dielectric constant of air is close to unity, the corresponding term $|K|_a^2$ is close to 0. Therefore it can be admitted from (2.53) that $K_m/\rho_m$ is constant for an air-ice mixture with any density $\rho_m$:

$$|K|_m^2 = (\rho_m)^2 \left( \frac{|K_i|}{\rho_i} \right)^2 \tag{2.54}$$

For $\rho_i = 0.92 \text{ g cm}^{-3}$ and $|K|_i^2 \approx 0.176$, we have

$$|K|_m^2 = 0.208 \rho_m^2 \tag{2.55}$$

with $\rho_m$ in g cm$^{-3}$.

Sometimes, for ice hydrometeors, it is preferable to consider the diameter of the liquid equivalent particles (of the same mass): the diameter ice spheres of the unit density would have; that is, $D_e$ with

$$D_e = D_i (0.92)^{1/3} \tag{2.56}$$

Thus, if $\rho_m$ is taken arbitrarily equal to 1 g $\cdot$ cm$^{-3}$, we have $|K|^2 = 0.208$.

*Small Melting Particles*

We can make the approximation that an ice particle in the course of melting consists of an ice sphere or a mixture of ice and air covered by a film of water whose thickness increases (except when the size of the particle is such that the water detaches as the melting progresses, as is the case for hailstones). To calculate the effective cross sections with the help of the Mie equations for such particles, two concentric spheres are considered with different refractive indices [Aden and Kerker, 1951]. Calculation shows that, whereas the thickness of the water film increases, the values of $\sigma$, $Q_s$, and $Q_t$ that at the beginning correspond to ice spheres (approximately one-fifth the liquid value), increase rapidly so that when the thickness of the water film reaches a tenth of the total radius (70% of ice), $\sigma$ and $Q_s$ have a value practically equal (90%) to that of an entirely liquid sphere of the same size. As soon as the melted fraction approaches 10% of the total volume, the attenuation reaches a maximum value that is double the effective cross section of the liquid sphere and then returns to the normal value while the melting continues. These variations apply to particles in the Rayleigh region. For large spheres, the variation of $\sigma$ as a function of the melted thickness is less rapid and the maximum of $Q_t$ is lower.

For large, low-density snowflakes, it is certainly preferable to consider a mixture of ice and water rather than two different concentric regions. This case is poorly known from the point of view both of the melting processes and scattering.

*Large Melting Spheres*

This is the case of hailstones. It is useful to have information on these to determine the criteria for detecting hail in thunderstorms (see Section 2.5.4). Such spheres lie outside the Rayleigh approximation range, in a region where the backscattering cross sections of ice and water are very different and fluctuate considerably as a function of $\alpha$. For example, observation by radar off the melting of an artificial hailstone suspended under a balloon reveals a variation in the value of $\sigma$ at the beginning of melting, then large oscillations as the diameter decreases (the excess melted part separates from the hailstone, the water film on the surface of the sphere having a maximum thickness of 1 mm [Atlas *et al.*, 1960]).

Extension of the Mie equations to concentric spheres shows that the cross sections of large melting spheres depend on the wavelength, the diameter of the sphere, and the thickness of the water film [Hermann and Battan, 1961b]. The results confirm that, for a small $\alpha$ (Rayleigh region), $\sigma$ increases when the particle is covered with water; whereas for $\alpha > 1$, the value of $\sigma$ as a function of $D$ oscillates between the theoretical curve for liquid spheres and that for solid spheres. (The few values situated a little outside this interval are explained by interference phenomena created by reflections at the surfaces of discontinuity of media with different refractive indexes.) The necessary water thickness for the melting sphere to scatter as a liquid sphere increases with the wavelength: it is higher than the maximum thickness of 1 mm for current radar wavelengths. Finally, although the theory is in agreement with the results of measurements on melting spheres, no simple relationships for this case give the different cross sections for large $\alpha$.

Under certain conditions, the hailstones consist of a mixture of water and ice; the liquid component is not limited to the outside film, sometimes constituting a significant fraction of the total mass. These particles of spongy ice with a large diameter have been the object of calculations for different values of the liquid fraction, either by considering spheres of mixed composition in their entirety or ice spheres covered by a layer of mixed composition [Battan and Hermann, 1962]. The experimental results are in good agreement with the calculations, if the ice spheres are considered to be surrounded by a water film corresponding to the amount of liquid entering the mixture [Atlas, Hardy, and Joss, 1964].

### 2.3.3 Nonspherical Particles

The problem of the diffusion of a plane electromagnetic wave by nonspherical hydrometeors was analyzed or discussed in many articles over the past 20 years in relation to the use of short wavelengths in telecommunication; that is, the depolarization of the hertzian beams of the terrestrial links or earth-space links and with the development of polarization diversity radars. When the ratio of the hydrometeors size to the wavelength is in the resonance region (i.e., size on the same order as the radar wavelength) the diffusion functions cannot be calculated analytically, and we must use various numerical techniques. Among those used for computations relative to hydrometeors, the following ones can be quoted: the point matching technique, in which the incident wave is developed in a Fourier series with coefficients determined by satisfying to the limit conditions, in the last squares sense, at several points; the extended boundary condition method or T-matrix method, which is based on the idea that the fields in the external region can be determined from considerations relative to the inside of the scatterer and the Fredholm integral equation method. A comparative presentation of these methods, their computational implementation, and their results is given in Oguchi [1981], Holt [1982], and Mon [1982].

When the ratio of the hydrometeors' size to the wavelength is small enough, which is the general case in radarmeteorology (except for the hail), Gans's extension [1912] of Rayleigh's theory to an ellipsoid, enables us to calculate simple functions sufficient for an approximate discussion of the phenomena (Labrum, 1952; Atlas, Kerker, and Hitschfeld, 1953; Shupiatskii, 1959; Seliga and Bringi, 1976, among others). In the Gans theory, the interaction of a plane wave with an oblate or prolate ellipsoid is described for three orthogonal dipole moments oriented following the three ellipsoid axes, independently excited by the components along the same axes, of the incident electric field. The multipole moments are neglected as in Rayleigh's theory. The backscattering cross section for horizontal ($H$) or vertical ($V$) polarization electric field is written [see, for example, Seliga and Bringi, 1976]

$$\sigma_{H,V} = \frac{16\pi^7 D_e^6}{9\lambda^4} \left| \frac{m^2 - 1}{4\pi + (m^2 - 1)P_{H,V}} \right|^2 \tag{2.57}$$

where $P_{H,V}$ are geometrical factors, such as

$$P_V = 4\pi - 2P_H = \frac{4\pi}{e^2}\left[1 - \left(\frac{1 - e^2}{e^2}\right)^{1/2} \text{arcsin}e\right] \tag{2.58}$$

for an oblate ellipsoid and

$$P_V = 4\pi - 2P_H = 4\pi\frac{1 - e^2}{e^2}\left[\frac{1}{2e}\ln\left(\frac{1 + e}{1 - e}\right) - 1\right] \tag{2.59}$$

for a probate ellipsoid. The term $e$ represents the eccentricity of the main elliptic section, such as

$$e^2 = 1 - \left(\frac{a}{b}\right)^2 \tag{2.60}$$

and $D_e$ is the diameter of the spherical equivalent volume of the hydrometeor. As in eq. (2.49) the backscattering cross section is proportional to $D_e^6$ and $\lambda^{-4}$.

When $(a/b) \rightarrow 1$, we have $P_H = P_V \rightarrow 4\pi/3$ and we find for $\sigma_{H,V}$ eq. (2.49). Equation (2.57) also indicates that the influence of the shape factor on $\sigma_{H,V}$ depends on the dielectric constant of the material. The result is that the effect of the sphericity deviations on the diffusion is maximum for liquid particles and negligible for low-density snowflakes, which reinforces the justification of considering the flakes as spheres for scattering. Similarly, because of the effect of the wavelength on the refractive index $m$, the shape effect increases with $\lambda$ for water particles and is independent of $\lambda$ (between 1 and 10 cm) for ice.

These various properties are summarized in Figure 2.12, which shows the calculated values for the ratio of the intensity of the field backscattered by particles with random orientation to the intensity that would be produced by spheres of the same volume as a function of the axial ratio [Atlas *et al.*, 1953]. This normalized scattering intensity is given by the water and ice, flattened or elongated ellipsoids with axial ratios lying between 0.1 and 10 for different values of $\lambda$. The values of the normalized attenuation effective cross sections under the same conditions also are shown in Figure 2.12. The backscattered intensity and the attenuation produced

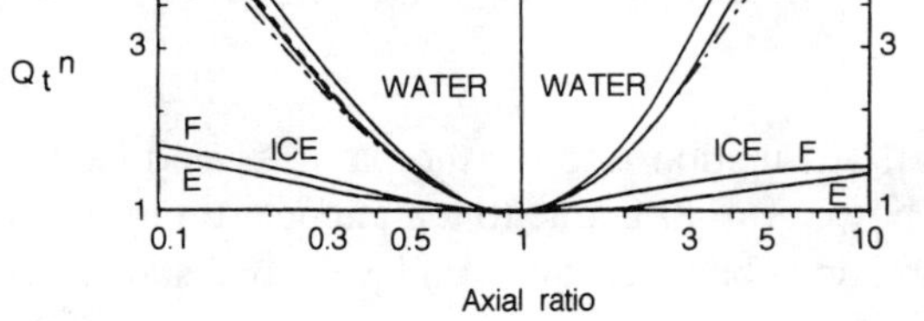


**Figure 2.12** Backscatter and attenuation by randomly oriented water and ice ellipsoids relative to water and ice spheres of equal volume (from [Atlas *et al.* 1953]).

by small water ellipsoids can be up to 10 and 30 times higher than those corresponding to spherical particles with flattened or elongated ellipsoids, respectively. In this connection, we recall that a small particle covered by water scatters is an entirely liquid particle. For ice particles, the intensity deviates little from the value corresponding to perfect sphericity even for extreme deviations in sphericity. Therefore, we can conclude that, in a general manner, the sphericity deviations with random orientation increase the backscattering and attenuation with respect to an equivalent spherical population, and this is true regardless of the orientation of the planes of polarization of the fields transmitted and received by the radar.

The situation is quite different if the particles have a predominant orientation. In this case, the scattering field depends on the orientation with respect to the plane of polarization of the incident wave: it is higher or lower than the field corresponding to spherical particles depending on whether the major axis of the ellipsoids is parallel or perpendicular to the plane of the transmitted wave.

## 2.4 ATMOSPHERIC ATTENUATION

Attenuation has been introduced in the radar equation (1.76) with the aim of taking into account losses by absorption and scattering between the radar and the target. In the meteorological atmosphere, two causes of attenuation are distinguished: gases and hydrometeors. In the hydrometeors, the attenuation for clouds is distinguished from that for precipitation, which are of the same nature but described by different equations.

The total attenuation rate $a_t$ is obtained from the terms $a_g$, $a_c$, and $a_p$, which are the corresponding attenuation rate for gases, clouds, and precipitation. These coefficients are additive, and we can write

$$\begin{aligned} a_t &= a_g + a_h \\ a_h &= a_c + a_p \end{aligned} \tag{2.61}$$

where $a_h$ is the total attenuation rate for the particles and excludes that of gases.

The incident power variation due to the attenuation by the hydrometeors, along the propagation direction, between the $r$ and $r$ + d$r$ distances is, equation (1.63),

$$\mathrm{d}P = -(\Sigma\, Q_t)P\, \mathrm{d}r \tag{2.62}$$

where the summation extends to an elementary volume of unit cross section and d$r$ depth. We thus can write

$$a_t = 0{,}4343 \int_{D_{\min}}^{D_{\max}} N(D) Q_t(D)\, \mathrm{d}D \qquad (2.63)$$

with

$$Q_t = Q_a + Q_s$$

In (2.63), if $Q_t$ is in cm$^2$ and $N(D)$ in m$^{-3}$, $a_t$ is in dB km$^{-1}$.

### 2.4.1 Attenuation by Gases

At wavelengths used by radars, among the various gases of the earth atmosphere, only oxygen and water vapor absorb significantly (Van Vleck, 1947a and 1947b). The water vapor presents a set of rotation lines around 1.348 cm (22.2 GHz), which affects the wavelengths ranging from about 1 to 2 cm, and around 0.164 cm (183.3 GHz). The oxygen presents a family of rotation lines around 0.5 cm (60 GHz), which, at atmospheric pressure, merge into a single band and a line isolated at 0.253 cm (118.8 GHz). Figure 2.13 shows the variation of the attenuation coefficients near the ground in the standard condition and for a humidity of 7.5 g m$^{-3}$. The coefficient values that depend on the temperature and pressure are given in Appendix 4.

The attenuation undergone between the radar and the target depends on the atmospheric layers crossed (it decreases with the density). It has its maximum value for horizontal aiming at sea level.

Globally, the attenuation for the gases, at the current radar wavelengths ($\lambda >$ 3 cm), is negligible only for the short-distance observations. At millimetric wavelengths, the $K_a$ band, centered around 0.8 cm benefits from a propagation window. The shorter wavelengths undergo prohibitive attenuations. The problem is particularly critical in humid tropical atmosphere.

### 2.4.2 Attenuation by Clouds

At radar wavelengths, the radioelectric dimensions of the cloud particles are sufficiently small that their attenuation cross sections satisfy the conditions of the Rayleigh approximation. In addition, for such particles the scattering is small compared to the absorption, and we can write $D \ll \lambda$, cf. eq. (2.48) and (2.50); and for ice or liquid particles, we can write $Q_t \approx Q_a$. Now $Q_a$, given by equation (2.50), is proportional to the volume of the water or ice particles assumed to be spherical. If

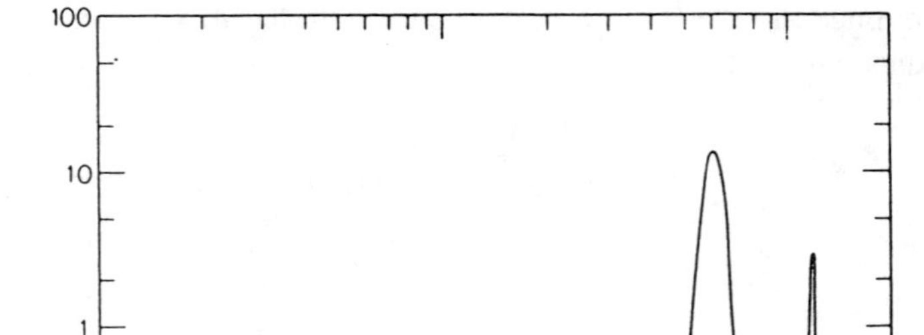


**Figure 2.13** One-way attenuation by oxygen (solid line) and water vapor (7.5 g $m^{-3}$; dashed line) at sea level for standard pressure $t°$ (from [Van Vleck, 1947]).

we denote the water content in the cloud by $M = (\pi/6)\rho \, \Sigma \, D^3$ in g $m^{-3}$ ($\rho$ is the density in g $m^{-3}$), we have

$$a_c = 0{,}4343 \, \frac{6\pi}{\rho} \left( \frac{\text{Im}(-K)}{\lambda} \right) \qquad M = k_c M \qquad (2.64)$$

The value of $\lambda$ is in cm, $\rho$ in g $cm^{-3}$, $M$ in g $m^{-3}$, and $a_c$ in dB $km^{-1}$. The term $k_c$ is the coefficient of attenuation by clouds per unit density in dB $km^{-1}$ (g $m^{-3})^{-1}$.

This expression underlines the fact that the attenuation in clouds is independent of the size spectrum of the particles and proportional to the water content. The coefficient $k_c$ depends on the wavelength and the temperature. Gunn and East [1954] have calculated the value of the attenuation in water and ice clouds as a function of these two parameters. Figure 2.14 was established from these results. It shows that the attenuation in clouds increases with the temperature and, above all, when the

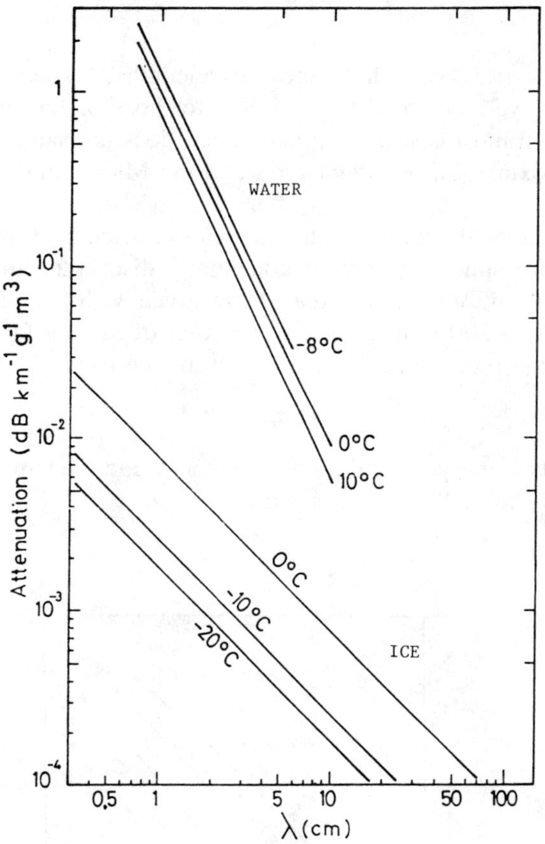


**Figure 2.14** One-way attenuation by clouds (term $C_n$ in equation 2.30) in dB km$^{-1}$ g$^{-1}$ m$^{-3}$ as a function of the wavelength and for different values of temperature.

wavelength decreases. Attenuation by ice clouds is very small and negligible in most cases. In water clouds, it is small only for decimetric wavelengths. Therefore, taking into account the small size of the constituents of clouds, they may not be detected but still create a nonnegligible attenuation.

### 2.4.3 Attenuation by Precipitation

The attenuation rate $a_p$ either can be calculated by assuming that the microphysical properties of the precipitation are known or determined experimentally by the simultaneous measurement of the attenuation and the rate of the precipitation.

### *Rain*

For values of $\alpha \ll 1$ (i.e., with the longest wavelengths, $\geqslant$10 cm, and precipitation with small drops), we have conditions similar to those of the previous section on clouds. However, in most cases, the precipitation does not satisfy the conditions of the Rayleigh approximation for attenuation and the Mie formula has to be used in eq. (2.63).

Figure 2.15 gives the value of the attenuation calculated as a function of the precipitation rate for spherical drops exponentially distributed for wavelengths between 0.83 and 15 cm. We can see that, for a given $\lambda$, for a rather broad interval of variation of $R$(1 to 200 mm $\cdot$ h$^{-1}$), the variation of $a_p$ as a function of $R$ can be adjusted with a good precision to an expression of the form

$$a_p = kR^{\gamma} \tag{2.65a}$$

where the constants $k$ and $\gamma$ depend on the frequency and the temperature; $a_p$ usually is expressed in dB km$^{-1}$. We also can write

$$a_p = k_p R \tag{2.65b}$$

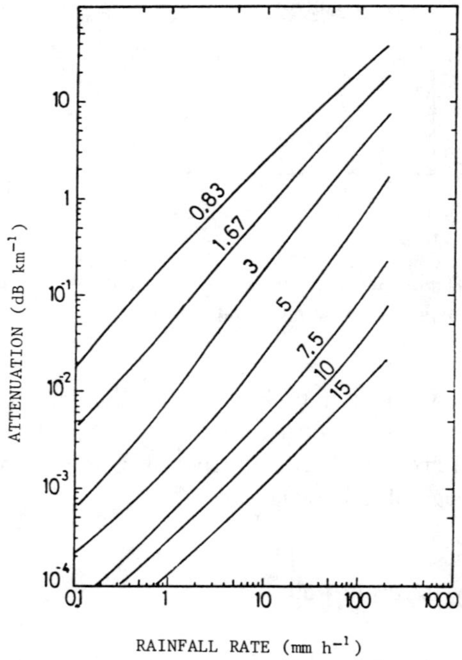


**Figure 2.15** One-way attenuation by rain calculated as a function of the intensity of precipitation for different values of the wavelength.

with $k_p = kR^{\gamma-1}$. In this way, a coefficient $k_p$ appears that depends on $R$ and is expressed in dB km$^{-1}$ (mm h$^{-1}$)$^{-1}$.

Many authors have concerned themselves with the value of these coefficients because of the importance of the attenuation phenomenon for hydrological measurements by radar and for radio transmission: among many others, Gunn and East [1954], Wexler and Atlas [1963], Medhurst [1965], Blevis [1974], Oguchi [1983]. The calculated or measured attenuation values show in agreement that $a_p$ in rain is approximately proportional to $R$. The most significant deviation lies around 10 GHz ($\lambda$ = 3 cm) (see Table A4.4, Appendix 4).

The temperature is involved in the attenuation via the refractive index of the drop (the term $I_m(-k)$ given in Appendix 4). Its effect on attenuation by rain at short wavelengths ($\lambda < 3$ cm) does not exceed ±20% between 10 and 30°C; it is usually much less (less than 5% for $\lambda$ = 8 mm). For increasing wavelengths, above 3 cm, the effect of temperature increases (it changes by a factor of 3 between 0 and 40°C for $\lambda$ = 5 cm) and gradually approaches the case of water clouds illustrated by Figure 2.14. However, Figure 2.15 shows that the attenuation decreases rapidly when the wavelength increases so that, qualitatively and from a practical point of view, the variations in the attenuation by rain with temperature can be considered of little importance at current radar wavelengths. Similarly, the size distribution chosen to represent precipitation (among the various types corresponding to natural precipitations) has little effect on the attenuation.

Figure 2.15 shows that $a_p$ becomes negligible at decimetric wavelengths. At shorter wavelengths ($\lambda < 5.5$ cm) the attenuation by rain must be taken into account in radar mesurements. However, when we consider the problems of attenuation by rain, it is important to remember that the duration and extent of precipitation zones with a given intensity vary inversely to their intensity.

The sphericity deviation of the rain drops (flattening, cf. Section 2.2.6) imply that the propagation conditions and notably attenuation depend on the polarization for oblique aimings: the $H$ polarization waves are more strongly attenuated than the $V$ polarization waves. The phase of the waves is also affected. Figure 2.16 shows the variation of the differential attenuation and differential phase shift according to the frequency for various values of the rain intensity over a horizontal propagation distance of 1 km for one way calculated by Oguchi (1974) with the Laws and Parson drop size distribution. We see that these "effects," which play a primordial role in the polarimetric measurements, are not negligible.

Table A4.5 in Appendix 4 gives the values of the parameters of relation (2.65a) for the $H$ and $V$ polarizations, for a horizontal propagation direction.

*Snow*

Attenuation by snow is due to the cumulative effects of absorption and scattering by the snowflakes. The attenuation rate $a_p$ therefore is given by (2.63) with $Q_t = Q_a +$

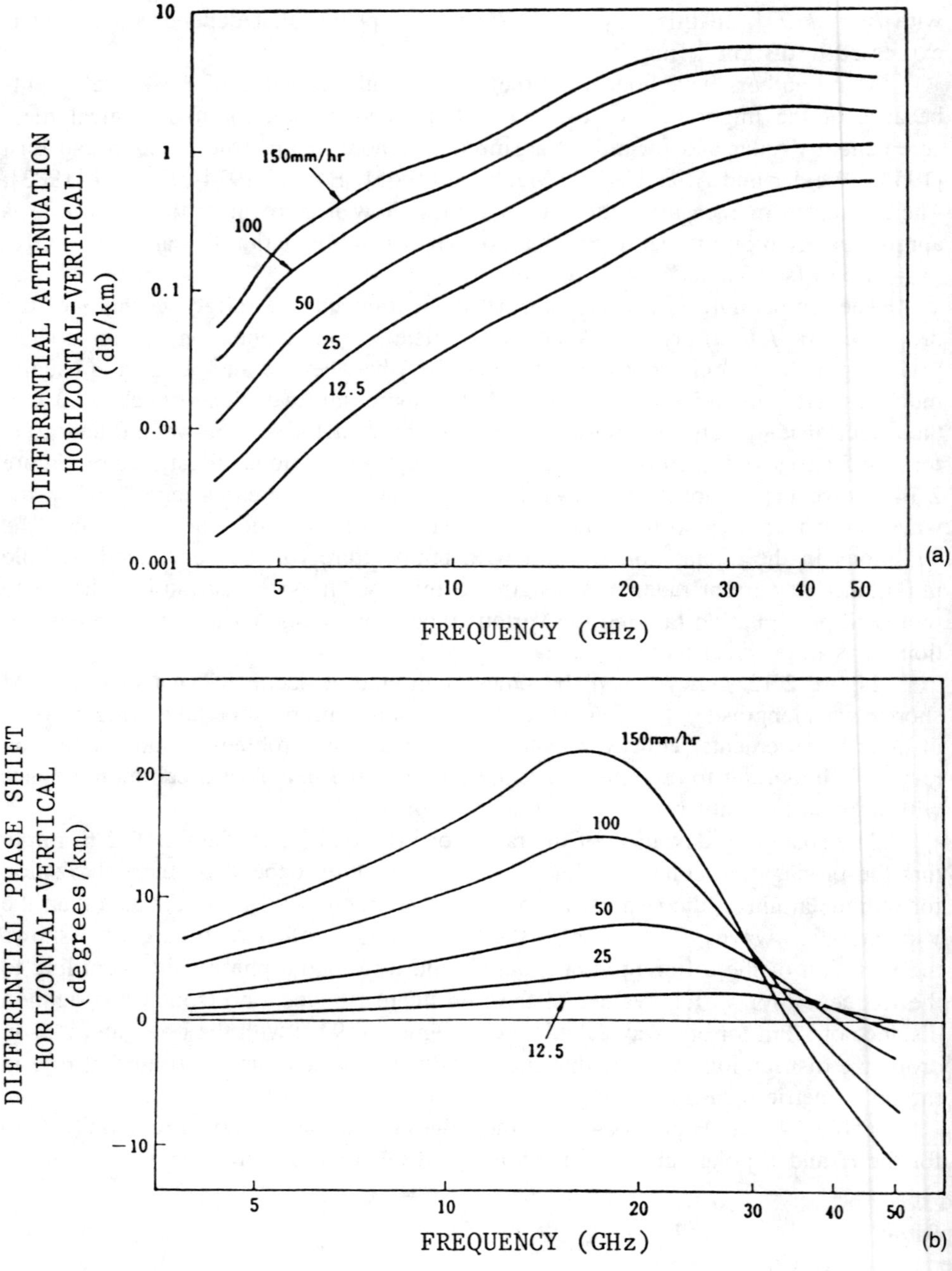


**Figure 2.16** (a) Differential attenuation and (b) differential phase shift as a function of frequency for rain. The parameter is rainfall rate (from [Oguchi and Hosoya, 1974]).

$Q_s$. If the corresponding values of $\alpha$ are sufficiently small, which is found in most cases, and if we consider compact spheres equivalent to the snowflakes, the cross sections $Q_s$ and $Q_a$ are given by (2.48) and (2.50), and their summation over unit volume is proportional to $\Sigma D^6$ and $\Sigma D^3$, respectively. If it is possible to select a functional relationshp between $\Sigma D^6$ and $R$ for the considered precipitation (see Section 2.5.2 on the relationships between $R$ and $\Sigma D^6$), these two terms can be expressed as a function of $R$ because according to (2.9), $\Sigma D^3 = 6R/\pi\rho v$, by assuming that all the particles fall at the same terminal velocity, which is an acceptable assumption. In fact, snow precipitation has a narrow spectrum of terminal fall velocity (typically around 1 ms$^{-1}$); furthermore, if it is assumed that below the convective regions that generate the precipitation (i.e., in calm air, the vertical velocity of the air is 0), we obtain an approximate relationship making it possible to calculate $a_p$ for snow as a function of temperature, $R$, and $\lambda$. For example, Gunn and East [1954] obtained, at 0°C,

$$a_p = 0.35 \times 10^{-2} \frac{R^{1.6}}{\lambda^4} + 0.22 \times 10^{-2} \frac{R}{\lambda} \tag{2.66}$$

where $R$ is in mm $h^{-1}$ of water, $\lambda$ is in cm, and $a_p$ is in dB km$^{-1}$.

Figure 2.17 shows some calculated values for attenuation by snow as a function of the rate of precipitation for four wavelengths. The values obtained are very small. We can see that the rate of snowfall is rarely higher than 4 mm h$^{-1}$ so that the corresponding attenuation is negligible except for large ranges at wavelengths less than 3 cm. However, these conclusions must be modified when melting snow is involved. Under these conditions, the combination of the effects of melting and deformation previously described lead to important attenuation, the value of which depends on the polarization of the radar wave (see Secton 5.5.2, the bright band).

*Hail*

Hail generally contains large particles ($D_{max} > 6$ mm) the dielectric properties of which are not homogeneous, the shape of which is not spherical, and the surface condition of which is not constant but changes rapidly, depending on the growth conditions and the temperature of the environment. The study of the influence of these different factors leads to the conclusion that it is easy to formulate a simple relationship defining the attenuation cross section (Section 2.3.2). Under these conditions, attenuation by hail can be calculated only if the physical properties of the population of hailstones are known with sufficient completeness, which is not generally the case.

By starting with the empirical size distribution of the hailstones described by eq. (2.26) for a condensed water content normalized to 1 g m$^{-3}$, the attenuation rates

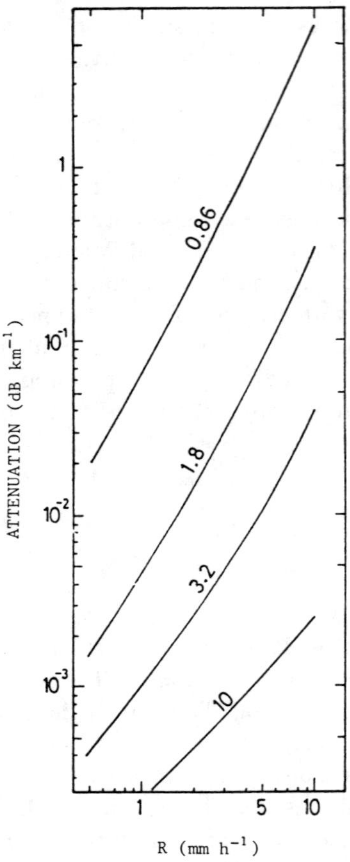


**Figure 2.17** One-way attenuation by snow calculated as a function of the intensity of precipitation for different values of the wavelength at 0°C.

have been calculated [Battan, 1971] as a function of the maximum size of the hailstones for dry ice spheres or spheres covered with a water film of variable thickness. The results reveal the earlier-mentioned influence of the surface of the hailstone: the attenuation is increased appreciably by the presence of a water film. At wavelengths less than a few centimeters, the hail produces a nonnegligible attenuation (see also Section 2.5.4). For example, with $D_{max}$ = 2.89 cm, the attenuation rate for hailstones, dry or covered with a water film, is 1.66 and 3.79 dB km$^{-1}$, respectively for $\lambda$ = 3.2 cm. These values are 0.33 and 2.60 dB km$^{-1}$ at $\lambda$ = 5.5 cm and 0.034 and 1.18 dB km$^{-1}$ at $\lambda$ = 10 cm.

## 2.5 BACKSCATTERING BY CLOUDS AND PRECIPITATION

### 2.5.1 Radar Reflectivity Factor

*Definitions*

We have seen in Section 1.3.2 that the average power backscattered by a population of particles homogeneously distributed in an elementary (pulse) volume is proportional to their radar reflectivity $\eta$; that is, to the sum of the backscattering cross sections ($\sigma_i$) of the individual particles over unit volume, eq. (1.72):

$$\eta = \frac{1}{V} \sum_V \sigma_i \tag{2.67}$$

or

$$\eta = \int_{D_{\min}}^{D_{\max}} \sigma(D) N(D) \, \mathrm{d}D \tag{2.68}$$

where $N(D)$ is the particle size distribution.

The radar reflectivity is a characteristic quantity of the target, and its definition does not imply any assumption on the nature of the scattering medium. Usually, at standard wavelengths of the meteorological radars, the conditions for the Rayleigh approximation concerning backscattering in clouds and precipitation (except hail) are satisfied. Therefore, $\sigma_i$ is given by (2.49), and we have

$$\eta = \frac{\pi^5}{\lambda^4} |K|^2 \int_{D_{\min}}^{D_{\max}} D^6 N(D) \, \mathrm{d}D \tag{2.69}$$

The integral is the radar reflectivity factor represented by the letter $Z$:

$$Z = \int_{D_{\min}}^{D_{\max}} N(D) \, D^6 \, \mathrm{d}D \tag{2.70}$$

For a discrete spectrum,

$$Z = \sum_i \Delta Z_i = \sum_i [N_i(D_i) \Delta D_i] D_i^6 = \sum_i n_i(D_i) D_i^6 \tag{2.71}$$

Just like $\eta$, the factor $Z$ is an average characteristic of the population of scatterers distributed in the pulse volume. It is independent of the wavelength, which makes it easier to compare measurements obtained with different radars.

The value of $Z$ must normally be expressed in $m^6\ m^{-3}$; however, as the size of the precipitation elements is on the order of a millimeter, this usually is given in $mm^6\ m^{-3}$ (with a conversion factor of $10^{18}\ mm^6/m^6$). Associated with this unit is a logarithmic scale defined by comparison with a reference level of 1 $mm^6\ m^{-3}$. The radar reflectivity factor thus is expressed in dBZ:

$$Z(\mathrm{dBZ}) = 10 \log[Z\ (\mathrm{mm}^6\ \mathrm{m}^{-3})] \tag{2.72}$$

If $Z$ is in $mm^6\ m^{-3}$, $\eta$ is in $cm^{-1}$, $D$ is in mm, and $\lambda$ is in cm, we have

$$\eta = \left(\frac{\pi^5}{\lambda^4} |K|^2 \times 10^{-12}\right) Z \tag{2.73}$$

When the observed scattering volume does not satisfy the conditions of the Rayleigh approximation or if there is any doubt that it does, it is convenient to characterize $Z$ by the equivalent reflectivity factor $Z_e$, which is equal to the reflectivity factor of a population of liquid and spherical particles satisfying the Rayleigh approximation and producing a signal of the same power. By definition,

$$Z_e = \frac{\lambda^4}{\pi^5 |K|_w^2} \eta \tag{2.74}$$

where $|K|_w^2$ is the dielectric factor of water taken by convention equal to 0.93 (average value for water). Thus,

$$\eta = 2.8 \times 10^{-10} \lambda^{-4} Z_e \tag{2.75}$$

with $\eta$ in $cm^{-1}$, $\lambda$ in cm, and $Z_e$ in $mm^6\ m^{-3}$.

For example, consider precipitation made of a uniform distribution of water drops of 1 mm in diameter. The concentration is one drop per liter of air, which is a value currently observed in nature. The water content is 0.52 $g\ m^{-3}$ and corresponds in still air to a precipitation intensity of 7.5 $mmh^{-1}$ for a terminal fall velocity of 4 $ms^{-1}$. The radar reflectivity factor $Z$ is $10^3\ mm^6\ m^{-3}$, or 30 dBZ. The back scattering cross section of a drop for a wavelength of 5 cm is $4.55 \times 10^{-7}\ cm^{-2}$, and the radar reflectivity for this same wavelength is $4.55 \times 10^{-4}\ cm^2\ m^{-3}$, or $4.55 \times 10^{-10}\ cm^{-1}$. Radars generally are calibrated with $|K|^2 = |K|_w^2 = 0.93$; that is, the $Z_e$ quantity is represented on the screen.

For snow and ice particles, the radar reflectivity factor $Z_i$ is defined by

$$\eta = \frac{\pi^5}{\lambda^4} |K|_i^2 Z_i \tag{2.76}$$

where $|K|_i^2$ is the dielectric factor of ice. With (2.74) and (2.76), we have

$$|K|_w^2 Z_e = |K|_i^2 Z_i$$

That is,

$$Z_i = \frac{|K|_w^2}{|K|_i^2} Z_e \tag{2.77}$$

or, with $|K|_w^2 = 0.93$ and $|K|_i^2 = 0.176$,

$$Z_i = 5.28 Z_e \tag{2.78}$$

and, taking the logarithm

$$Z_i(\text{dBZ}) = Z_e(\text{dBZ}) + 7.2(\text{dB}) \tag{2.79}$$

If $\rho_i$ is taken equal to 1 g m$^{-3}$, when the dimension of the melted equivalent particles to $|K|_i^2 = 0.208$ (cf. Section 2.3.2) is considered for the computation of the ice reflectivity factor, we have

$$Z_i = 4.47 Z_e \tag{2.80}$$

$$Z_i(\text{dBZ}) = Z_e(\text{dBZ}) + 6.5 \text{ (dB)} \tag{2.81}$$

which is the range of validity of the Rayleigh approximation for the backscattering by precipitation.

In Section 2.3.1, the range of validity of the Rayleigh approximation for the computation of the backscattering and attenuation cross sections for individual hydrometeors was discussed. It is interesting to examine again this point for a distribution of large hydrometeors, such as rain. Figure 2.18 shows the variation of $Z$ calculated for an MP distribution by the exact formulas according to $R$ up to 100 mm h$^{-1}$ for a set of wavelengths ranging from 0.62 and 10 cm [Wexler and Atlas, 1963]. The relations between $Z$ and $R$ for 3.2 and 10 cm calculated with the Rayleigh approximation do not differ appreciably from those resulting from the use of Mie's theory.

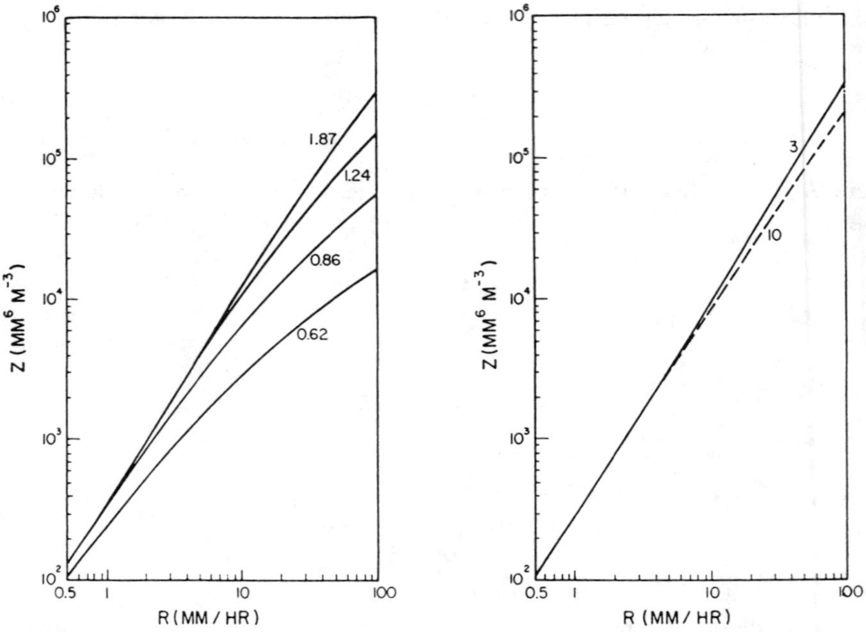


**Figure 2.18** Plots of $Z$ as a function of $R$ computed using Mie scattering (solid lines) or Rayleigh scattering (dashed line) for MP drop-size distributions (from [Wexler and Atlas, 1963]).

At 10 cm, the deviation is smaller than 0.4 dB for $R < 100$ mm $h^{-1}$. At short wavelengths, except for weak rain rate, the deviation can be considerable. The temperature does not influence the error at 10 cm and remains weak at 3.2 cm. For $\lambda <$ 3.2 cm, the deviation is negligible only at weak rainfall rates.

### 2.5.2 Z and X Relations

Natural clouds and precipitations, except hail, are seen to satisfy the conditions of the Rayleigh approximation for current radar wavelengths. The radar thus enables the measurement of the reflectivity factor $Z$. In order to interpret this measurement in terms of useful physical quantities $X$ (with $X = R, M, D_0, A$, *et cetera*), we endeavor to determine the relations between $Z$ and these various variables. Numerous works were devoted to the study of such relations for their validity depends on that of the radar important application in the meteorological measurements field, more particularly hydrology. The relations have the general form:

$$Z = aX^b \tag{2.82}$$

where $a$ and $b$ are coefficients.

Just like the microstructure of clouds and precipitation, on which they depend, these coefficients vary in space and time. We therefore can consider instantaneous or local values and average values.

The coefficients can be experimentally determined by measuring a set of $Z$ and $X$ corresponding value pairs and performing least square adjustment. There are two possible ways to obtain the data relative to $Z$ and $X$:

1. A set of values of $N(D)$ is measured with a spectrogranulometer[5] [Joss and Waldvogel, 1967 and 1969; Campistron, Despaux, and Lacaux, 1987; Gertzman and Atlas, 1977]; from these $Z$ and $X$ are calculated by numerically integrating the relations of Section 2.2.3.
2. $Z$ and $X$ are measured independently in space and time synchronization, $Z$ by radar and $X$ by any technique (for example, for $R$ with a pluviometer).

"Theoretical" values of the $a$ and $b$ coefficients can also be obtained by computation with sufficiently restrictive hypotheses on the distribution $N(D)$.

*Precipitation*

The measurement of precipitation is an important application of the meteorological radar. In Section 2.6 we will see various methods proposed for this measurement. The most obvious and immediate one consists of using a simple relation between $Z$ and $R$. That is why there is a voluminous literature on this type of relations. The coefficients of the $Z$-$R$ relation are found to present a great natural variability. This variability can be associated to numerous fators that can be tentatively put together in two groups.

1. The place factors, geographic and climatic: local peculiarities of the atmosphere and its circulation (latitude, humidity, and available instability; depth of the troposphere; orographic and coastal effects; *et cetera*) that, according to the place and season, the dynamical, thermodynamical, and microphysical processes, are responsible for the precipitation development change and, together with them, $N(D)$.
2. Factors linked to the cloud structure: at a same place, $N(D)$ varies from one type of cloud to another and, for the same type, with the evolution of some processes. For example, coefficient $a$ is seen to increase and exponent $b$ to decrease when the convection intensity acting on the precipitation development increases.

It is thus doubtful that we obtain sensibly better results to measure $R$ if we use, for relation $Z$-$R$, coefficients adapted to the site and to the type of precipitation; or

[5] Spectrogranulometers make possible a direct measurement of the size spectrum of the precipitations or clouds, mainly by using optical or electrostatic procedures or capture followed by counting.

that the representativeness of a relation improves when the size of the sample considered increases (i.e., when it is averaged in space and time). A large number of experimental relationships (more than 100) have been published [among others, Marshall and Palmer, 1948; Gunn and Marshall, 1958; Fujiwara, 1965; Cataneo and Stout, 1968; Stout and Mueller, 1968; Sekhon and Srivastava, 1970; Carbone and Nelson, 1978].

*The Relations*

The most frequently quoted relation is by Marshall and Palmer [1948], which represents stratiform rain

$$Z = 200R^{1.6} \tag{2.83}$$

For convective rain (thunderstorms), Sekhon and Srivastava [1971] find from doppler radar observations in vertical aiming

$$Z = 300R^{1.35} \tag{2.84}$$

These coefficients are close to those found for convective rains in other regions of the earth. For snow, Sekhon and Srivastava [1970] obtain by reprocessing a set of observations coming from different places,

$$Z = 1780R^{2.21} \tag{2.85}$$

In these expressions, $Z$ is in mm$^6$ m$^{-3}$ and $R$ is in mm h$^{-1}$. In (2.85) $Z$ is the measured value (i.e., $Z_i$) and $R$ the equivalent height of rain. For a same precipitation rate, the snow reflectivity factor is much higher than that for rain; this difference reflects the difference of the terminal fall velocities. Equations (2.83)–(2.85) are shown in Figure 2.19. The formulas apply only when the conditions of the Rayleigh approximation are satisfied.

Relations between $R$, $M$, and $D_0$, as well as distribution parameters can be established in the same way. The relations corresponding to the three cases mentioned earlier are given in Table 2.2. The distributions are exponential. $D_0$ for the snow [Sekon and Srivastava, 1970] is the melted equivalent diameter. All these expressions work only for vertical air in which velocity is negligible in comparison with termina fall velocity; for the snow, this is a relatively restrictive condition.

When, because of the considered wavelength or the size of the scatterers (hail), the Rayleigh conditions are not fulfilled, if $N(D)$ is known, we can establish these relations by computing the useful sections of Mie. $Z$ is then replaced by $Z_e$. Table 2.3 shows a set of such relations established for a Marshall and Palmer distribution.

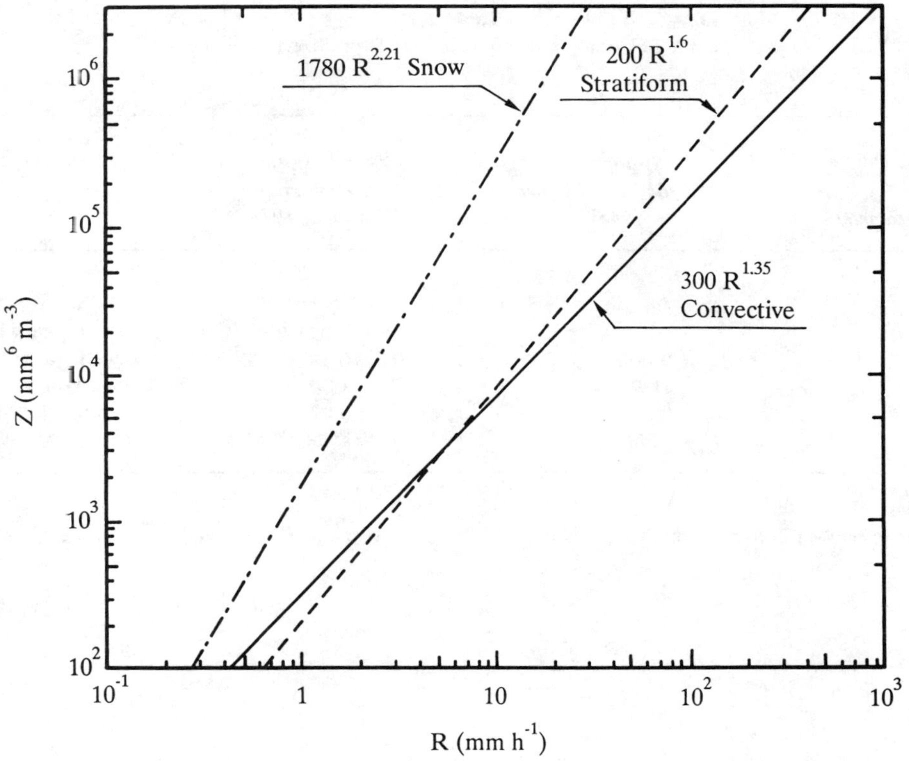


**Figure 2.19** *Z* and *R* relations from eqs. (2.83)–(2.85).

Their interest is in the way they illustrate the influence of the value of $\lambda$ on $Z$ in rain. The relations corresponding to the attenuated wavelengths have no practical use if the attenuation is not corrected.

The preceding relationships, between parameters describing precipitations and their radar properties, are redundant (owing to their dependency on $N(D)$) and can be combined in several ways. We can show [Atlas and Ulbrich, 1974] that, for rain, the six main parameters are the precipitation rate $R$, the water content $M$, the median volume diameter $D_0$, the attenuation $a_p$, the reflectivity factor $Z$, and the optical extinction coefficient $e$. These form the base of a diagram so that knowledge of two of these quantities allows us to derive the other four. An example of this type of diagram, called *RAINPAD* (Rain parameter diagram) is shown in Figure 2.20 for a precipitation with a Marshall and Palmer distribution. The RAINPAD, mainly insensitive to the form of the distribution, can be used for the discussion of problems associated with the measurement of precipitation.

**Table 2.2**
$Z$ and $X$ Relations in Precipitation for Drop-Size Distributions

| *Parameter* | *Stratiform [Marshall and Palmer, 1948]* | *Thunderstorm [Sekhon and Srivastava, 1971]* | *Snow [Sekhon and Srivastava, 1970]* |
|---|---|---|---|
| $Z$-$R$ | $Z = 200\ R^{1.6}$ | $Z = 300\ R^{1.35}$ | $Z = 1780\ R^{2.21}$ |
| $M$-$R$ | $M = 0.072\ R^{0.88}$ | $M = 0.052\ R^{0.94}$ | $M = 0.250\ R^{0.86}$ |
| $D_0$-$R$ | $D_0 = 0.09\ R^{0.21}$ | $D_0 = 0.13\ R^{0.14}$ | $D_0 = 0.14\ R^{0.45}$ |
| $\Gamma$-$R$ | $\Gamma = 41\ R^{-0.21}$ | $\Gamma = 38\ R^{-0.14}$ | $\Gamma = 22.9\ R^{-0.45}$ |
| $N_0$-$R$ | $N_0 = 0.08$ | $N_0 = 0.07\ R^{0.37}$ | $N_0 = 0.025\ R^{-0.94}$ |
| $\Gamma$-$M$ | $\Gamma = 22\ M^{-0.24}$ | $\Gamma = 25\ M^{-0.15}$ | $\Gamma = 11\ M^{-0.52}$ |

*Note*:
$Z$ is in mm$^6$ m$^{-3}$, $R$ is in mm h$^{-1}$, $M$ is in g m$^{-3}$ m$^{-3}$, $D_0$ is in cm, $\Gamma$ is in cm$^{-1}$, $N_0$ is in cm$^{-4}$.

**Table 2.3**
$Z_e$-$R$ Relations Computed Usind Mie Scattering at 0°C for Marshall-Palmer Distributions (from Wexler and Atlas, 1963)

| *λ (cm)* | *R (mm h$^{-1}$)* | $Z_e$ | *λ (cm)* | *R (mm h$^{-1}$)* | $Z_e$ |
|---|---|---|---|---|---|
| | 0–5 | $240\ R^{1.1}$ | | 0–20 | $330\ R^{1.54}$ |
| 0.62 | 5–20 | $345\ R^{0.9}$ | 1.87 | 20–50 | $500\ R^{1.4}$ |
| | 20–200 | $540\ R^{0.75}$ | | 50–100 | $750\ R^{1.3}$ |
| | 0–5 | $350\ R^{1.32}$ | 3.21 | | $275\ R^{1.55}$ |
| 0.86 | 5–20 | $450\ R^{1.15}$ | 4.67 | | $280\ R^{1.45}$ |
| | 20–100 | $780\ R^{0.95}$ | 5.5 | | $280\ R^{1.45}$ |
| | | | 10 | | $295\ R^{1.45}$ |
| | 0–5 | $356\ R^{1.5}$ | | | |
| 1.24 | 5–20 | $460\ R^{1.35}$ | | | |
| | 20–100 | $820\ R^{1.15}$ | | | |

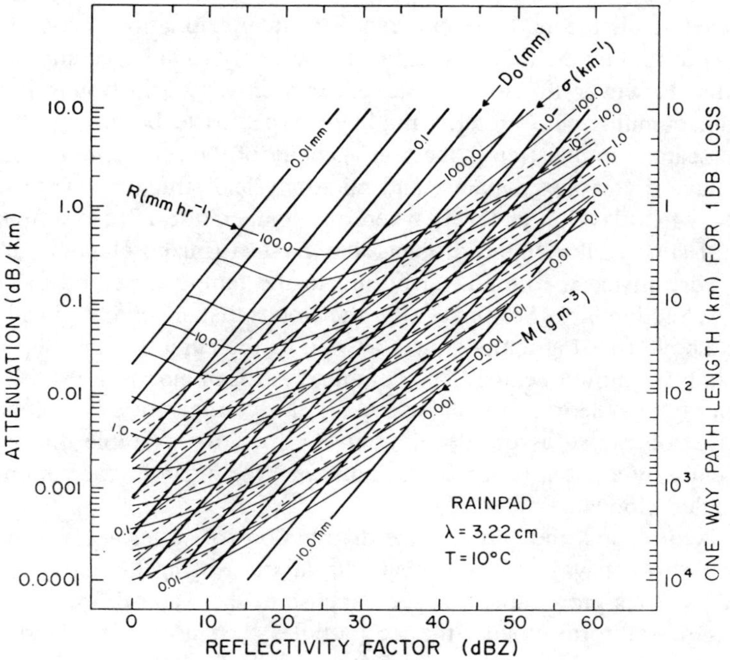


**Figure 2.20** Rain parameter diagram or RAINPAD for $\lambda = 3.22$ cm at 10°C calculated for MP drop-size distribution. Measurement of any two of the six parameters lets us obtain all six. A similar RAINPAD can be developed for other wavelengths (from [Atlas and Ulbrich, 1974]).

## *Clouds*

Quantitative observation of clouds is of great scientific and operational usefulness particularly in cloud modeling and for icing, radiative, and electromagnetic propagation problems. Using radar for hydrological remote sensing in clouds implies the knowledge of reliable $Z$-$X$ relationships for interpreting the radar reflectivity factor $Z$ in terms of significant cloud physical parameters. In the past, few theoretical and experimental works have faced this problem, mainly because available radars were not suitable for nonprecipitating cloud observation. More recently the situation has evolved favorably with improvement in the sensitivity of the equipment (high power and short wavelength radars) [Atlas, 1954; Gossard and Strauch, 1983].

The problem of the quantitative interpretation of radar reflectivity factor measurements comparatively is much more complex for clouds than for precipitation. The first problem is the scatter from clouds. In calculating the reflectivity of clouds and precipitation, we assume that the particle concentration is not correlated. If the distribution is partially ordered, the scatter will be partially coherent and less than that expected if all the particles are random. In precipitation, coherent scatter is negligible, but for clouds, this problem is not perfectly clear especially in very stable cloud forms. However, for the wavelength currently used in tropospheric weather observation, cumulus cloud returns might be expected to be largely dominated by incoherent scatter. In question is the homogeneity of the scattering medium. Clouds generally have a complex dynamic and microphysical structure (cf. Sections 2.2.1 and 2.2.4). Most develop at levels where the temperature is below freezing. They are made of supercooled liquid droplets and crystals (mixed clouds) but in proportions that vary inside the cloud, according to the temperature and the vertical air motions (cf. Section 5.5). Schematically, the clouds that are entirely in a liquid phase are clouds above freezing temperatures as well as the cumuli whose top is at a temperature slightly below freezing. The completely frozen clouds are stratiform clouds of the upper troposphere, of the cirrus type, whose temperature is well below freezing (when a convective ascent develops in a cirrus, for example due to instability created locally by a latent heat release, however, there can be a creation of patches of supercooled clouds).

In mixed clouds, the particle size distributions of water and ice must be considered in a distinct way. We saw that, the mixture being thermodynamically unstable, ice particles grow rapidly at the expense of the water droplets. The result is that in mixed stratiform clouds, the ice particles, because of their more important size (and despite the weak values of $|K|_i^2$) generally bring the main contribution to reflectivity, that of the droplets being negligible in relative value.

In single-phase clouds, when a precipitation develops, we saw that the size distribution becomes bimodal; one of the modes represents the small drops of the cloud, the other one the bigger elements of the nascent precipitation. In this case, the reflectivity is determined by the particles that have a precipitating size. The *Z-X* relations corresponding to the precipitation component therefore must be used. In convective clouds whose development reaches the congestus stage, we always find particles of precipitable size. Sometimes, the convective clouds are embedded in the stratiform clouds.

When a precipitation component develops inside a cloud, the biggest particles cross the cloud basis (condensation level) so that even if the precipitation evaporates before reaching the ground, the lower limit of the echoes is determined by the precipitation component. Radar therefore is not very reliable for measuring the height of the cloud basis.

In nonprecipitating clouds, the physical parameters that can be used to interpret the reflectivity measurements are the water content $M$ and the median volume diameter $D_0$. Sauvageot and Omar [1987], from granulometric measurements by instrumented aircraft, studied the relation between *Z-M* and $D_0$ in warm cumuli with size distributions intermediary between the maritime and continental types. By selectively treating the samples according to the maximum diameter $D_M$, they found that a clear correlation between $Z$, $M$, and $D_0$ appears only for samples that contain droplets with a diameter smaller than about 200 $\mu$m (Figure 2.21). This limit corresponds to that generally admitted between precipitating and nonprecipitating droplets. They showed that the corresponding reflectivity threshold is around $-15$ dBZ (or 0.032 mm$^6$ m$^{-3}$). The obtained relations for $D_M \leqslant 200$ $\mu$m are

$$Z = 6.8 \times 10^{-2} M^{1.9} \quad \text{with} \quad \rho = 0.74 \tag{2.86}$$

$$Z = 3.63 \times 10^{-7} D_0^{3.2} \quad \text{with} \quad \rho = 0.83 \tag{2.87}$$

$$D_0 = 46 M^{0.6} \tag{2.88}$$

For $Z$ (mm$^6$ m$^{-3}$) $> 1$, they are

$$Z = 240 R^{2.1} \quad \text{with} \quad \rho = 0.98 \tag{2.89}$$

where $Z$ is in mm$^6$ m$^{-3}$, $R$ is in mm h$^{-1}$, $M$ is in gm$^{-3}$, and $D_0$ is in $\mu$m; $\rho$ is the correlation coefficient.

Ice clouds present a level of additional complexity because the shape of the crystals (and their falling speed) vary according to temperature. However, we saw that the backscattering cross section of the small ice hydrometeors is not affected much by their shape. For ice clouds with a reflectivity ranging between $-25$ and 3 dB$Z_i$, Sassen [1987] proposed the following experimental relations:

$$Z_i = 28.85 R^{1.24} \quad \text{with} \quad \rho = 0.91 \tag{2.90}$$

$$Z_i = 114 M^{1.44} \quad \text{with} \quad \rho = 0.90 \tag{2.91}$$

with the same units as earlier.

### *Manufactured Clouds*

The industrial era saw the birth of a new species of meteors, manufactured clouds. There is a great variety of them: aircraft contrails, various pollution fogs and stratus clouds, dry or wet plumes linked to localized thermal releases, to name a few. These clouds, which can interact with the surrounding natural clouds, have become a significant component of the climate in some industrial areas; even if their effects are generally not very intense, because of their persistence or their occurrence frequency, they can have a nonnegligible local climatic impact: periods of sunshine, humidity, precipitation, and so on. Radar provides a potential for watching and measuring such clouds on the condition that they are detectable.

The radar observation of the manufactured clouds follows the same physical principles as those of the natural clouds. The differences bear essentially on the microphysical structure and on the form of $N(D)$. Thus, the wet plumes of the cooling

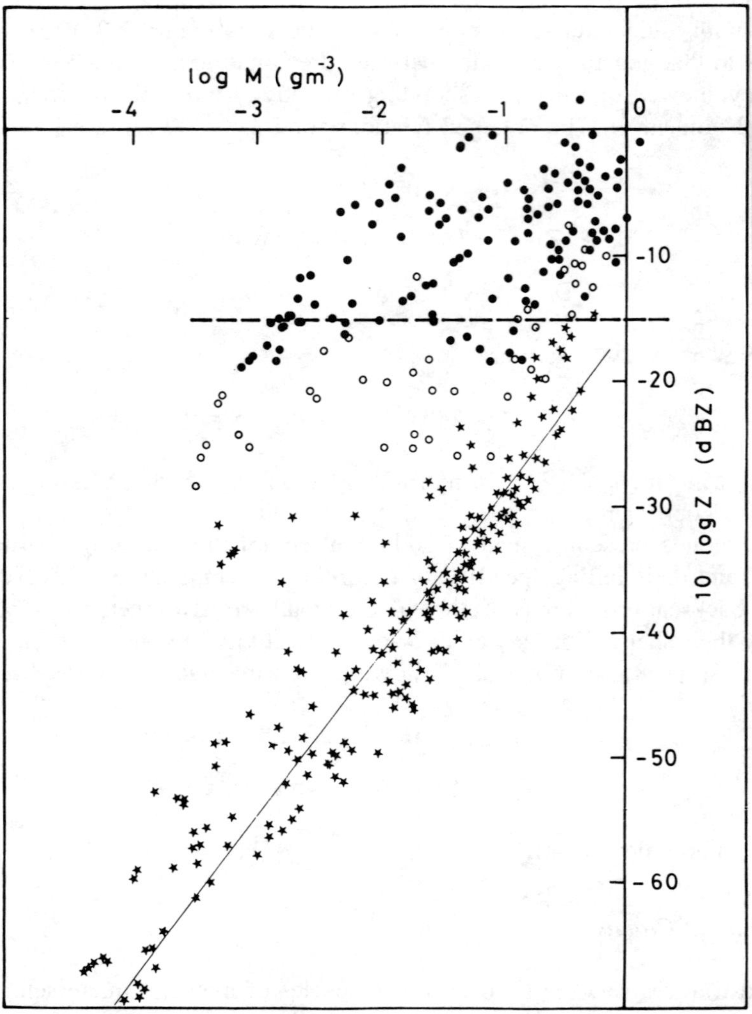


**Figure 2.21** Plot of $Z$ and $M$ values measured by aircraft. Stars, open circles, and solid circles are for droplet-size distribution, respectively, with $D_M \leqslant 100$ $\mu$m, 100 $\mu$m $< D_M \leqslant 200$ $\mu$m, and 200 $\mu$m $< D_M < 300$ $\mu$m. The horizontal line is the proposed limit of $-15$ dbZ. The regression line for $D_M \leqslant 100$ $\mu$m is $Z = 0.03M^{1.31}$ (from [Sauvageot and Omar, 1987]).

towers of a nuclear power plant showed that, because of the absence of big droplets, the radar reflectivity factor of a recondensation cloud is for a same value of $M$, about 20 dB weaker than for a cumulus (Sauvageot, 1989). The relation is

$$Z = 7.9 \times 10^{-4} M^{1.8} \qquad \text{with} \quad \rho = 0.94 \tag{2.92}$$

with the same units as (2.86). The curve slope of (2.92) is almost the same (1.8) as that of eq. (2.86). To the recondensation cloud are added the "drift droplets" of cooling water entrained out of the tower in the exhaust air flow. The drift droplets contain dissolved salts. They are polluting and do not evaporate totally in clear air (because of the presence of salts). Consequently, the cloud seen by the radar is more spread out than the visible cloud. The radar enables the measurement of the precipitating rate associated to the drift droplets whose reflectivity is

$$Z_D = 21 R_D \tag{2.93}$$

where $Z_D$ is in mm$^6$ m$^{-3}$ and $R_D$ is in mm h$^{-1}$.

Radar also let us show the relation between cooling towers and a snow precipitation reaching the ground 10 km downward of the exhaust and spreading over 12 km [Sauvageot, 1987].

### 2.5.3 Polarization Measurements

An essential characteristic of the "meteorological target" is its microstructure; that is, the nature, shape, and orientation of the hydrometeors that compose it. Most meteorological radars emit and receive a single fixed linear polarization; they give no direct information on the nature of the scatterers. Because the polarization depends on physical parameters such as the shape and orientation of the hydrometeors, it can measure these parameters. This is the reason why several methods and techniques were devised to extract the "polarization signal" of the radar echo and quantitatively interpret it. In what follows, the hypothesis is that the radar observes Rayleigh scatterers.

*The Backscattering Matrix*

Generally, the polarization of a wave can be considered elliptical, the circular and linear polarization corresponding to extreme and particular aspects of the general shape. The polarization ellipse covered during the time by the electric field vector, in the plane perpendicular to the Poynting vector, is described by the dimension of its two axes and by its orientation, or else by the amplitude and relative phase of its two linear or circular components. In linear polarization, the horizontal component

($H$) and the vertical one ($V$) are distinguished. In circular polarization, considering the propagation direction, a right circular (RC) and a left circular (LC) are distinguished. RC revolves clockwise when we look at a wave moving away from us.

If we consider the linear components, the scattering field is defined by the backscattering matrix $S_L$:

$$S_L = \begin{bmatrix} S_{HH} & S_{HV} \\ S_{VH} & S_{VV} \end{bmatrix} \tag{2.94}$$

The diagonal terms $S_{HH}$ and $S_{VV}$ represent the amplitudes of the linear copolarized scattering components or main components; that is, transmission $H$ and reception $H$, transmission $V$ and reception $V$, respectively. The terms off-diagonal $S_{HV}$ and $S_{VH}$ correspond to the orthogonal linear components; that is, $H$ transmitted and $V$ received and *vice versa*. The terms of the scattering matrix are complex and thus the amplitude and the phase of each one must be obtained to completely characterize the scattering medium.

If we consider circular components, we write them in the same way, with a matrix $S_C$:

$$S_C = \begin{bmatrix} S_{RR} & S_{RL} \\ S_{LR} & S_{LL} \end{bmatrix} \tag{2.95}$$

With, between the $S_L$ and $S_C$ terms, the relations are

$$S_{RR} = \frac{1}{2}(S_{VV} - S_{HH} - 2iS_{HV}) \tag{2.96}$$

$$S_{LL} = \frac{1}{2}(S_{VV} - S_{HH} + 2iS_{HV}) \tag{2.97}$$

$$S_{RL} = S_{LR} = \frac{1}{2}(S_{VV} + S_{HH}) \tag{2.98}$$

Between the components of the backscattered electric field $E^r$ and those of the incident field $E^t$, we have the matricial relation

$$E^r = (S)E^t \tag{2.99}$$

*Differences between Linear and Circular Polarization*

Consider a radar transmitting $H$ or $V$ polarization. For a spherical hydrometeor or an ellipsoid with a symmetry axis parallel or perpendicular to the polarization plane, like a raindrop falling in still air with zero canting angle, the off-diagonal terms of $S_L$ equal zero. The received wave has the polarization of the transmitted wave; there is no depolarization. The diagonal terms are proportional to the axes of the polarization ellipse. When the axes of ellipsoidal hydrometeors form an angle not equal to zero with the polarization plane of the incident wave, off-diagonal components appear. The polarization ellipse associated with the backscattered wave has the same orientation as the scattering ellipsoid.

For a wave transmitted in circular polarization, like right (RC), the wave backscattered by a perfectly spherical scatter is left circular (LC, because of the change in the propagation direction) and cannot be received by the transmission channel. The diagonal terms equal zero. For ellipsoidal hydrometeors, whether or not the symmetry axes are common, the return wave is elliptically polarized. It can be solved in its two components RC and LC (Figure 2.22). The relative phase of the two components, determined by the angular orientation of the ellipse, is identical to the hydrometeor orientation (i.e., to the canting angle). The transmission channel is called a *direct channel*. The component it receives (here RC), which for a hydrometeor scattering is the weakest of the two, is the copolar component. The opposite component (here LC), the strongest one, is cross polar. The opposite channel that receives it is called the *main channel* or *orthogonal channel,* it is the one that gives the reflectivity in circular polarization. In the particular case of a linear scatterer, such as chaff (cf. Section 4.6) the two components are equal.

The property of single circular-polarization antennas to reject the opposite circular is used in some nonmeteorological radar (for example, air traffic control radar) to reduce the contribution of precipitation echoes and to improve the detection conditions of the target. The level of rejection depends on the ellipticity of the hydrometeors and the quality of the antenna (see later).

Finally, we must note that in the literature on polarization, the symbols and definitions are not always homogeneous, particularly for the circular polarization. A comparison of the results thus requires a peculiar care.

### *The Polarization Parameters*

To discuss the radar observable parameters and the techniques associated to them, we use the coherence matrix $J$ defined by

$$J = \begin{bmatrix} \langle E_1 E_1^* \rangle & \langle E_1 E_2^* \rangle \\ \langle E_2 E_1^* \rangle & \langle E_2 E_2^* \rangle \end{bmatrix} \tag{2.100}$$

Where $E_1$ and $E_2$ are the complex orthogonal components of the backscattered electric field, eq. (2.99). If the incident wave has the polarization of $E_1$, the diagonal

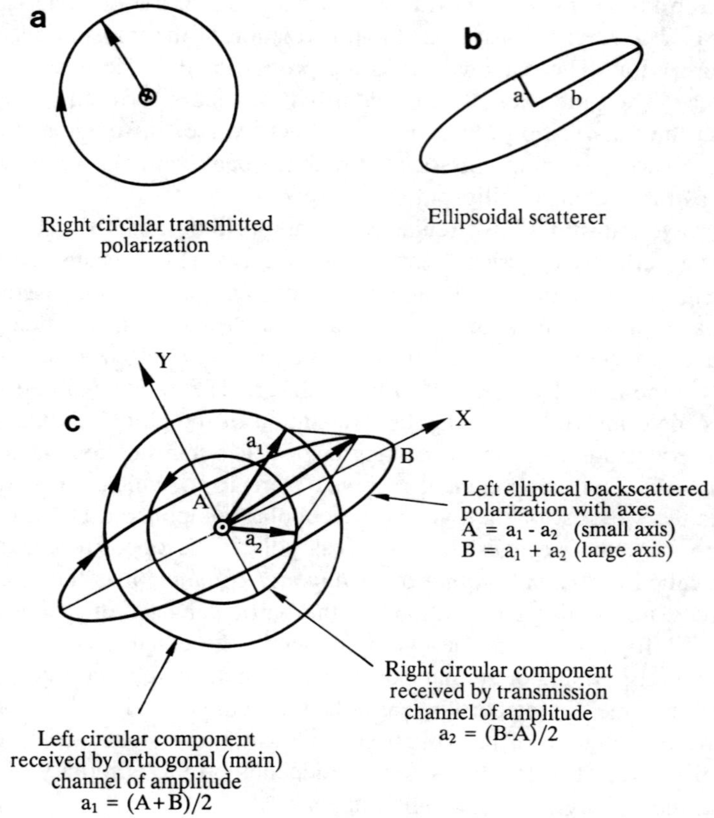


**Figure 2.22** (a) Right circular transmitted wave; (b) ellipsoidal scatterer; (c) left elliptical backscattered wave resolved in two circular component of opposite senses. We can see that (1) the axial ratio of the elliptical wave is $A/B = (a_1 - a_2)/(a_1 + a_2)$—if $a_1 = a_2$ polarization is linear; (2) the circular component of largest amplitude (orthogonal channel) has the sense of the backscattered elliptical wave; (3) if the phase of one of the circular component is retarded by $\phi$ with respect to the other component, the orientation of the resultant elliptical wave shift without deformation by $\phi/2$ in the inverse sense of the rotation of the retarded component.

terms represent the power of the copolar and cross-polar signals. The off-diagonal terms represent the complex correlation between the two components of the signal.

The first studies on depolarization, which began with the invention of meteorological radar, bore only the level of the orthogonal component relative to the main component, called the *linear depolarization ratio* (LDR) or *circular depolarizaton ratios* (CDR). By associating the corresponding reflectivity factor to each measurable term, we can write

$$\mathrm{LDR}_H(\mathrm{dB}) = 10 \log\left(\frac{Z_{HV}}{Z_{HH}}\right) \tag{2.101}$$

or the similar relation in $V$, and

$$\mathrm{CDR}(\mathrm{dB}) = 10 \log\left(\frac{Z_{RR}}{Z_{RL}}\right) \tag{2.102}$$

we also use, although more rarely, the *cancellation ratio* (CR). This term was first defined by Newell *et al.* [1958] as a combination of linear and circular polarization:

$$\mathrm{CR}(\mathrm{dB}) = 10 \log\left(\frac{Z_{RL}}{Z_{HH}}\right) \tag{2.103}$$

The cancellation ratio was redefined by McCormick and Hendry [1975] as the opposite of the CDR; that is,

$$\mathrm{CR}(\mathrm{dB}) = 10 \log\left(\frac{Z_{RL}}{Z_{RR}}\right) \tag{2.104}$$

Later, Seliga and Bringi (1976) defined the differential reflectivity $Z_{\mathrm{DR}}$, which is

$$Z_{\mathrm{DR}}(\mathrm{dB}) = 10 \log\left(\frac{Z_{HH}}{Z_{VV}}\right) \tag{2.105}$$

Interesting information also can be obtained considering the correlation at zero time lag between the two orthogonal components of the received wave. McCormick and Hendry [1975] show that for a circular polarization transmission the received signal as defined by $J$, contains four real quantities designated by $W_1$, $W_2$, $W_3$, and $W_4$. The first two, $W_1$ and $W_2$, are the intensities, respectively, of the copolar and cross-polar circular components, and $W_3 + iW_4 = W$ designates the complex correlation between the amplitudes of those two components. From these four quantities, they draw the reflectivity ($Z_e$) in circular polarization, the circular depolarization ratio ($W_1/W_2$), the correlation $\rho$, and the phase difference $\phi$ between the two components, defined by

$$\rho e^{i\phi} = \frac{W}{(W_1 W_2)^{1/2}} \tag{2.106}$$

$\rho$, which is also designated ORTT with

$$\rho = \text{ORTT} = \frac{|W|}{(W_1 W_2)^{1/2}} \tag{2.107}$$

can be approximately interpreted as expressing the nonspherical hydrometeor fraction having a common preferential orientation in the considered medium. However $\rho$ also depends on the shape effects. The value of $\phi$ is equal to twice the apparent mean orientation angle with the vertical. This angle, designated ALD, is such that

$$\text{ALD}_{\pm} = \pm\frac{1}{2}(\arg W_{\pm} - 180°) \tag{2.108}$$

where the plus and minus signs correspond to RC and LC, respectively.

This interpretation of $\rho$ and $\phi$ is based on the hypothesis that the propagation medium is isotropic; that is, their value is determined by backscattering and the propagation medium does not intervene. In the intense rain of short wavelengths, this condition is not fulfilled, and both the differential attenuation and the differential phase shift affect the ellipticity of the wave. These effects make the interpretation of $\rho$ and $\phi$ more difficult. McCormick and Hendry show that they can be detected by studying the dependency in distance (penetration depth of the medium) of the complex quantity $W/W_2$.

In the rain, $\rho$ is found to increase with $Z_e$ by about 10% at 10 dBZ and up to 80% at 50 dBZ; $\rho$ is weak in snow. Hendry and McCormick [1976] observed rapid changes of $\rho$ associated with lightning. Hendry and Antar [1984] suggested that $Z_e$, $\rho$, and CDR can be used to distinguish the various types of precipitation, although in a somewhat unprecise way in some cases (e.g., snow and weak rain).

The quantity $\phi$ is of lesser meteorological interest and influenced by the propagation effects. In addition, it can be useful to evaluate the importance of such effects.

Generally, hydrometeors (notably rain, which is the most often species observed by radar) tend to have a horizontal orientation, which has the following consequence: the propagation effects degrade the measurements more strongly in circular polarization than in linear polarization. At short wavelengths, the differential attenuation effects can be important, even prohibitive. At large wavelengths (i.e., when the attenuation is weak), the main cause of depolarization is the differential phase shift (cf. Section 2.4.3).

*The Measurement*

To measure the depolarization ratio, we must transmit a polarization and receive the direct and the orthogonal components either simultaneously or alternatively. LDR depends on the orientation of the polarization plane of the transmitted wave, on the hydrometeors orientation, on their shape, on their dielectric factor, and on their degree of common orientation. Comparatively, CDR is not very sensitive to the particles orientation and depends principally on the shape and the dielectric factor; however, it is about 3 dB higher if the hydrometeors have a common orientation than if they are equioriented. The values of the depolarization ratio have a span of about 30 dB. The typical levels are about −30 dB in a light rain (0.6 mm $h^{-1}$) and in snow, −25 to −20 dB in a moderate rain, −15 dB in an intense rain (50 mm $h^{-1}$), and −10 dB in the melting layer (under 0°C), which corresponds to the maximum value.

Among the limits of the measurement of the depolarization ratio, one is due to the fact that this parameter is measurable only if the orthogonal component remains detectable (the signal-to-noise ratio of the main component must be higher than the expected value of the parameter), which excludes weak echoes and thus nonprecipitating or slightly precipitating clouds.

To measure the four terms that describe the backscattering completely, we can, for example, transmit alternately in $H$ and $V$ polarization and receive simultaneously the $H$ and $V$ components in a dual-channel receiver. In practice, the four terms generally are not measured, because of the technical difficulties of the operation and because it is not always necessary to obtain this information, considering the nature of the observed medium.

Thus measuring the differential reflectivity $Z_{DR}$ can be relatively simple by alternately transmitting and receiving $H$ and $V$ in linear polarization. $Z_{DR}$ depends on the shape, on the common orientation degree, and on the dielectric factor of the hydrometeors, but is independent from the number of particles. Snow, graupels, and hail correspond to weak values of $Z_{DR}$, whereas for intense rain, it can reach 4 dB (cf. Section 2.6.7).

The measurement of the terms describing the backscattered wave polarization must be made very quickly, before the target geometry is modified; that is, inside the autocorrelation radius of the target in the sense of (1.104) because the decorrelation during the interpulse deteriorates the information. This is generally the case for measuring $Z_{DR}$ with a single-channel radar by commutation.

For a random medium, such as meteorological targets, $Z_{DR}$ is obtained by averaging the ratio of two correlated variables:

$$Z_{DR}(\mathrm{dB}) = \frac{1}{N} 10 \sum_{i=1}^{N} \log\left(\frac{p_H}{p_V}\right)_i \tag{2.109}$$

where $N$ is the number of independent pairs of the power value in $H$ and $V$ polarization; that is, $p_H$ and $p_V$, respectively. The precision of the measurement depends on the number of independent samples and the cross-correlation degree of the terms used to calculate the ratio [Chandrasekar and Bringi, 1988]. The quick polarization commutation enables the conservation of a high degree of correlation (Figure 2.23). A few tens of independent samples of the ratio thus are sufficient to obtain $Z_{DR}$ with a precision of 0.2 to 0.3 dB (Figure 2.24); comparatively, to obtain $Z_{DR}$ with the same precision from $Z_H$ and $Z_V$ estimated by uncorrelated measurements, several

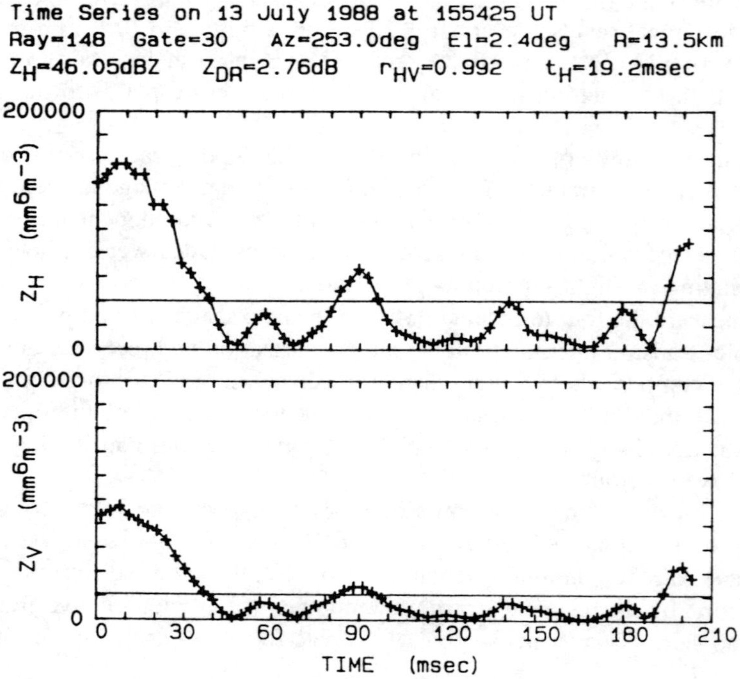


**Figure 2.23** Time series of $Z_H$ and $Z_V$ in heavy rain taken on with the 10 cm Chilbolton radar. Successive values of $Z_H$ and $Z_V$ fluctuate as the particles rearrange in space; the decorrelation time of each time series is related to the width of the doppler spectrum of the scattering particles. The series correspond to about 10 independent samples ($\tau_i \approx$ 20 ms) (from [Caylor and Illingworth, 1989]).

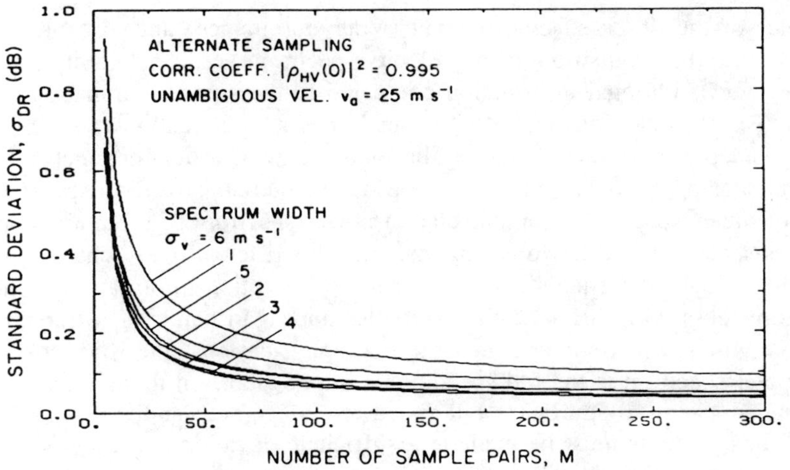


**Figure 2.24** Standard deviation of $Z_{DR}$, $\sigma_{DR}$ in a function of the number of sample pairs $M$, for alternate sampling method. Correlation at zero time lag between sample pairs is $|\rho_{HV}(0)|^2 = 0.995$. Spectrum width $\sigma_V$ is shown as a parameter and pulse repetition time is 1 ms so that, at a wavelength $\lambda$ of 10 cm, the unambiguous velocity $v_a = 25$ m s$^{-1}$. Increasing $\sigma_V$ decorrelates the sample pairs for a given acquisition time. $|\rho_{HV}(0)|^2 = 1.0$, that is, for perfectly correlated pairs, $\sigma_{DR} = 0$ and is independent of $\sigma_V$ (from [Sachidananda and Zrnic', 1985]).

thousand independent samples would be necessary. The estimator of $Z_{DR}$ using log ratio estimators typically has twice the standard error as those using square law estimators.

For polarization measurements, the quality of the antennas (illumination functions, reflectors, *et cetera*) is a decisive factor. For example, in circular polarization, if the transmitted wave presents an ellipticity, because of the technical shortcomings of the antenna, an "instrumental" orthogonal component is generated. For an ellipticity of 0.94 axial ratio, the rejection of the orthogonal component is 24 dB, which represents a good-quality antenna. The very good meteorological polarimetric radar antennas have a rejection that can reach 40 dB. If the causes for error are known and quantified, in some cases we can correct for their effects (Metcalf and Ussailis, 1984; Johnson, 1984; Pointin, Raymond, and Fournet-Fayard, 1988).

The value of the polarization parameter depends, of course, on the elevation angle of the beam, as the effects linked with a horizontal common preferential orientation are maximum for horizontal aiming and almost equal to zero for vertical aiming. Usually the polarization measurements are made at a low-elevation angle.

### *Polarization and Doppler Measurements*

The association of polarization and doppler measurements first presents the advantage of data complementarity. The doppler measurements bring information on the dynamics of the observed environment (wind, turbulence) and also on the microphysics, with the acquisition of the velocity spectra in vertical pointing (cf. Chapter 3). However, additional information compared to the simple summation of the polarization and doppler information (i.e., dual-channel polarization and single-channel doppler independently acquired) can be obtained with a dual-channel polarimetric doppler radar by analyzing the polarimetric characteristics of each spectral band in the frequencies space. We can calculate a power spectrum of the main channel and the cross spectrum of the two orthogonal channels [Metcalf and Echard, 1978; Metcalf, 1986]. Such a frequency sorting of the polarization information is interesting if hydrometeors with different characteristics appear in bands of different frequencies. We will see in Chapter 3 that the microphysical characteristic most affecting the doppler spectrum is the fall speed. This implies that, on the one hand, only the hydrometeors with a significant fall speed are concerned; and, on the other hand, that the observations must be made at a sufficient elevation angle ($> 30°$) (at weak elevation angles, the spreading of the doppler spectrum reflects principally the air movement and the contribution of the hydrometeors with peculiar microphysical characteristics is distributed throughout the spectrum).

*Hydrometeor Identification from Polarimetric Observations*

The problem of the radar identification of hydrometeors, in which the complete quantitative solution would be equivalent to a microphysical description of the scatterers, has yet to find a general solution. Multiparameter methods using doppler measurements at an elevation angle close to the vertical and measurements with several wavelengths offer some partial and indirect solutions. The polarimetric observations, with measurements with the shape and orientation of hydrometeors, present an important potential in this field. The exhaustive analysis of this question, combining the various radar observables and considering various hypotheses on the structure of the observed medium, lead to many theoretical possibilities and have been discussed by numerous authors [among others, Hendry and Antar 1984; Metcalf, 1986; Jameson, 1985 and 1989].

Rain (below melting level) is the simplest case because there is a simple relation between the size of the drops and their axial ratio, a stable state of common orientation, and a density (i.e., a dielectric factor) known and almost constant. It was proposed, in rain, to use the polarimetric measurement of the shape of the drops to obtain one or several parameters of the DSD [for example, Jameson, 1983a and 1983b; Illingworth and Caylor, 1989]. The main application consists in using this information to overcome the influence of the DSD variability on the radar estimation of the rain rates (cf. Section 2.6.7). The parameters used are $Z_{DR}$ or $C_{DR}$.

Compared to the polarimetric measurements of liquid hydrometeors, the ice hydrometeors important differences are:

1. an extremely variable density (from 0.9 to 0.1 g $cm^{-3}$ or less);
2. a very complex shape with equivalent axial ratios not linked to size;
3. modes of fall in which the orientation of the particles can go from strict horizontality, with simple shape crystals, to an equioriented state.

The result is that the indications given by the polarimetric radar on the ice hydrometeors are qualitative and ambiguous. Figure 2.25 shows, as an example, the influence of the density on $Z_{DR}$ according to the axial ratio. Schematically, the anisotropy of shape and the common horizontal degree of orientation of the hydrometeors decreases as the size increases. In the ice precipitation, an anticorrelation thus is observed between the polarization parameters and reflectivity. For example $Z_{DR}$, the most often measured parameter, is close to zero in hail (cf. Secton 2.5.4) and heavy snow. It increases in moderate precipitation. In light precipitation, it depends on the type of crystals concerned. In the nonprecipitating or slightly precipitating clouds of stratiform character, a dependence is observed between the temperature and the value of $Z_{DR}$ linked to the influence of the temperature on the shape of the crystals (cf. Section 2.2.6); the polarization parameters are horizontally stratified (Sauvageot, Kouadio, and Etty, 1986, and Section 5.5.2). In the melting layer,

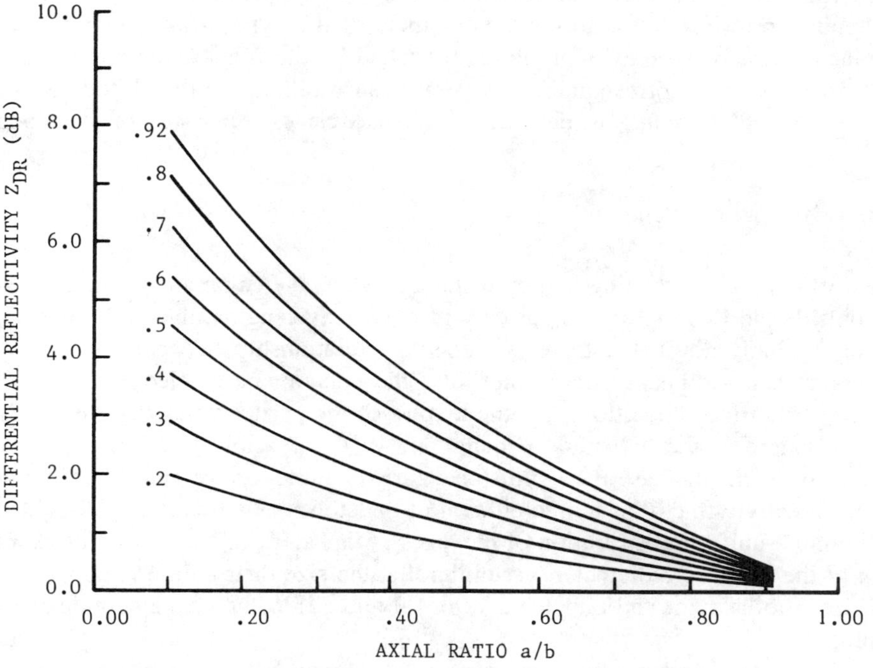


**Figure 2.25** Differential reflectivity of an oblate ellipsoid of ice (size 400 μm) as a function of axial ratio with particle density (in g $cm^{-3}$) as a parameter (courtesy K. Kouadio).

polarimetric parameters such as $Z_{DR}$ have a high value because the melting hydrometeors have a very flattened shape, the melted fraction spreading around the still solid fraction. In the mixed clouds and precipitation (mixture of solid and liquid hydrometeors) the polarimetric measurement cannot enable the estimation of the fraction corresponding to each phase. Generally, in a mixture of different size and phase hydrometeors, the polarimetric response of each component is pondered by its reflectivity; the characteristics of the small hdyrometeors are hidden by those of the big ones. Abnormally high values of $Z_{DR}$ were observed much higher than the melting level in convective clouds and were attributed to the presence of big drops or hydrometeors coated with water [Caylor and Illingworth, 1987]. Such observations remain exceptional. The combination of several polarization parameters (for example, CR and ORTT [McCormick and Hendry, 1975]) and the reflectivity enables us to make finer observations.

### 2.5.4 Hail Precipitation Detection

Starting with the first attempts to detect hail by radar by Byers and Braham [1949], numerous works have tried to describe the significant properties of the structure of a thunderstorm in relation to the development of this type of precipitation. Of particular interest is the maximum altitude reached by the echoes, the maximum value of their power and corresponding temperatures (in relation to the altitude of the 0°C and −35 to −40°C isotherms, between which water can exist in a state of supercooling).

*Single-Wavelength Reflectivity*

The difficulties in detecting hail by using a single wavelength are because hailstones lie outside the Rayleigh region at current radar wavelengths; the study of backscattering by hail shows that there is no simple relationship between the size of the scatterers and the backscattered intensity; the phenomena of electromagnetic resonance, the surface condition, and the texture of the particles but also the small horizontal extent of the hail trails [Changnon, 1970], possibly less than the width of the radar beam, increase the confusion. Various authors have calculated the equivalent reflectivity factor of a precipitation of hailstones as a function of the diameter, assuming a uniform distribution of the sizes [Atlas and Ludlam, 1961] or as a function of the maximum diameter, assuming that the size distribution is known [Sulakvelidze, Bibilashvili, and Lapcheva, 1965]. Figure 2.26 shows the complexity of the results.

Investigations have been made on a correlation between the radar reflectivity in the thunderstorm and the properties of the hail observed on the ground. For example, identification criteria have been suggested for the presence of hail in the

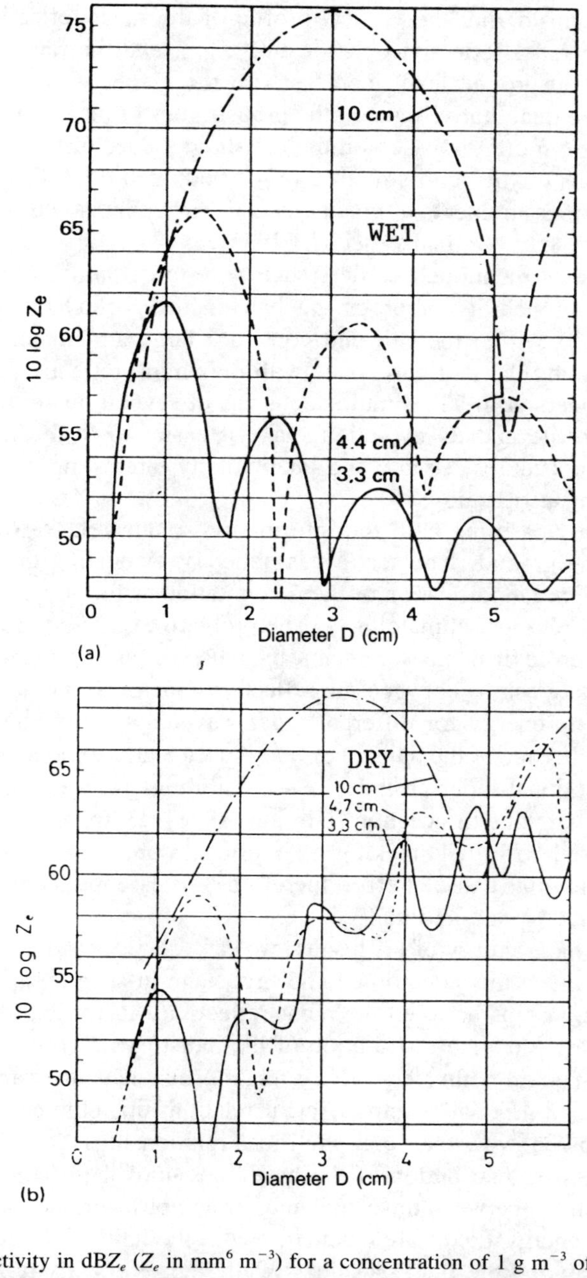


**Figure 2.26** Reflectivity in $dBZ_e$ ($Z_e$ in $mm^6\ m^{-3}$) for a concentration of 1 g $m^{-3}$ of spherical hailstones of uniform size as a function of the diameter assuming that the hail is (a) "electromagnetically humid" and (b) dry for different values of the wavelength specified on the curves. The equivalent precipitation rates are based on the relationship $Z_e = 4.86\ R^{1.37}$ (from [Atlas and Ludlam, 1961]).

thunderstorm by considering the vertical profile of the radar reflectivity: the maximum value of the radar reflectivity factor aloft $Z_{e,\max}$ and the ratio of this quantity to the value of $Z_e$ at ground level increase with the intensity of the thunderstorm. Taken together, the quantitative data on the probabilities of detection and false alarm obtained for hail from criteria based on the threshold values of the reflectivity or on the particular forms of the echo structure (e.g., hook echo) confirm the usefulness of the reflectivity measurements at nonattenuated wavelengths with conventional radars for identifying hail [Donaldson *et al.,* 1975].

For the reasons mentioned earlier, such criteria certainly provide overall information on the probability of observing hail on the ground but do not make it possible to locate it within the cell and contain a large amount of uncertainty. In fact, zones containing the hail and zones with maximum reflectivity do not always coincide in a thunderstorm. The simultaneous use of several quantities derived from the reflectivity profile and its evolution make it easier to deal with the problem. However, these procedures, simple and qualitatively interesting, remain empirical, with little physical justification.

Often, when measuring hail, the information sought relates to the quantity of hail arriving at the ground, whether this is aimed at evaluating the efficiency of a technique for artificial modification of the precipitation or for any other reason (evaluation of damage to crops, climatology). The problem expressed in this way is similar to that of hydrological measurements by radar discussed in Section 2.6. The analysis of the relationships between the reflectivity factor of the hail and the fluxes of mass and kinetic energy for different radar wavelengths between 3 and 10 cm, by assuming that the size of the hailstones obeys a truncated exponential distribution, shows that conventional radar provides real possibilities for a precise remote determination of the quantities in question [Ulbrich, 1978]. However, the possible presence of rain, mixed with hail, making a significant contribution to the reflectivity factor, is a nonnegligible cause of error; therefore, it is essential to be able to separate the components due to rain and hail.

To improve the accuracy of the results, we can use radar and an array of ground-based measurements at the same time. The first stage in such an approach involves studying the relationships between the radar reflectivity factor and the mass and kinetic energy fluxes from size distribution of the hailstones measured on the ground with an array of spectral hailmeters. These relationships have the same general form as the *Z-R* and the *Z-M* relationships. Actual relationships obtained from measured spectra are compared with the semi-empirical relationships based on various assumptions on the size distribution of the hailstones until there is an approximately satisfactory agreement between the actual and semiempirical relationships. Then, the mass and kinetic energy fluxes are calculated with the help of the relationship "adjusted" in this way and the field of values of the radar reflectivity factor measured in the vicinity of the ground. Testing the method by comparing the results derived from the radar measurement with actual values obtained from the data of hailmeters leads to a relatively good agreement (better than approximately 20% for the total kinetic energy during the same hailfall, according to [Waldvogel *et al.*, 1978a and 1978b]).

*Dual-Wavelength Reflectivity*

Another radar method consists of exploiting the dependence of the reflectivity on the wavelength in the Mie scattering region to detect the hail and obtain information on the size of the hailstones from simultaneous observations at several wavelengths. Figure 2.26 was established for wavelengths of 3.3, 4.7, and 10 cm, a uniform distribution of the hailstone size, a water concentration of 1 g $m^3$, and "electromagnetically" humid or dry particles. It shows that for hailstones with size between 1 and 4.5 cm, the equivalent reflectivity factor at 3.3. and 10 cm (i.e., $Z_{e3.3}$ and $Z_{e10}$) differs appreciably. For the pair of values $Z_{e4.7}$ and $Z_{e10}$ the differences are appreciable between 2 and 4.5 cm. Although the use of the pair $Z_{e3.3}$ and $Z_{e10}$ has resulted in satisfactory detection attempts [Sulakvelidze *et al.*, 1965], it is unreliable because the attenuation by the rain experienced by the shortest radiation is considerable. The observation of the target across intense convective rain leads to attenuation differences that can reach more than 10 dB and these affect the measured reflectivity differences. Thus, in the absence of any attenuation correction, a pair of wavelengths containing radiation too strongly attenuated should be prohibited. The pair 4.7 and 10 cm from this point of view is more favorable.

The validity of the method depends on a large number of other factors; in particular, the validity of the interpretation of the cause of the measured reflectivity deviation, taking into account the uncertainty of the actual physical properties of the observed precipitation (shape, distribution, texture), the homogeneity of the targets with respect to the size of the corresponding beams at the two wavelengths, and so forth. Also based on measurement with two wavelengths, the method suggested by Eccles and Atlas (1973) is simpler to use and lets us eliminate certain defects associated with the direct use of $Z_e$. It uses the derivative with respect to distance of the logarithm of the ratio of the reflectivities for the two wavelengths. Let us consider precipitation observed at the same time by two radars with wavelengths $\lambda_1$ and $\lambda_2$. The average power from the range $r$ is given by, see eqs. (2.114d),

$$\bar{P}_i = C_i \frac{Z_{ei}}{r^2} 10^{-0,2\int_0^r a_i \mathrm{d}r} \qquad (2.110)$$

where $i$ (= 1 or 2) refers to the two wavelengths, $C_i$ is the radar constant, $Z_{ei}$ is the equivalent reflectivity factor, and $a_i$ is the attenuation rate. The logarithm of the ratio of the received powers, expressed in decibels, is

$$\begin{aligned} y(r) &= 10 \log(P_1/P_2) \\ &= 10 \log(C_1/C_2) + 10 \log(Z_{e1}/Z_{e2}) - 2 \int_0^r (a_1 - a_2)\, dr \end{aligned} \tag{2.111}$$

If $\lambda_1 = 10$ cm and $\lambda_2 = 3$ cm, $a_1$ is negligible compared to $a_2$. Furthermore, if the observed volume contains only liquid precipitation or precipitation sufficiently small to belong to the Rayleigh region, $Z_{e1} = Z_{e2}$ and the function $y$ increases in a monotonic manner with $r$ because of the stronger attenuation undergone by the shortest wavelength. On the other hand, if hail precipitation with a diameter larger than 1 cm is present in the beam, $y$ increases more rapidly at the front of the hail volume and decreases at the rear because the hailstones have a much larger backscattering cross section at 10 cm than at 3 cm ($Z_{10} \gg Z_3$). Figure 2.27 illustrates this situation for the case of a parabolic distribution of rain in a region containing a hail front also with a parabolic distribution. One then considers the function $dy/dr$; in the absence of hail, this function expresses the attenuation coefficient at 3 cm. In the presence

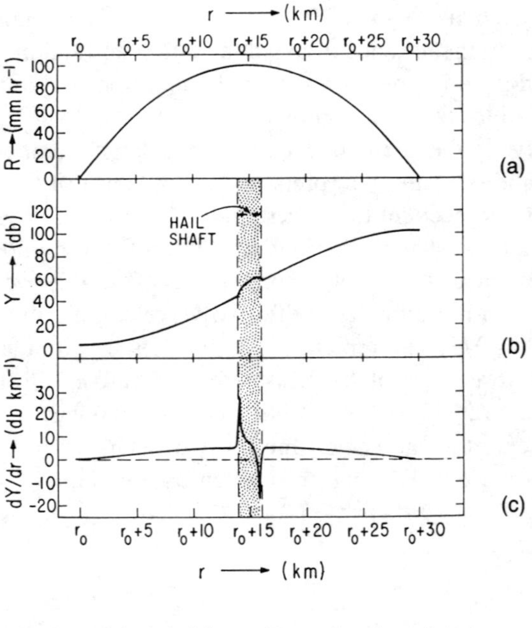


**Figure 2.27** Parabolic distribution of the rainfall rate $R$ as a function of the range $r$ and corresponding variation of $y$ and $dy/dr$, where a hail trail of 2 km depth (depending on $r$) is centered around $r = 15$ km (from [Carbone *et al.*, 1973]).

of hail, it has a sharp variation, positive at the front of the hail zone, small or negative at the rear of the hail zone. Whereas the positive variation of $dy/dr$ can be attributed to an attenuation by a region of high water content, the negative value can be due only to hail of large size. The authors therefore suggest considering the negative value as an indicator that unequivocally locates the rear boundary of a hail zone. They conclude that this technique makes it possible to detect smaller hail concentrations within precipitation than only taking into consideration the ratio of the reflectivities at two wavelengths; because only a small ratio of radar reflectivity is required at two wavelengths, the technique is not affected by the attenuation and is independent of the absolute calibration of the radars. However, the method presents difficulties concerning the coincidence of the beams relative to each wavelength [Rinehart and Tuttle, 1982], attenuation differences [Tuttle and Rinehart, 1983], and measurement simultaneousness. The validity of this method is considered to be insufficiently established. A negative value of $dy/dr$ indicates the presence of hail with size larger than 1 cm; however, it is no sure indication of the furthest boundary of the hail zone and can be produced within a region where the size of the hail or the thickness of the water film surrounding the hailstone varies with distance [Jameson and Srivastava, 1978]. In conclusion, the inconveniences of this method are technical difficulty in obtaining measurements having the necessary precision and a lack of clarity in the hail signal.

### *Circular Polarization*

The measurement of polarization combined with that of reflectivity has been suggested as a means of distinguishing different types of precipitation, in particular detecting the presence of hail [Barge, 1974; McCormick and Hendry, 1975; Hendry and Antar, 1984]. The comparison on thunderstorms between measurements by radar with low elevation and observations of the precipitation at ground level shows that, when the reflectivity in altitude is less than 30 dBZ, hailstones larger than small shot are not observed. On the other hand, when the reflectivity is larger than 50 dBZ, large hailstones are observed from the ground. In the transition region, the probability of observing a hailfall increases with the circular depolarization ratio. The relation between a high CDR and the presence of hail is due to the usually irregular shape of the hailstones. Figure 2.28 summarizes the results of Barge [1974] on the combined study of the two quantities $Z_e$ and CDR. The method suffers the effects of depolarization occurring during the propagation in rain between the radar and target, which are unduly attributed to the backscattering.

### *Linear Polarization*

A clearer hail signal can be obtained with the joint measurement of $Z_e$ and a parameter of linear polarization. Indeed, the hailstones generally tumble and gyrate, so

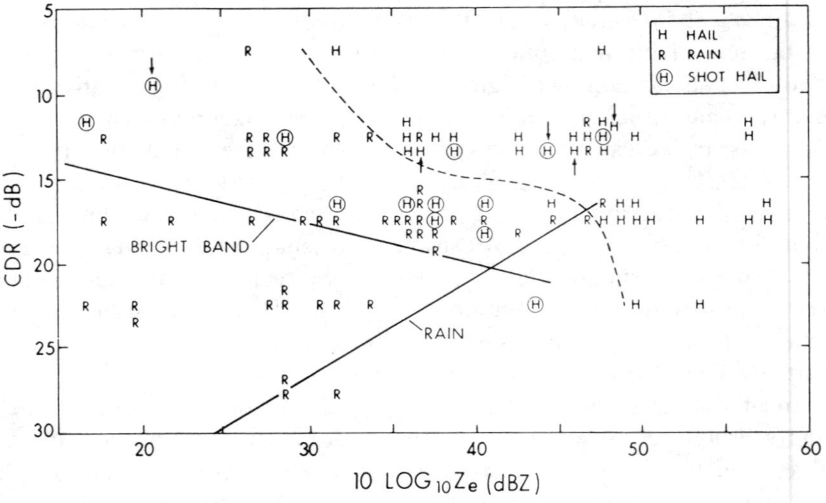


**Figure 2.28** Values of the circular depolarization ratio *versus* the equivalent radar reflectivity factor $Z_e$ (in dBZ) measured for elevation angles of the radar beam between 1° and 2°. The dashed line separates rain and shot-sized hail from larger-size hail. The two solid lines represent the depolarization and reflectivities for rain and bright band, respectively (from [Barge, 1974]).

that, in spite of their sphericity defects (responsible for their response in circular polarization), hail precipitation presents no preferential orientation and the values of $Z_{DR}$ they produce are close to zero. They are even slightly negative for hailstones of peculiar shape falling with their larger dimension vertically oriented. The small melting hailstones (under about 10 mm) seem to correspond to slightly positive values of $Z_{DR}$. The rain is characterized by $Z_{DR} > 0.5$ dB and a reflectivity factor that can reach 60 dBZ. The anticorrelation of $Z_e$ and $Z_{DR}$ that comes from the statistical relation valid for the rain, defines the signature in the $Z_{DR}$ of hail. Thus, in a region of high reflectivity associated with the intense precipitation of storms, the high values of $Z_{DR}$ are due solely to the rain, whereas the $Z_{DR}$ weak or equal to zero shows the presence of hail. When there is a mixture of rain and hail, the contribution of the hail is supposed to prevail over that of the rain, because of the size of the hailstones. The $Z_{DR}$ signature of the hail is easily identifiable below the melting level. Aydin, Seliga, and Balaji [1986] propose an $H_{DR}$ parameter combining $Z_e$ and $Z_{DR}$ to detect the presence of hail. The measurements at $\lambda = 10$ cm with ground control show the validity of using $Z_{DR}$ to detect the presence of hail [Bringi, Seliga, and Aydin, 1984; Illingworth, Goddard, and Cherry, 1986]. Bringi, Vivekanandan, and Tuttle [1986] present observations that show that the localization of the hail signature from $Z_{DR}$

below the melting level, from LDR above the melting level, and from the dual-frequency ratio are in agreement and enable us to see the vertical structure of the hail shaft.

### 2.5.5 Lightning Detection

Localizing and observing lightning poses considerable problems because of the small volume it occupies and its brevity, its suddenness and the unpredictability of its trajectory. The complexity of the problem is increased by the uncertainty as to its relationships with the properties and structure of the generator cloud and the processes responsible for the formation of charged zones.

Although the directivity characteristics of the beam hamper the observation of lightning in some respects, radar is one of the main instruments for studying atmospheric electricity and related problems dealing with both the causes and the consequences of the phenomenon.

The electrification of thundercloud is due to the charging of graupels and ice crystals by collision and their separation by differential vertical velocities. Lightning is a transitory electrical discharge of high intensity produced when the electric field associated with a distribution of charge reaches the limiting break-down value of air (several hundred kilovolts per meter within the cloud). It is commonly observed in cumulonimbus clouds but can also be produced in snow, sand storms, or volcanic eruption clouds [Uman, 1969].

*Sferics: Passive Detection of Lightning*

An effect of electric discharge in a thunderstorm is the transmission of an electromagnetic field proportional to the variation in time of the intensity of the discharge current (more precisely, proportional to the variation of the total dipole moment associated with the charges present in the thunderstorm). This emission, called *sferic,* can be easily detected up to very large distances with the help of ordinary radio receivers at frequencies on the order of tens of kilohertz. It gives rise to observations of the radiogoniometric type, making it possible to localize the discharges by triangulation. At current radar wavelengths, the energy is sufficient for the (radar) receivers to detect sferics at ranges on the order of 10, even 100 km, when the discharge is produced in the aiming direction of the antenna. Such an observation cannot give any information on the distance of the lightning but only its direction and possibly its duration, its intensity, and fine structure (counting pulses, *et cetera*). Observations of this type have been described by Ligda [1956] and Oetzel and Pierce [1968] for wavelengths between 10 and 50 cm. They show that the sferics have a duration on the order of 100 $\mu$s and a discontinuous structure associated with the development of the precursor. The distribution of the intensity as a function of the frequency is approximately proportional to $f^{-1}$, where $f$ is the frequency. The power of the signal in the radar band has been measured at a level of approximately $10^{-11}$W (that is, $-80$ dBm).

*Active Detection*

Radar can be used to detect lightning from the backscattering by the ionized high-temperature channel of a few centimeters in diameter that reveals the trajectory of the energy. Attempts have been made in this direction [Ligda, 1956; Browne, 1951; Atlas, 1958; among others]. It seems that the echo cannot be explained by the sharp variation in the refractive index created by heating the air around the ionized channel but rather by the scattering of the free electrons produced by the ionization of the air. The duration of the observed echoes is on the order of a few tenths of a second followed by an exponential decrease associated with the disappearance of the ionized channel by recombination of the electrons and the ions. The critical point for detecting the ionized channel is the concentration of free electrons, which governs the intensity of the backscattered signal and its duration. Various estimates of these quantities have been made. A value of approximately $10^{12}$ $cm^{-3}$ was found, leading to an initial concentration of $10^{17}$ $cm^{-3}$ with a decrease to $10^{10}$ $cm^{-3}$ in approximately $10^{-2}$ s [Uman and Orville, 1964; Mazur, Zrnic', and Rust, 1985].

The radar backscattering cross section of cloud-to-ground lightning has been calculated for low-density and high-density plasma corresponding to two extreme cases [Dawson, 1972]. The results confirm that the cross section of the dielectric discontinuities due to the warm air of the channel is negligible [Browne, 1951].

For high-density (or superdense) plasma, the channel can be considered a perfect reflector. The backscattering cross section then depends only on the ratio of the radius of the channel ($a$) to the wavelength and the polarization. For two $2\pi a/\lambda >$ 1, the cross section $\sigma$ is determined by geometrical optics and is independent of the polarization

$$\sigma = 2\pi ra \tag{2.112}$$

where $r$ is the radar-target distance.

For $2\pi a/\lambda < 1$, we find that the cross sections $\sigma_\perp$ and $\sigma_\parallel$ for polarization perpendicular or parallel to the channel, respectively, are given by

$$\sigma_\perp = 8r\pi^5 a^4/\lambda^3 \tag{2.113a}$$

$$\sigma_\parallel = \frac{2r\lambda}{\pi|1 - i(2/\pi)\ln(\gamma ka/2)|^2} \tag{2.113b}$$

where $k = 2\pi/\lambda$ is the radar wave-number and $\gamma$ is a constant equal to 1.781.

With the help of a model of the ionized channel in decline (i.e., ions in course of recombination) and by interpolation between these limiting cases, curves of time variation of the radar cross section have been obtained for wavelengths between 3.2 and 150 cm. The results show that radars using wavelengths greater than or equal to 10 cm can detect lightning for periods of approximately 100 ms after the first discharge. At shorter wavelengths, less than a few centimeters, the cross section decreases very rapidly with decrease in the concentration of electrons in the channel so that the lightning probably can be detected for only a few milliseconds. Figure 2.29 shows the results obtained for parallel and perpendicular polarization and a range of 3.66 km. Note that the cross section depends on the polarization only for channels with a small radius and large wavelengths.

However, the observation of lightning by direct radar detection may be confronted by the presence of echoes of precipitation with a level comparable to that expected of the lightning alone. Because the echoes of precipitation are proportional

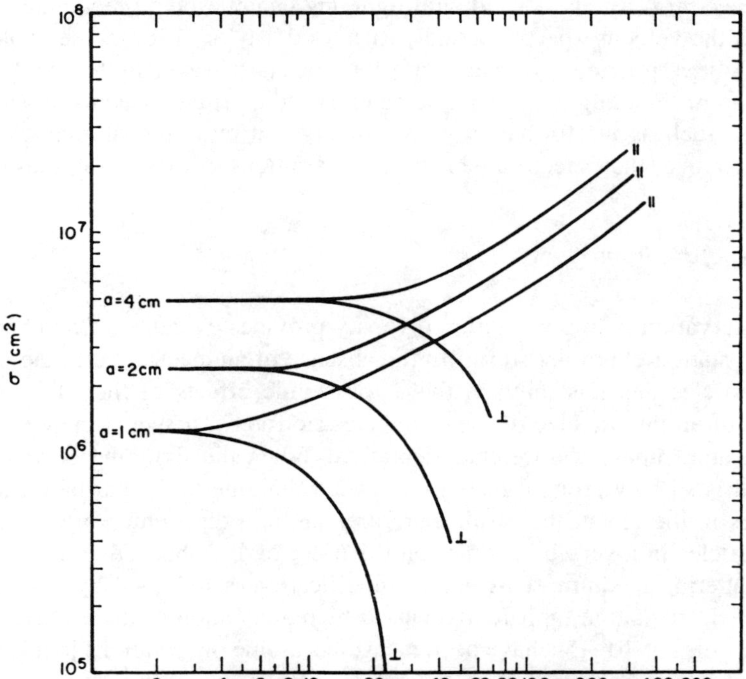


**Figure 2.29** Backscattering cross section of a perfectly reflecting lightning channel as a function of the wavelength. The channel radius is 1, 2, or 4 cm; the range is 3.66 km. The parallel (‖) and perpendicular (⊥) polarizations are indicated (from [Dawson, 1972]).

to $\lambda^{-4}$, we must use relatively long wavelengths for detecting lightning inside the precipitation. For a wavelength of 10 cm, detecting lightning in rain only seems feasible with circular polarization or the help of cross-linear polarization between transmission and reception aimed at reducing rain echoes. For normal linear polarization, wavelengths above 10 cm must be used (for example, $\lambda > 50$ cm) [Cerni, 1976].

A classification of "intracloud" discharge echoes has been proposed [Ligda, 1956, from observations with a wavelength of 23 cm]. This kind of lightning is sometimes very long, a few tens of kilometers (150 km for the longest) with a large number of branches. Four types of lightning have been distinguished depending on the position of the charge centers and the discharge mechanism.

Radar observations show that, in certain cases, echoes from successive discharges constituting the same vertical lightning involve higher and higher levels in the cloud until the heights may exceed 10 km; the most intense and prolonged echoes are returned by the lower part of the cloud. Investigations with doppler radar show that the spectral peaks associated with lightning channels and precipitation are often distinct in the velocity spectra (Zrnic', Rust, and Taylor, 1982). The exploration of the possibilities provided by radar with long or short wavelengths for the physical observation of lightning is far from being exhausted; progress seems possible in several areas, such as the formation of precursors and returning discharges, their velocity, their size, their electron density, and variations of these quantities over time.

### *Alignment of Hydrometeors*

Radar observation with polarization diversity provides evidence of another effect of the electrostatic field of the storm. In the absence of an electric field, the orientation of the particles depends only on the aerodynamic effects of their fall. Under the influence of an intense electric field, the electric forces, it seems, may become predominant and impose a different orientation. When the lightning is produced, the local field is sharply brought back to a lower value due to the partial disappearance of charges in the cloud; the result therefore can be a disorientation or reorientation of the particles in a very brief time (on the order of 1 s) that reduces the anisotropy of the scattering medium. This abrupt modification is followed by a slower return to the initial orientation, apparently caused by regeneration of the electrostatic field. Recovery times of 10–15 s have been measured. Some observations in this way show the existence of two stable states separated in time: one with a vertical orientation of the particles assumed to be due to the influence of the electrostatic field, the other with a horizontal orientation of the particles under the influence of air resistance [Hendry and McCormick, 1976; Hendry and Antar, 1982].

At centimetric wavelengths, these changes in orientation obviously are accompanied by a modification of the conditions of propagation of parallel or perpendicular polarization waves with respect to the anisotropy axes. These modifications are revealed by observing the changes that occur in the backscattered signals in each polarization as a function of the penetration depth.

Various authors have also considered the consequences of thunderstorm discharges on the freezing of the cloud elements or precipitation elements undergoing supercooling, the assumed effect of the discharge being to promote the freezing. Under these conditions, the discharge must be accompanied by a reduction in the reflectivity corresponding to a change in the value of the factor $|K|^2$ from 0.93 to 0.176 until the increases in the size of the solid particles in the mixture compensate and exceed this effect. However, this process remains hypothetical.

### 2.5.6 Artifacts

The measurement of the clouds, and precipitations, reflectivity can be affected by artifacts associated to peculiarities of the target. The most well-known cause is linked to the air refractivity, considered in Section 1.2.7. We saw that the refractivity distribution can lead to modifications in the shape and direction of the beam. In the storms, a quasi-specular reflection of the beam can occur on surfaces presenting a strong refractivity gradient.

Another type of artifact was recently described: flare echoes detectable as anomalies in the reflectivity and velocity fields in the storm. The echo has a long shape; it is radially lined up behind a very high reflectivity core (rain or hail). The process proposed to explain the flare echo (Zrnic', 1986; Wilson and Reum, 1987) appeals to

1. A diffusion of the incident beam by the hydrometeors toward the ground;
2. A scattering from the ground to the hydrometeors;
3. A scattering of the hydrometeors toward the radar.

In the velocity field, the signature of the flare echo is an anomaly occupying the same region as the reflectivity anomaly (the velocity of the flare echo depends on the radial velocity and the fall speed of the scattering hydrometeors).

### 2.5.7 Particular Meteorological Forms of the Radar Equation

*Expression for the Received Power*

To relate the power of the received signal to the physical properties of the medium observed with the help of the radar equation, the radar reflectivity is substituted by the factor $Z$, which can be defined by the preceding formulas as a function of significant quantities in meteorology. By substituting (2.69) in (1.76a) we obtain

$$\overline{P_r} = \left(\frac{c\pi^3}{1024 \ln 2}\right)\left(\frac{P_t\tau G_0^2\theta_0\phi_0 L_r}{\lambda^2}\right)\left(L^2|K|^2 \frac{Z}{r^2}\right) \tag{2.114a}$$

In this formula, all the terms are expressed in the same system of units and the angles $\theta$ and $\phi$ are in radians.

If we use the units currently used in practice in (2.114a) we obtain

$$\overline{P_r}(\text{mW}) = \left(\frac{\pi^5 \times 10^{-19}}{2^{10} \times 1.08 \ln 2}\right)\left(\frac{P_t(\text{W})\tau(\mu\text{s})G_0^2\theta_0(\text{deg})\phi_0(\text{deg})L_r}{\lambda^2(\text{cm})}\right) \cdot \left(L^2|K|^2 \frac{Z(\text{mm}^6\ \text{m}^{-3})}{r^2(\text{km})}\right) \tag{2.114b}$$

The constant coefficient between parentheses is approximately $3.99 \times 10^{-20}$. We can also write

$$\overline{P_r} = C_3 L^2|K|^2 \frac{Z}{r^2} \tag{2.114c}$$

where $C_3$ is a constant that depends on the radar used and the units.

Remember that eqs. (2.114) apply to a distributed target when the following asumptions are satisfied:

1. The target occupies the entire volume of the pulse;
2. It consists of spherical particles or particles that can be approximated by spheres;
3. The size of the particles is small with respect to the wavelengths and satisfies the conditions of the Rayleigh approximation;
4. The dielectric ($|K|^2$) and granulometric ($Z$) properties are homogeneous in the volume considered;
5. The antenna beam is approximately Gaussian;
6. Its polarization is linear;
7. The effects of multiple scattering are negligible.

When condition 1 is not satisfied (2.114) cannot be used, unless it is possible to determine a corrective coefficient for the occupation of the beam expressing the fraction of the pulse volume occupied by the target, weighted by the relative value of the antenna gain in the corresponding solid angle. When conditions 2, 3, and 4 are not satisfied, $Z$ is replaced by the equivalent reflectivity factor $Z_e$ with $|K|^2 = 0.93$. In this case and for an attenuated wavelength, (2.114c) can be simply written

$$\overline{P_r} = C_4 L^2 Z_e / r^2 \tag{2.114d}$$

where $C_4 = 0.93 C_3$ is a constant. If the attenuation is negligible, (2.114d) lets us obtain $Z_e$ by a radar measurement giving $\overline{P_r}$ and $r^2$. In a general manner, any useful quantity that can be related to $\eta$ or $Z$ by an analytical expression can be made explicit in these equations; in particular, the intensity of precipitation $R$, water content $M$, or the median volume diameter $D_0$.

### *Logarithmic Form*

Various quantities expressing the properties of the signal usually are given by using a logarithmic scale: received power in dBm, radar reflectivity factor in dBZ. It follows that it is useful to write the radar equation in a logarithmic form. To do this, we define a reference level so that the signal can be written in terms of a dimensionless quantity. Expressing the powers with respect to the reference level $P_m$ (often taken equal to the minimum detectable power) and by taking the logarithm, we obtain, from (1.76b),

$$10 \log \eta = 20 \log r - 10 \log C_5 + S + 2 \int_0^r a \, dr \tag{2.115}$$

in which $r$ is the range, $C_5 = C_2 / P_m$ is a constant that depends on the radar and the selected reference level, and $S$ is the value of the received signal in dB with respect to the reference level.

A particular signal reference can be introduced into the equation, such as a value obtainied by calibration (for example, from a precipitation with known property); that is, $Z_c$. Starting from equation (2.114c) we have

$$Z(\text{dBZ}) = Z_c(\text{dBZ}) + 20 \log r/r_c + P_r(\text{dBm}) - P_{r_c}(\text{dBm}) + 10 \log \frac{|K_c|^2}{|K|^2} + 2 \int_{r_c}^{r} a \, dr \tag{2.116}$$

where the subscript $c$ refers to the signal used as a reference. For a nonattenuating medium, by regrouping known terms, the equivalent reflectivity factor is

$$Z_e(\text{dBZ}) = P_r(\text{dBm}) + 20 \log r + C_c \tag{2.117}$$

*Propagation in an Attenuating Medium*

When, because of the wavelength used and the nature of the medium observed, the attenuation is not negligible, the last term of equation (2.116) must be integrated along the beam trajectory. This "attenuation correction" is useful essentially in the field for which the radar data are exploited quantitatively This field is generally limited to radial distances less than 100–150 km. It is necessary to examine separately the three terms of equation (2.61); that is, $a_g$, $a_c$, and $a_p$.

The attenuation rate for gases generally is weak, often negligible, except at millimetric wavelengths and in humid tropical air. It is estimated empirically and approximately from the humidity vertical profile by assuming that the medium is stratified.

For $a_c$ and $a_p$ the spatial distribution of the hydrometeors, being very heterogeneous, the attenuation rate depends on $r$ and must be evaluated for each distance. When the attenuating environment is detected (i.e., gives an echo) and when there is a relation between the attenuation rate and the radar reflectivity factor $Z$, the integral of eq. (2.116) can be calculated by iteration step by step [Hildebrand, 1978]. The relation between $a$ and $Z$ is obtained by combining the $a$-$X$ relations (where $X = R$ or $M$, cf. Section 2.4) and the $Z$-$X$ relations (cf. Section 2.5.2). If $a = \alpha X^{\beta}$ and $Z = cX^d$, we have

$$a = \alpha \left( \frac{Z}{c} \right)^{\beta/d} \tag{2.118}$$

where $\alpha$, $\beta$, $c$, and $d$ are the empirical coefficients of the relations that must be chosen according to appropriate hypotheses concerning the propagation medium. This medium is supposed to be composed of only one type of hydrometeor: rain, snow, cloud.

Let us call $Z_a(r)$ the attenuated reflectivity factor measured by the radar at the distance $r$. By putting $Z_a$ in (2.118), we obtain an estimation of the attenuation rate, $a(r)$. With equation (2.116) and $a(r)$, a first estimate of the corrected reflectivity factor is calculated; that is $Z'(r)$:

$$Z'(r) = Z_a(r) + 2 \sum_{x=1}^{r-1} a(x) \tag{2.119}$$

where $Z$ is in dBZ and $a$ in dB km$^{-1}$ for one path.

By introducing $Z'$ in (2.118) a new, more exact estimation is obtained of the attenuation rate, $a'(r)$, which enables the computation of a second estimation of the corrected reflectivity factor:

$$Z''(r) = Z_a(r) + 2 \sum_{x=1}^{r-1} a'(x) \tag{2.120}$$

The same process is repeated until the successive iterations no longer produce significant modifications of $Z$ and $a$ and until the cumulative attenuation (i.e., second term on the right in (2.120)) $a'(x)$, changes from the previous iteration by less than 1 dB. The attenuation correction starts when entering the attenuating medium and stops at the maximum distance for which a signal is observed. Of course, we must exclude the areas perturbed by the ground echoes, notably the area close to the radar.

When, because of a calibration error or a wrong choice in the coefficients of relation (2.118), the correction produces an overestimation of the attenuation, the calculation diverges. In this case, it must be stopped after some iterations; the convenient number of iterations has to be determined empirically. Hildebrand [1978] shows that the errors made on the coefficients of (2.118) can significantly alter the results but that these alterations are weak compared to the effects of the radar calibration error. The computation must be limited to the cases where the radar calibration error is smaller than the maximum attenuation (dB $km^{-1}$). For a radar wavelength of 5 cm, the method gives bad results when the maximum values of $Z$ are higher than 60 dBZ. This threshold decreases when the wavelength considered decreases. Moreover, it cannot be reliably corrected if the attenuating medium is not homogeneous (for example, when there is a melting layer) or if it is composed of a mixture of different types of hydrometeors.

Usually, the attenuation correction is made, considering only the effect of rains. However, we must keep in mind that for $\lambda < 5$ cm, the attenuation by the water clouds, possibly present in the radioelectric path, can be nonnegligible, even if their contribution to the reflectivity is negligible, and, if the case arises, they are not detected. For example, for an observation at a distance of 50 km through a cloud at 0° C, whose liquid water content is equal to 1 g $m^{-3}$, and whose reflectivity factor is $-12$ dBZ, the attenuation is about 9 dB and 3 dB, for $\lambda = 3.2$ cm and 5.7 cm, respectively. Such an attenuation is equivalent to that of a rain having intensities of 6.5 mm $h^{-1}$ and 8.6 mm $h^{-1}$, respectively. For the airborne radar, which in general is in the $X$ band, the attenuation by the clouds is always an important term.

*Correction of the Attenuation as a Function of the Range*

The radar equation applied to a meteorological target filling up the beam shows that the power of the received signal is proportional to the reciprocal of the square of the range of the target. The correction of this variation is obtained by forming the product $r^2\overline{P}_r(r)$, which is independent of the range. This operation is "range normalization," by means of which the signals with the same reflectivity are represented with equal levels regardless of range.

Such a correction can be made directly in the receiver to obtain a normalized display of the echo levels on the screen, by a variation of the receiver gain proportional to the square of the transit time of the energy between the radar and the target so as to "cover" the interval of the useful ranges between $r_{\min}$ and $r_{\max}$. The value of $r_{\max}$ depends on the maximum gain of the receiver. The correction also can be obtained by a reduction in the receiver gain (or attenuation) proportional to $(r_{\max}/r)^2$ in the same interval. In this case, the correction is accompanied by a loss in sensitivity of the radar, which hampers the observation of weak signals (for this reason the correction circuit can always be switched on or off by the operator).

## 2.6 PRECIPITATION MEASUREMENTS

### 2.6.1 Introduction

The most frequent quantitative application of the radar observation is to measure precipitation fields. Compared to an array of conventional pluviometers, the measurement of precipitation by radar offers important advantages:

- Spatial continuity of the observations (the mesh of the pluviometer networks usually is larger than the scale of the severe convective cells, giving intense precipitation that can thus "escape" from the network or induce an error by nonrepresentaive peaks);
- Improved access to the observation of the variability of the precipitation;
- The possibility to observe the three-dimensional structure of the system generating the precipitation;
- The possibilitiy of real-time surveillance over a wide area from a single measurement point;
- The facility for data acquisition, storage, and processing.

This method lends itself to a large number of applications, such as the climatology of precipitation for statistics for radio propagation and agriculture; the prediction of floods and their surveillance during critical periods; the study of soil erosion, the evaluation of water resources in regions with difficult access and sparse populations, thus lacking routine measurement points; the evaluation of snow in relation to the prediction of avalanches, or for the measurement of available reserves in the catchment area of hydroelectric dams. To discuss the various methods proposed, the expression of the precipitating rate $R$, eq. (2.9), can be considered, in the most simple case of an exponential distribution (2.15):

$$R = \frac{\pi}{6} N_0 \int_{D_{min}}^{D_{max}} D_e^3 \, e^{-\Lambda D_e} \, [V(D_e) - W] \, dD_e \tag{2.121}$$

$D_e$ is the equivalent spherical diameter of the liquid hydrometeor.

This equation shows that $R$ depends on the two parameters of the distribution $N_0$ and $\Lambda$ and the vertical velocity of the air $w$. The measurements are made at a low level, as close as possible to the ground, and $w$, which is unknown, is supposed to be equal to 0. If this hypothesis is not satisfied, the measurement of $R$ (for $w$ = 0) overestimates or underestimates the real rate according to whether the precipitation falls in ascending or descending air, respectively.

The evaluation of $R$ by radar is based on measurements of the reflectivity $Z$ at one or two polarizations, on measurements of attenuation $A$, or on a combination of these two types of measurement. When only one radar observable is measured, $R$ is computed by an analytical relation of the $R$-$Z$ or $A$-$Z$ type, ignoring the spatiotemporal variations of $N(D)$. When two radar observables are measured, we determined two parameters of $N(D)$ and, from it, $R$. Some of these methods are notably improved if we have calibration means at the ground. Such calibration lets us correct instrumental errors and some propagation effects and reduce the variance of the results due to the measurement insufficiencies concerning $N(D)$.

The quality of the radar measurements depends on a large number of meteorological or hardware factors. Among the latter, the antenna beamwidth is of fundamental importance because it determines the spatial resolution of the radar; in fact, the dimensions of the pulse volume fix the minimum size of the small-scale variations accessible to the radar and therefore its capacity to measure the exact value of local extremes. More generally, the resolution arises in problems associated with the homogeneity of the target mentioned in Section 2.5.1 as a condition of the application of the meteorological radar equation. In fact, a beam with a width at the current value of 1° corresponds to a resolution volume whose lateral dimension is 2.6 km at 150 km. The condition for regularity of the precipitation concerns the horizontal homogeneity of the size distribution (as a result of the same physical process of formation) but also the vertical homogeneity. Other meteorological factors are involved, including modification of the precipitation during its fall to the ground within the layer in which the measurement is made.

Other major problems in hydrological measurement by radar are those of parasite echoes (ground echoes, in particular) and, above all, screening that can obscure part of the beam. These are related to the configuration of the ground, the location of the radar, and the shape of the beam. In summary, the main sources of error that can arise in hydrological measurements by radar are

1. Homogeneity defects of the target:
   - Different types of precipitation (for example solid and liquid);
   - Anomalies in the size distribution—multimode distribution or presence of particles situated outside the Rayleigh region (hail);
   - Beam occupied in a nonuniform manner;
   - Presence of the melting layer;
   - Presence of parasite targets.
2. The modification of the precipitation between the ground and the region where the radar measurement is made:
   - Evaporation or growth;
   - Vertical movements of the air.
3. Propagation problems:
   - Attenuation (precipitation clouds, gases, radome);
   - Abnormal propagation;
   - Screening.
4. Radar calibration faults and measurement errors.
5. The periodicity of the measurements maladjusted to the type of precipitation observed.
6. Errors in data processing, in particular in the procedures for measuring the average power of the radar signal and calculating the time and space averages.

These causes of error can be corrected or avoided if suitable experimental precautions are taken.

### 2.6.2 Single-Wavelength Reflectivity

The most current method for measuring $R$ by radar simply consists of using a conversion formula $R$-$Z$ (cf. Section 2.5.2). By using (2.82) with $X = R$ in (2.114c), we obtain

$$\overline{P_r} = C_3 L^2 |K|^2 \frac{aR^b}{r^2} \tag{2.122}$$

After correcting the attenuation if necessary, we obtain

$$R = \left( \frac{r^2}{aC_3 |K|^2} \overline{P_r} \right)^{b-1} \tag{2.123}$$

The weakness of the method of course is in the choice of the coefficients $a$ and $b$ of the conversion relation $Z$-$R$. We saw in Section 2.5.2 that these coefficients vary in the same way as the parameters of $N(D)$ they depend upon. The use of constant coefficient can lead to considerable error that can reach 300% on instantaneous local values.

When only information provided by radar is available, attempts can be made, based on climatological or circumstantial information, to make the best choice of the particular relationship that can be applied to the observed precipitation; for example, warm front or cold front precipitation, snow, or rain. In this way, some improvement is made with respect to the systematic use of a single relationship for all cases, which is the simplest procedure. However, the possibilities offered by this method seem limited. In fact, the variations remain that can exist within the same type of precipitation.

With the aim of refining the choice of the coefficients to be used, an empirical study has been made of possible correlations between the structure of the echoes (considered from the point of view of their classification by form recognition and their physical properties, such as the average rate or variance) and the values of the coefficients in the conversion formula; the correlations observed are weak [Puhakka, 1974, among others].

Happily, the influence of the variability of $a$ and $b$ on the precision of the results decreases when the measurements are averaged in time or space. We find that, for averages done on a few hours and on condition that the radar is properly calibrated, the correspondence between radar measurement and control measurement by ground raingauge are in good agreement. Errors due to strong attenuation for a precipitation measurement by backscattering are very difficult to correct and involve excessive uncertainties on the measurements. Therefore, we must use relatively long wavelengths for these measurements. A wavelength of 10 cm is accepted as a good compromise for measurements of precipitation by backscattering. The observation range also has an effect on accuracy; this remains approximately constant up to a range of about 100 km (depending on the beamwidth, the wavelength, and so on) and then deteriorates rapidly. Despite their mediocre accuracy, measurements of this type, by virtue of their flexibility and their simplicity in implementation, offer useful applications whenever an approximate measurement is sufficient; for example, in systems for a general hydrological surveillance or an early flood warning.

### 2.6.3 Radar and Rain Gauge

The precision of the precipitation measurements from the radar reflectivity can be improved considerably if the radar is calibrated by using independent measurements of $R$ at the ground. Usually, these measurements are made with rain gauges, which implies that this type of instrument is considered a viable reference.

*Rain Gauge Networks*

Current rainfall measurements use intensity rain gauges giving a reading every time that a certain "quantum" of precipitation has been collected (there are a large number of variants and different degrees of complexity in the measurement, recording, or data transmission techniques). This simple and robust equipment measures the average rate of precipitation during the time for collecting a quantum (the quantum can vary, depending on the equipment, between a calibrated drop and a few deciliters, just as the collection area can vary) with an error whose main cause is the air flow around the sensor; the error is an underestimation of approximately 10% for a wind of 5 m $s^{-1}$, 20% at 10 m $s^{-1}$, 20–40% between 10 and 30 m $s^{-1}$.

The relative error made in the measurement of the space-time distribution at the ground of precipitation by sampling with the help of a network of rain gauges depends on the structure of the observed precipitation and the characteristics of the network used. It increases with the spacing between the measurement points (mesh) of the network and decreases when the rainfall rate, the extent of the network, and the duration of the event considered increase [Huff, 1971].

Huff [1970] shows that the average sampling error $E$ can be estimated from the general equation

$$\log E = a + b \log P + c \log G + d \log T + e \log A \qquad (2.124a)$$

where $P$ is the area's mean precipitation in mm, $G$ is the gauge density in $km^2$ per gauge, $T$ is the storm duration in hours, $A$ is the area of the gauge network in $km^2$, and $a$, $b$, $c$, $d$, and $e$ are coefficients. The sampling error $E$ (in mm) represents the absolute difference between the best estimate of the true mean, $P$, obtained from the maximum density of rain gauges and the sample mean precipitation calculated from the gauge amounts for a given value of $G$.

After 10 years of observation with a network of 49 rain gauges distributed on 1000–1500 $km^2$ in Illinois, Huff obtained

$$\log E = -0.49 + 0.68 \log P + 0.94 \log G - 0.01 \log T - 0.75 \log A \qquad (2.124b)$$

The rain-gauge networks used for testing radar measurements on average have a density of one measurement point per 10–20 $km^2$. Under such conditions and except in the case of small local thunderstorms or in the presence of high values of the air flow velocity, the error in the measurement of precipitation quantity with a network of rain gauges is less than 5%.

*Radar Calibrated by Rain Gauges*

In this method, the radar data ensure the density and continuity of the observations. The sampling network is optimized at the maximum number of rain gauges necessary to reach the wanted precision.

The representativeness and the validity in time and space of such a calibration depend on the particular conditions of the observation and especially on the homogeneity of the precipitation in the observed zone. In fact, the validity of the radar calibration decreases as the distance between the scattering volume considered and the calibration point increase; therefore, it is preferable to spread the calibration points over the entire observed region. The optimum characteristics of this network depend on the desired accuracy and spatio-temporal resolution as well as on the type of precipitation observed and the special features of the terrain.

Measurements made both by radar and by a rain-gauge network are carried out by sampling. Because of this, structures whose scale is less than twice the sampling mesh size fall outside the resolution. When they are sampled differently by the radar and the array, they tend to decorrelate the measurements.

Comparison between the average values of the measurements of precipitation made by radar and those carried out by a rain-gauge network reveal important differences during the same event or from one precipitation event to another: the ratio $(\overline{G/H})$, where $G$ and $H$ are the heights of water measured by the rain-gauge network and by the radar, respectively, frequently varies by a factor of two from one storm to another. It therefore is important to find an improvement by adjusting the factors used to convert the radar data in terms of rainfall.

*Adjustment from One Precipitation Event to Another.* In the simplest calibration procedure, the average difference between the radar estimates and the rain-gauge network estimates is determined and a correction is applied uniformly to the radar estimates. The multiplicative adjustment factor ($F$) is given by

$$F = \sum_{i=1}^{N} G_i \Big/ \sum_{i=1}^{N} R_i \tag{2.125}$$

or by

$$F = \frac{1}{N} \sum_{i=1}^{N} G_i / R_i \tag{2.126}$$

With (2.125), the data are weighted proportionally to the height of the rain, whereas with (2.126) all the pluviometer-radar values have an equal weight. In each case, the corrections concern the error associated with the calibration of the radar and the error due to the $Z$-$R$ relationship.

*Spatial Adjustment*. The differences between the ground measurements and estimates obtained by radar have physical causes that vary over space and time. Therefore, a more precise adjustment than that resulting from the use of an average factor may be obtained by referring to a sufficiently close rain gauge (as one moves away from the station in question, the quality of the adjustment deteriorates until it becomes less than the overall adjustment). The utilization of calibration rain-gauge network data thus has led to development of relatively sophisticated interpolation procedures derived from multivariate objective analysis [Krajewski, 1987; Creutin, Delrieu, and Lebel, 1988].

The method proposed by Brandes [1975] presents the following stages:

- The values of the reflectivity factor $Z$ measured by the radar (2.114) are converted in precipitation rate with a $Z$-$R$ relation (the choice of the relation used is not crucial for the results; it seems logical however to use the relation that is the climatically best adapted).
- The rainfall rates are converted in incremental precipitation depth for the time intervals of the radar measurements then summed in each resolution volume during the chosen time duration (duration of the rainy event or hourly or daily sequences, *et cetera*). The field obtained is converted into Cartesian coordinates then filtered slightly to decrease the small-scale fluctuations.
- The $G/R$ ratios are computed for each rain gauge having at least 2.5 mm of rain, using the radar data located in a given radius around the rain gauge.
- From the $G/R$ data, we determine a field of adjustment factor that is applied to the values of rain estimated by radar to produce a first radar-corrected rain field $H_e$.
- The differences between the radar corrected rain depths $R_C$ and the rain-gauge measured rain depths are used to sharpen the adjustment factors so that the corrected field agrees with the rain data at gauge location.

Figure 2.30 is an example of the result obtained by adjustment of the radar data from a rain-gauge network.

Taken as a whole, the error in the measurement of the quantity of water reaching the ground decreases when the surface considered and the duration of integration increase. For example, for surfaces less than 50 km$^2$ and integration durations less than 10 min, the error in the measurement of precipitation is greater than 35%. In current cases of convective rain, a radar calibrated by two rain gauges for an area of 1000 km$^2$, allows the same accuracy as an array of 50 rain gauges distributed over the same area; that is, approximately 20%. Over 1000 km$^2$, the quantity of precipitation integrated over 1 hr is estimated at 40% with 10 rain gauges; if a calibrated radar is added, an accuracy greater than 15% is obtained.

According to the results in agreement of the various authors cited, it seems that, for long integration durations (> 24 hours), the measurement of the precipitation by combining radar data and a rain-gauge network leads to average errors of

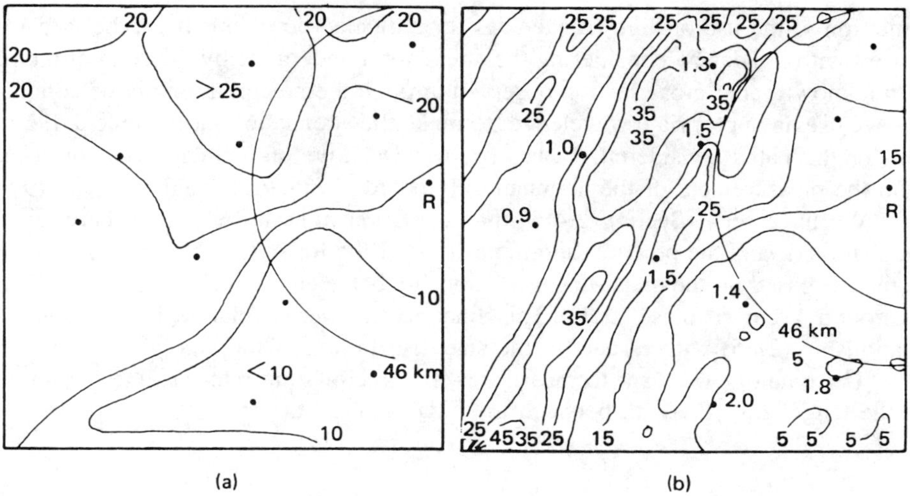


**Figure 2.30** Precipitation field (a) at the ground derived from measurements of a rain-gauge network and (b) derived from adjusting the estimates of the radar with the rain-gauge network data. The isohyets are labeled in mm of rain. The points locate the rain gauges. Plotted values are calibration factors. The letter *R* refers to the position of the radar. In the region between the radar and the marker at range 46 km, the radar data are distorted by ground echoes and not analyzed (from [Brandes, 1975]).

approximately 10–30% for a rain-gauge network with one measurement point per 1000 to 2000 km$^2$. Whereas the error decreases when the spacing between the rain gauges decreases, the value of the radar contribution also vanishes because a rain-gauge network by itself, if it is sufficiently dense, leads to the same accuracy as the combined data. For integration durations longer than 24 hours, the equality between the accuracy of the network by itself and combination of the network and radar is reached with a network consisting of one rain gauge for an area of 250–300 km$^2$.

In all cases, the accumulation of rain data over long periods leads to results in agreement with the "standard" relationship $Z = 200\ R^{1.6}$ to better than a factor of approximately 2. The accuracy of the estimates for snow with the help of a calibrated radar is similar to that obtained for rain measurements using the same technique [Collier and Larke, 1978].

### 2.6.4 Single-Wavelength Attenuation Measurements

We saw that precipitation attenuates the electromagnetic radiations. This attenuation can be measured on the microwave link that has the transmitter and the receiver at each extremity, considering the variations of the signal received for a constant transmission (this problem is well known in telecommunication as it constitutes the main limitation to the use of high frequencies for radioelectric links). It can be measured as well with a radar by considering the attenuation undergone by the echo of a target with a known radar cross section observed through the precipitation on the condition that we use an appropriate wavelength to make the attenuation measurable at the rate and for the paths considered. It was proposed to give an estimate of $R$ on a path from the measurement of the attenuation [Godard, 1965; Atlas and Ulbrich, 1977].

We have seen (Section 2.4.3) that a relationship can be written between the attenuation $a_p$ and the precipitation intensity $R$ of the form $a_p = kR^{\gamma}$, eq. (2.65). The method is based on the quasi-linear relationship between $a_p$ and $R$ with a wavelength of around 1 cm ($\gamma$ close to 1). This linearity implies a relative independence of relationship (2.65) with respect to the size distribution of the precipitation.

The quantity we want to measure is the precipitation rate, at time $t$, averaged on the length $L$ of the path between the radar and the target:

$$\bar{R}(t) = \frac{1}{L}\int_0^L R(t, l)\, \mathrm{d}l \tag{2.127}$$

The attenuation measured along the path is

$$A_p = k\int_0^L R^{\gamma}(l)\, \mathrm{d}l \tag{2.128}$$

The distribution of $R$ along the path is unknown. However, we can interpret the measurement of $A_p$ by referring to an eqivalent average rate, $R_e$, as

$$A_p = LkR_e^{\gamma} \tag{2.129}$$

or

$$R_e = \left(\frac{A_p}{Lk}\right)^{\gamma-1} \tag{2.130}$$

Substituting (2.128) into (2.130), we obtain

$$R_e = \left[\frac{1}{L}\int_0^L R^{\gamma}(l)\, \mathrm{d}l\right]^{\gamma-1} \tag{2.131}$$

When comparing (2.127) and (2.131) we see that, if $\gamma = 1$, we have $R_e = \bar{R}$. The condition $\gamma = 1$ is necessary and sufficient for the same quantity of precipitation on the radar-target path to create the same attenuation, whatever the distribution of $R(l)$. The term $\gamma$ for wavelengths between 0.86 cm and 3.2 cm increases from 1.04 to 1.16 whereas, in a corresponding manner, the average error of the estimate of $R$

from $a_p$ increases from 9% to 21%. These increases reflect a lesser independence of $\gamma$ with respect to the size distribution when the wavelength used increases [Atlas and Ulbrich, 1977]. However, even at 3.2 cm the error remains small. The method therefore is of value because of its precision.

It presents various inconveniences, however. It is a global measurement that gives no spacial distribution of $R$. It requires as many reflectors on the field as the radials we want to measure. The values of the attentuation coefficients are such (cf. Appendix 4, Table A4.3) that the measurement dynamics at a given wavelength is limited (with the same $\lambda$, it is impossible to measure precisely the weak and strong intensities on a path of a given length). Therefore, it is rarely used for the meteorological measurement of precipitation by terrestrial radar. We will see in Section 2.9 that on the other hand, it is the main method used to measure precipitation from space.

### 2.6.5 Dual-Wavelength *a-R* Method

To avoid resorting to a ground target and determining $R$ according to distance, Eccles and Mueller (1971) proposed measuring the attenuation, considering the variation of the signal between two distances $r$ and $r + \Delta r$ with two wavelengths, at least one of them being attenuated. We suppose that in the dual-wavelength radar the beams corresponding to each $\lambda$ are colinear and identical (collimated). The mean power of the echo for the wavelength $i$ (with $i = 1, 2$) at distance $r$ is given by equation (2.110). If $\lambda_1$ and $\lambda_2$ are such that the hydrometeors are Rayleigh scatterers ($Z_{e_1} = Z_{e_2}$), we can write

$$y(r) = 10 \log\left(\frac{P_1(r)}{C_1} \Big/ \frac{P_2(r)}{C_2}\right) = -2 \int_0^r (a_1 - a_2)\, \mathrm{d}r \tag{2.132}$$

For the radial distance interval $\Delta r$, we have

$$y(r + \Delta r) - y(r) = 10 \log\left[\frac{P_1(r + \Delta r)}{P_2(r + \Delta r)} \Big/ \frac{P_1(r)}{P_2(r)}\right] = 2 \int_r^{r+\Delta r} (a_2 - a_1)\, \mathrm{d}r \tag{2.133}$$

If $\bar{a}$ designates the average attenuation rate on the interval $\Delta r$, we have

$$2\bar{a}\Delta r = 2 \int_r^{r+\Delta r} (a_2 - a_1)\, \mathrm{d}r \tag{2.134}$$

that is,

$$\bar{a} = \frac{10}{\Delta r} \log\left[\frac{P_1(r + \Delta r)}{P_2(r + \Delta r)} \Big/ \frac{P_1(r)}{P_2(r)}\right] \tag{2.135}$$

The wavelengths are chosen to make $a_1$ negligible *versus* $a_2$ (for example, $\lambda_1 = 10$ cm, $\lambda_2 = 3.2$ cm). We saw (Section 2.4) that the attenuation rate and the precipitation rate are linked by a relation of the type $a_p = kR^\gamma$, where $k$ and $\gamma$ are coefficients depending on the wavelength. We thus have

$$\bar{R} = \left(\frac{\bar{a}}{k}\right)^{1/\gamma} \tag{2.136}$$

This method can be used even if $a_1$ is not negligible and likewise enables the estimation of the water content $M$. Moreover, it does not require radar calibration. However, the results obtained are mediocre for the reasons mentioned in Section 2.5.4, concerning the hail detection by the dual-wavelength method.

### 2.6.6 Dual-Wavelength *N*(*D*) Method

Another approach to the measurement of $R$ with a dual-wavelength radar consists of using the data to determine the two parameters of the distribution $N(D)$ [Goldhirsh and Katz, 1974]. To simplify, we suppose that the distribution is exponential, even if the method can be applied to other forms of dual-parameters distribution, and that only $\lambda_2$ is attenuated. The measurements of $P_1$ and $P_2$ at two radial distances separated by $\Delta r$ let us computate the average on $\Delta r$ of the attenuation rate $\bar{a}$ and the radar reflectivity factor $\bar{Z}$, supposing that, on the $\Delta r$ interval, the rain is uniform. From (2.135) we have

$$\bar{a} = \bar{a}_2 = N_0 \int_0^{D_{\max}} Q_t(D)\, \mathrm{e}^{-\Lambda D}\, \mathrm{d}D \tag{2.137}$$

where $Q_t$ is the attenuation cross section.

The value of $Q_t$ is not an analytical function. However, in the field of drop sizes that significantly contribute to the rain, we can replace $Q_t$ for the approximation [Atlas and Ulbrich, 1974]

$$Q_t \approx CD^n \tag{2.138}$$

where $C$ is a coefficient empirically determined.

Substituting (2.138) into (2.137) and integrating, we obtain

$$\bar{a}_2 = CN_0\Gamma(n + 1)/\Lambda^{n+1} \tag{2.139}$$

The reflectivity factor is, eq. (2.5) with $\mu = 0$,

$$\bar{Z} = N_0(6!)\Lambda^{-7} \tag{2.140}$$

We thus have two equations, from which the two unknowns $N_0$ and $\Lambda$ can be computed. This method works only if the hypotheses concerning the form of the distribution $N(D)$ and the local uniformity of the rain rate are satisfied. It also suffers from the constraints mentioned in Section 2.5.4.

### 2.6.7 Dual Polarization

The method proposed by Seliga and Bringi [1976] is as follows. If we suppose that the rain observed is correctly represented by the exponential distribution $N(D_e) = N_0 \exp(-\Lambda D_e)$, then $R$ is given by equation (2.121), which is a function of $N_0$ and $\Lambda$. From (2.68), (2.69), and (2.15), we have

$$Z_{H,V} = \frac{\lambda^4}{\pi^5 |k|^2} N_0 \int_0^{D_{\max}} \sigma_{H,V}(D_e)\, e^{-\Lambda D_e}\, dDe \tag{2.141}$$

where $\sigma_{H,V}$ is given by (2.57). We thus can write

$$Z_{DR} = 10 \log \left| \frac{\int_0^{D_{\max}} \sigma_H(D_e)\, e^{-\Lambda D_e}\, dDe}{\int_0^{D_{\max}} \sigma_V(D_e\, e^{-\Lambda D_e}\, dDe} \right| \tag{2.142}$$

$$\frac{Z_H}{N_0} = \int_0^{D_{\max}} \sigma_H(D_e)\, e^{-\Lambda D_e}\, dDe \tag{2.143}$$

These two equations show that $Z_{DR}$ and $Z_H/N_0$ depend only on $\Lambda$. The simultaneous measurement of $Z_{DR}$ and $Z_H$ by the radar lets us determine $\Lambda$ and $N_0$ and thus compute $R$.

In these equations, $\Lambda$ can be replaced by $D_0$, taking into account relation (2.19). Figure 2.31 shows the variation of $Z_{DR}$ and of $Z_H/N_0$ in the function of $D_0$ for an exponential rain. The values of $D_0$ put in abscissa correspond to precipitation rates ranging between 0.16 and 300 mm h$^{-1}$. This method applies to rain because the relation between $D_e$ and the axial ratio $a/b$ of the drops is determined (2.29). The method just described also can be applied in the hypothesis of a drop size distribution of log normal or gamma modified form on the condition that we link two of the

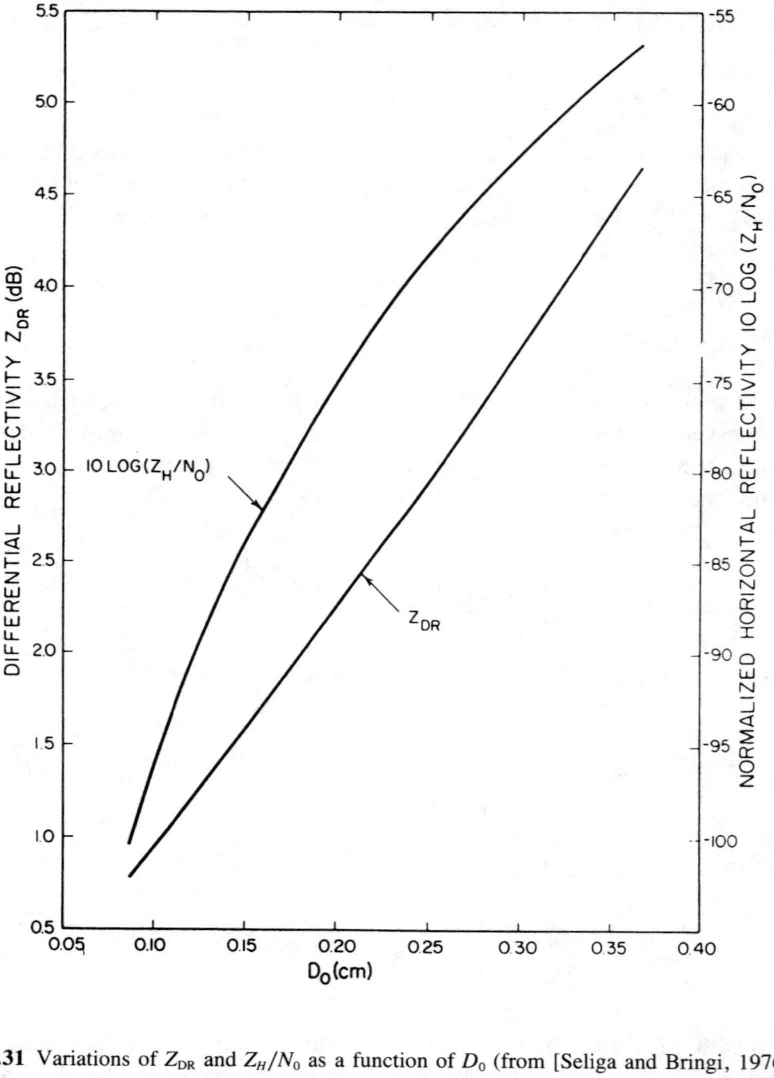


**Figure 2.31** Variations of $Z_{DR}$ and $Z_H/N_0$ as a function of $D_0$ (from [Seliga and Bringi, 1976]).

three parameters of these distributions as indicated in Section 2.2.5 [Ulbrich, 1983; Feingold and Levin, 1987].

From these various relations (2.121, 2.142, and 2.143) we can write an analytical relation expressing $R$ as a function of $Z_{DR}$ and $Z_H$. For intensities under 200 mm h$^{-1}$ and exponential distributions with $30 \leq N_0 \leq 30{,}000$ (mm$^{-1}$ m$^{-3}$) and $1.0 < \Lambda < 4.5$ (mm$^{-1}$), Sachidananda and Zrnic' find

$$R = 6.84 \times 10^{0.1(Z_H - 30 - 4.86 Z_{DR})} \tag{2.144}$$

with $R$ in mm h$^{-1}$, $Z$ in dBZ, and $Z_{DR}$ in dB. This empirical formula matches the theoretical values at less than 3%.

From measurements of drop size distribution in the intense rains of Central Illinois, Seliga, Aydin, and Direskeneli, [1986] propose the following empirical relations:

$$\begin{aligned} R &= 1.95 \times 10^{-3}\, Z_H Z_{DR}^{-1.04} \quad \text{with } 0.2 \leq Z_{DR} \leq 0.7 \\ R &= 1.59 \times 10^{-3}\, Z_H Z_{DR}^{-1.67} \quad \text{with } 0.7 \leq Z_{DR} \leq 2.6 \end{aligned} \tag{2.145}$$

$$\begin{aligned} M &= 1.04 \times 10^{-4}\, Z_H Z_{DR}^{-1.29} \quad \text{with } 0.2 \leq Z_{DR} \leq 0.7 \\ M &= 7.81 \times 10^{-5}\, Z_H Z_{DR}^{-2.04} \quad \text{with } 0.7 \leq Z_{DR} \leq 2.6 \end{aligned} \tag{2.146}$$

$$D_0 = 1.68\, Z_{DR}^{0.64} \tag{2.147}$$

with $R$ in mm h$^{-1}$, $M$ in g m$^{-3}$, $Z_H$ in mm$^6$ m$^{-3}$, $Z_{DR}$ in dB, and $D_0$ in mm. Analytical expressions among $R$, $Z$, and $Z_{DR}$ were also proposed by Illingworth and Caylor [1989].

The principle of the method also is applicable using the measurement of CDR. Indeed, in the rain, because there is a common orientation of the scatterers, $Z_{DR}$ and CDR bring the same information. The superiority of $Z_{DR}$ over CDR is because the linear polarizations are less sensitive than circular ones to the propagation effects. As shown by the previously mentioned relations, the precision of the measurement of $R$ depends on that of $Z_H$ and $Z_{DR}$. The precision of the measurement of $Z_{DR}$ depends on the cross correlation of the terms of the $Z_H/Z_V$ ratio (cf. Section 2.5.3). It is necessary, to have a high degree of cross-correlation, to have an $S/B$ ratio superior to about 20 dB; below that, the precision deteriorates. With a very good polarimetric radar, we can measure $Z_{DR}$ with a precision of 0.1 dB, which is equivalent to a change in the axial ratio of the drops of about 0.01.

Error [Sachidananda and Zrnic, 1987; Chandrasekar and Bringi, 1988] comes from two sources: errors on the measurement of $Z_{DR}$, notably due to sidelobe contamination; and errors because the target does not match the considered hypotheses in the theory, notably DSD moving away significantly from the exponential or gamma forms, drop average axial ratios moving away from the equilibrium, or the presence of non-Rayleigh scatterers for the used wavelength. As indicated in Section 2.2.2 the analytical distributions correspond to averages whose instantaneous distributions in one point often draw noticeably apart. It is necessary therefore to average over a sufficient area. Applied in bad conditions, the methods of estimating $R$ by dual polarization can lead to results inferior to those using the simple relation $Z$-$R$. An uncertainly remains concerning the quantitative interpretation of the polarization parameters $Z_{DR}$, $L_{DR}$, or CDR for determining the drop-size distribution. Consequently, the methodology of estimating precipitation intensity from the polarization measurement and the quality of the results (i.e., the "attainable" improvement compared to other methods) has not yet led to firm and definitive conclusions.

### 2.6.8 Area Integral Methods for Convective Rainfall

Common experience quite obviously shows that, for convective cells, there is a certain correlation between the area occupied by a cloud, its height, its life duration, and the volume rain rate. However, if we consider instantaneous measurements on a small number of individual cells, the correlation is very loose because of the important variance among these parameters. By integrating in time or space, this variance is reduced. The tighter correlations that appear then can be exploited in the measurement. Such is the case of the relation between the area occupied by the convective rains and the rain volume at the ground.

Doneaud *et al.* [1984] observed that the volume of rainfall $V$ from individual storms can be estimated with surprising accuracy simply from the so-called *area-time integral* (ATI) of the radar echo in excess of a specified reflectivity threshold over the life of the storm, without worrying about the internal structure of the system. Following a similar process, Chiu [1988] noticed that 98% of the instantaneous area-wide rainfall variance can be explained by the fractional area covered by the radar echoes higher than a given level of intensity. He thus proposed to estimate the average instantaneous rain rate on wide areas from such a measurement of the fractional area; a single snapshot is sufficient to obtain the measurement, and it is not necessary to determine in detail the distribution of the rainfall rate in the considered convective field.

Atlas, Rosenfeld, and Short [1990] rediscussed these methods and showed that their validity is founded on the existence of a well-behaved probability density function of rain rate either from the many storms at one instant or from a single storm considered during its whole time of activity. The method can be formulated as follows.

*Instantaneous Areawide Method*

Consider a precipitation field covering, at moment $t$, a horizontal area $A_0$, and we call $P(R)$ the percentage of the area where the precipitation rate is between $R$ and

$R + \mathrm{d}R$. $A(\tau)$ is the area with $R > \tau$. The volume of rain in the area $A(\tau)$, in which $R > \tau$, is

$$V(\tau = A(\tau)\, R_C(\tau) \tag{2.148}$$

where $R_C(\tau)$ is the average rain rate in the $A(\tau)$ area:

$$R_C(\tau) = \int_\tau^\infty RP(R)\, \mathrm{d}R \Big/ \int_\tau^\infty P(R)\, \mathrm{d}R \tag{2.149}$$

$V(\tau)$ corresponds to the fraction $\phi$ of the total volumetric rate $V_t = V(\tau = 0)$:

$$\phi = \int_\tau^\infty RP(R)\, \mathrm{d}R \Big/ \int_0^\infty RP(R)\, \mathrm{d}R \tag{2.150}$$

and, according to (2.148), we have

$$V_t = \frac{V(\tau)}{\phi} = A(\tau)\frac{R_C(\tau)}{\phi} = A(\tau)\, S(\tau) \tag{2.151}$$

with $S(\tau) = R_C(\tau)/\phi$. The average rain intensity in the area $A_0$, that is, $V_t/A_0$ can thus be written

$$\langle R\rangle = S(\tau)\, F(\tau) \tag{2.152}$$

where $F(\tau) = A(\tau)/A_0$ is the fractional area occupied by the rain that has a rate of $R > \tau$. Its measurement from a snapshot of the precipitation field by the radar enables the deduction of $\langle R\rangle$ on the condition that $S(\tau)$ is known. $S(\tau)$ is determined for each value of $\tau$ by the probability density function of $R$ because

$$S(\tau) = \int_0^\infty RP(R)\, \mathrm{d}R \Big/ \int_\tau^\infty P(R)\, \mathrm{d}R \tag{2.153}$$

The probability density function $P(R)$ and thus also the terms deduced from it, notably $S(\tau)$, are climatological characteristics of the convective fields considered. For convective rains, $P(R)$ has a log normal distribution. If $P(R)$ exists, $S(\tau)$ for a specified value of $\tau$ is a constant. Thus, $\langle R\rangle$ is a linear function of $F(\tau)$. $S(\tau)$ can be obtained either from $P(R)$ by (2.153) or by a regression on all the measured values of $\langle R\rangle$ and $F(\tau)$.

In practice, the determination of the isopleths of $R$ (or isohyets) and the area with $R > \tau$ is done from the value field of $Z$ observed by the radar, using the relation $Z$-$R$, which is the most climatologically adapted (i.e., we consider a threshold of $Z$). The condition of validity of the method is that the precipitating field considered is wide enough for the population of convective cells included in $A_0$ observed at the moment $\tau$, each in a different stage of evolution, to constitute a representative sample of the probability density function of $R$. A typical value of $A_0$ is $10^4$ km$^2$. The value of $\langle R \rangle$ thus determined at a given moment is representative of the intensity of rain generated by the convective field considered during its autocorrelation time; that is, for $A_0 \geqslant 10^4$ km$^2$, a typical duration of 5–6 hr.

*ATI Method*

Likewise, if an individual storm or a small group of storms is considered, observed by the radar during a $T$ duration, we can write that the total volumetric rain $V$ generated by the storm, after (2.151), is

$$V = \int_T V_t \, \mathrm{d}t = S(\tau) \int_T A(\tau) \, \mathrm{d}t \tag{2.154}$$

The integral of $A(\tau)$ on the time $T$ is the ATI. The volumetric rain $V$ thus is a linear function of the ATI. The validity condition, here again, is that the ATI is sufficiently wide (i.e., the time $T$ sufficiently long) for the successive stages of the storm, observed during the time $T$, to be statistically representative of $P(R)$. We thus make an hypothesis of ergodicity of the field of $R$. The conditions of the "instantaneous method" on $A_0$ and the ATI are the same.

For example, for a radar reflectivity threshold of 25 dbZ (i.e., about 1 mm h$^{-1}$), Donneaud *et al.* [1984] find that, for North Dakota storms, we have

$$V(\text{in km}^2\ \text{mm}) = 3.7\ [\text{ATI (in km}^2\ \text{h)}] \tag{2.155}$$

with a correlation coefficient of 0.98 and a standard deviation of +45% and −31% (the two values differ because the regression is done from the log).

Rosenfeld, Atlas, and Short [1990] proposed an improvement on the instantaneous areawide method, consisting of taking into account the height and the temperature of the precipitating clouds; that is, the available precipitable water. This method, called HART (or height area rainfall threshold), leads to estimates of $\langle R \rangle$ with a standard deviation between only 5 and 10%.

These methods of rainfall measurement ($V$ or $\langle R \rangle$) from the integral areas present much fewer advantages for the applications using radars aboard planes or satellites. The principle of these methods also is relevant for a transposition to the precipitation estimate from the infrared satellite imagery (cf. Section 2.9). The principle is the same as that of the two methods presented earlier, considering, instead of the isopleths of $R$, isopleths of the infrared brightness temperature ($T_B$) of the cloud tops. We thus observe that $\langle R \rangle$ is proportionate to the areas where $T < T_B$; that is, to the areas where the cloud tops are colder (and thus higher) than the isotherm corresponding to $T_B$.

## 2.7 RADAR NETWORKS

The use of a single ground-based radar enables local observations to be made, the quality of which, as has been seen previously, is degraded with increasing distance from the radar. The association of several radars spread over the territory to be covered for the observations provides for overcoming this limitation. The term *network* is used for such a system when the data collected by each radar are combined to produce an integrated description of the meteorological fields on the scale of a region, a country, or a group of countries.

Small networks are sometimes organized in a temporary fashion over small areas within the framework of specific meteorological research operations. These arrays are defined as a function of the specific objectives and elude any generalization. The arrays in which we will be interested are those of operational meteorology organized in a permanent fashion with a view to providing information in real time or with a short time lag. This information is intended for users with highly varying concerns such as air traffic, agriculture, civil engineering, *et cetera*. They therefore must be as complete as possible.

The organization of such arrays is essentially a technical problem, the solution of which depends on the computing and telecommunication resources for digital processing and data transmission. As with most radar applications, the choices concerning the geographical and technical characteristics of the arrays are a compromise between often competing sets of requirements.

### *Geographical Distribution*

As far as the distribution of the stations is concerned, the optimum solution in terms of homogeneity of the observation density is a triangular grid array because each radar covers a circular area. However, it often is necessary to take various constraints into account. Constraints may be associated with the terrain and topographical features: presence of natural obstacles (orography) or artificial obstacles, contour of emerged land. Constraints of interest may be involved, demarcation of zones meriting a particular meteorological or hydrological surveillance, or political and administrative constraints, like frontiers. Opportunity constraints can include demarcation of already existing installations that can facilitate the operating conditions of the radar. All these factors mean that actual networks often depart quite appreciably from the optimum distribution and may have blind spots (i.e., gaps in coverage).

The size of the mesh (i.e., the spacing between radars) depends on the information required. We have seen that, for high-quality hydrological measurements, the effective range of the radar is more limited than for a simple qualitative localization of the precipitating zones. Similarly, doppler measurements with two or three radars (Section 3.5) do not tolerate large spacing.

### *Technical Aspects*

The choice of the technical characteristics of the network depends on the information expected and the climatology of the region, which are obviously related. It also depends on cost as a function of the demand and the economic efficiency of the service expected. The NEXRAD network, the most powerful one, which planned to handle large destructive storms among others, uses doppler radars with a wavelength of 10 cm and so little attenuated (see characteristics in Appendix 4).

The European network consists of noncoherent radars some of which existed before the network was set up. They therefore do not have homogeneous characteristics, which is not a drawback in itself. Most of the equipment uses a wavelength of 5 cm (for example, the RODIN radar; see Appendix 4).

The choice of scanning frequency affects the information collected and the complexity of processing. Some networks used mainly for measuring rainfall are limited to panoramic scanning (PPI) at a low and constant angle of elevation. The use of sequences with variable angles of elevation for each azimuthal revolution lets us observe the three-dimensional structure of the precipitation fields. The transmission of digitized, preprocessed, and compressed data uses earth-based lines or direct or satellite radio links.

Data processing can be carried out at several sites and at several levels of intervention or concentrated at one site. For example, the elimination of ground echoes and spurious data, the calibration and conversion to intensity of precipitation, the interpolation and the conversion of the systems of coordinates all can be done at a local level. The combination of the data from different radars, the combination with other data, in particular satellite data, can be made at the central site. The same is true for the distribution of information produced by the system; information of local interest can be left at the local level. In fact, the configuration of the operating system of the network is related to the product under consideration.

### *Products*

The products of a radar network are different and varied; they are forced to adapt to meet the needs expressed by the users and therefore are subject to change. Their main objective is short-term forecasting (cf. Section 2.8). However, the data can be stored, wholly or in part, in crude form or compressed to create data banks that can be used for long-term objectives, for example, in climatology or for the evaluation of water resources. The proposed products are fields of physical data and the results of analysis and processing carried out using automatic procedures.

The terminals that can be used to receive and display the products of the network usually have other useful information available to describe and understand the meteorological situation and predict its evolution in the short term. This supplementary information is essentially the animation of images from geostationary satellites (Meteostat in the European network) and the output of large- and medium-scale models [Gilet, 1984]. Radar networks are an essential component of the detection and warning systems for the prevention of natural risks: hail, tornadoes, flash floods, wind shear, and microbursts.

## 2.8 SHORT-TERM FORECASTING

The more complex and technology-dependent human activities become, the greater are the needs of weather forecasting on all time scales and also the greater is their value in terms of economic viability. The forecast is the result of implementing a very broad range of observation and calculation techniques. It may concern the weather in the broadest sense of the term or a particular parameter. However, the users are most interested in clouds, precipitation, and the violent and intense phenomena that may be associated with them. Conventionally, *short term* and *very short term* are taken to mean a forecast that is valid from a few minutes to a few hours, including a detailed description of the local weather. Such a forecast is sometimes close to a simple description of the current state of the weather, and because of this it is also called *nowcasting*. It lies below the resolution of digital forecast models used to describe the general evolution of the large-scale and medium-scale characteristics of the meteorological field.

Short-term forecasting uses all the information available on the state of the atmosphere and the trends of its evolution as well as all forms of analysis and processing of these data both objective and subjective. The base data for short-term forecasting are the results of forecast models that give the large-scale "framework," the local radio sounding data for analyzing the vertical structure of the atmosphere (in particular, the stability), the data of the ground measurement networks, and especially, for the characteristics of the cloud and precipitation field, the data from radar and geostationary satellites that allow the local situation to be monitored over a short-term interval.

The preparation of the short-term forecast involves combining all these data into a consistent summary, conceptualizing a model of the local situation, and extrapolating this model over time. The difficulty of the operation clearly depends on the complexity of the situation, the characteristic scale, and the lifetime of the phenomenon involved. For example, a forecast involving the precipitation field associated with the convection line of a winter depression system at moderate latitude moving above the sea can be made over several hours with a good accuracy (intensity, duration of the rain at a point) by a simple linear extrapolation of the movement observed by radar during the preceding period. This is no longer true for forecasting tornadoes or microbursts in a region of complex orography in the presence of a highly unstable air mass and high-speed winds. In this situation, the complexity of the forecast, due to the nonlinear nature of the phenomenon, is increased by the problem of making a rapid decision on the warning that is one expected outcome of the forecast.

At the present time, short-term forecasting still relies mainly on subjective analysis: its quality and its efficiency, especially in complex situations, depend on the value of the information at the forecaster's disposal but also on his or her experience and intuition [see, for example, McGinley 1986]. Assimilation and objective analysis procedures as well as expert systems (shape recognition, *et cetera*) are gradually being introduced into short-term forecasting. They have not yet replaced the forecaster but they do facilitate more complete utilization of the available data, in particular the combined processing of radar and satellite data [see, for example, Austin and Bellon, 1982; Doswell, 1986]. With the deployment of large radar networks, these methods stand to undergo swift development.

## 2.9 RADARS AND SATELLITES

Radar networks provide a satisfactory solution to the problem of measuring ground rainfall and tracking pluviogenic systems on a local and regional scale. Thus, for a region equipped with radar resources providing a homogeneous cover, satellite data only contribute supplementary information concerning the surrounding region of larger scale than the network. They facilitate the identification of meteorological structure by presenting an overall view with a spatial continuity that may be masked or not observable by the network due to its configuration (for example, the network on the Japanese archipelago) or blind spots and the effects of the terrain (orography, coastal effects) on the precipitation field. They also allow extrapolation of the short-range forecast deduced just from the network.

However, a radar network, just like conventional networks for meteorological measurements, in particular pluviometric measurements, is highly inhomogeneous and covers only a tiny part of the planet. Oceans cover 75% of the surface of the earth, to which have to be added the inhospitable regions of the continents (cold and hot deserts, jungles), which are completely unsuitable for measuring the distribution of precipitation on a global scale. At present, the liberation of latent heat associated with the formation of precipitation is the main cause of the heating of the atmosphere. Because of this, the precise observation and the tracking over time at short intervals of the distribution of precipitation over the entire globe is essential for understanding the global dynamics of the atmosphere, for modeling these dynamics, and for meteorological and climatic forecasting. Remote satellite detection of clouds and precipitation may provide a satisfactory answer to the problem [see, for example, Barrett and Martin, 1981]. This is why considerable research has been carried out on estimating precipitation from satellites, which are emerging as the solution of the future. We point out in passing that a similar statement can be made in relation to measuring other essential variables for understanding the atmosphere (temperature, humidity, pressure, aerosols, *et cetera*), the global measurement of which also depends on satellite techniques.

Several methods can be used for the satellite estimation of precipitation. Technical considerations (antenna, power requirements) predominate in making a choice. Methods currently used are based on passive detection with the help of radiometers. However, meteorological radar will be placed in orbit in the very near future.

### 2.9.1 Technical Aspects

Satellites observing earth's environment mainly are in geostationary orbit and describe an almost circular polar orbit. Geostationary satellites are at a height of approximately 36,000 km in the plane of the equator and move in the same direction as the earth with the same angular velocity (so-called geosynchronous period of one day). They therefore are fixed with respect to the surface of the earth so that they always look at the same area. They make observations of this area at very short time intervals, usually 30 min. Such an observation period (or sampling period) is less than the lifetime of medium-scale and large-scale structures in the atmosphere; therefore, they can follow the structures' evolution and provide very short-range forecasting.

Polar satellites occupy a much nearer orbit, usually between 400 and 1,500 km above the surface of the earth, and their orbital period is only 1 hr 30 min to 2 hr. Generally, they cut the equator at a large angle of inclination so that their trajectory brings them close to the earth's poles. These satellites are usually synchronous with the sun and fly over the same region of the earth twice every 24 hours, once at night and once during the day but always at the same time. However, other choices can be made; for example, for observing the tropical atmosphere, inclinations on the order of 30° are more favorable (in this case, the high latitudes are not flown over). From one orbit to the next, the orbital plane rotates about the earth and several orbits must be traversed for all the earth's surface (at observable latitudes) to be seen.

The radiometers' on-board satellites observing the earth's atmosphere or meteorological satellites operate mainly

- In the *visible* (VIS) region between approximately 0.5 and 0.7 $\mu$m; in this band by day they "see" the distribution and appearance of the clouds.

- In the thermal *infrared* (IR) region. The bands used are those of the two main windows of atmospheric air transparency, between approximately 3.5 and 4.2 $\mu$m and between 10.5 and 12.5 $\mu$m. In this region, they measure the temperature of the upper surface of the clouds, which can also be interpreted in terms of the altitude.
- In the *microwave* (MW) region around 1.55 and 0.81 cm and below 0.5 cm. Whereas, in the VIS and IR clouds are opaque, in the MW region they are partially transparent. Therefore, it is possible to see precipitating zones included in the clouds. For technical reasons involving the size of the antennas necessary to obtain a sufficient resolution of the footprint of the beam on the ground, up till now MW radiometers have been used only in low-orbit satellites (stealth).

### 2.9.2 Estimation of Precipitation with Visible and Infrared Data

Such estimation involves indirect methods. The general principle exploits the relationships between one or more physical characteristics of the clouds that can be deduced from satellite data (for example, the temperature at the top, which for convective clouds is a measure of the vertical expansion) and the ground rainfall. To establish these relationships (i.e., calibrate the measurement), rain data are obtained on the ground at a calibration site with the help of a rain-gauge network, with or without a radar. Proposed estimation methods vary in complexity. They can be classified into five types:

1. Cloud indexing methods based on a statistical relationship between the rainfall rate and the type of cloud classified as a function of the altitude of the top and the extent of the cloud;
2. Methods based on parameterization schemes for the convection or on cloud models;
3. Methods based on following the variation of parameters such as the temperature at the top or the area of the cloud;
4. Methods based on the occurrence of clouds with cold tops;
5. Shape recognition methods.

These methods often are used in conjunction [Griffith *et al.*, 1978; Negri and Adler, 1981; Wu, Weinman and Chin, 1985; Motell and Weare, 1987; Adler and Negri, 1988; Flitcroft, Milford, and Dugdale, 1989; Garand, 1989; among others]. They give quite precise results for cumulative estimations over relatively long periods, on the order of 20% for rain estimation on a monthly scale. However, the quality deteriorates for estimations over short time intervals (a few hours to one day). Furthermore, they are effective mainly for convective clouds and are hardly effective for systems with a stratiform structure whose flat top does not reveal the dynamics of its lower layers. Very fortunately, the main rain producing clouds, at tropical latitudes in particular, are convective clouds with cold tops. However, although all heavy rainfalls are associated with cold tops, the opposite is not true.

For estimates made at short time and space intervals, using the bihourly data of geostationary satellites, the ground data provided by rain-gauge networks are not sufficiently precise by themselves for the calibration. In this case, the use of isolated radars or in networks not only lets us refine the estimates of ground rainfall but also provides information on the extent and the shape of the rain areas, which are quantities that can be correlated with satellite data [Tsonis and Isaac, 1985; Bellon and Austin, 1986].

In conclusion, these methods of estimating precipitation have shortcomings arising from the need to base them on indirect methods (the top of clouds), and so are imprecise. They also have advantages, because they can be made from geostationary satellites; if suitable ground-based calibration resources are available, they allow high-resolution estimates in terms of time and space, useful for a large number of applications.

### 2.9.3 Rain Estimation by Passive Microwave Methods

For the estimation of rain, microwaves have potential importance because of their power to penetrate clouds and precipitation and because various cross sections of hydrometeors grow with their dimension (cf. Section 2.3). The proposed estimation methods are based on the emission (and therefore also absorption) or the scattering of the microwaves by the precipitation, following the frequency of the radiometers. These frequencies obviously are chosen to lie outside the absorption regions of water vapor and hydrogen. A broad distinction can be made among three cases.

At low microwave frequencies, that is, below the resonance frequency of water vapor at 22.235 GHz (precisely, the frequency used is 19.35 GHz), clouds and ice precipitation situated in the upper part of the clouds are not involved much due to weak absorption and emission and the minor contribution of scattering. The radiometer "sees" mainly the rain and we try to relate the radiation emitted by the drops, that is, the brightness temperature of the rain, to its intensity on the ground. However, this method, called the *emission* or *absorption-emission method,* can be applied only above a cold and homogeneous backcloth so that the rain appears with sufficient contrast, and there is no ambiguity in the cause of the variations in brightness temperature $T_B$ (with $T_B = \varepsilon T$, where $\varepsilon$ is the emissivity and $T$ the physical temperature of the material). This is the case above oceans, where the rain appears warmer than the background because the surface of the ocean has a low emissivity, approximately 0.5, whereas the rain drops have an emissivity close to 1. Over land, in which emissivity is variable and close to 0.9, the method cannot be used (unless a radiometer is used pointing upward from the surface). It has other limitations and presents difficulty in interpretation. First of all, although rain makes the major contribution to the observed brightness, the contribution of water vapor and water clouds cannot be negligible, in particular at low intensities of precipitation (< 5 mm $h^{-1}$). However, a correction can be made for the contribution of water vapor if a channel at 22 GHz is available to measure its value. Then, the brightness temperature of the precipitation depends on the height of the rain column, and so on the height of the freezing level (isotherm 0° C) with, under certain circumstances, deviations with respect to this level (for example, growth by warm processes at negative temperatures). Finally and most important, the relationship between $T_B$ and $R$ is not linear. Figure 2.32 illustrates this problem, which concerns the value of $T_B$ calculated by Wilheit *et al.* [1977] by neglecting the ice layer for precipitation of variable intensity and thickness. Starting from a value of 15–20 mm $h^{-1}$ the relationship is inverted because the rain layer becomes opaque, and $T_B$ decreases with rain rate because of the cold cosmic background scattering by large drops and the emission at lower temperatures. The method therefore can be used only between approximately 1 and 20 mm $h^{-1}$ with an accuracy of a factor of 2.

At high microwave frequencies (i.e., above the oxygen complex lying between 50 and 70 GHz), the main contribution of the precipitation to the brightness is due to *scattering* by frozen hydrometeors situated in the upper part of the precipitation column (in the case of clouds with large vertical extension). In fact, at these frequencies, the radiation has relatively little penetrating power; furthermore, for frozen hydrometeors, the effective absorption cross sections (and therefore emission cross

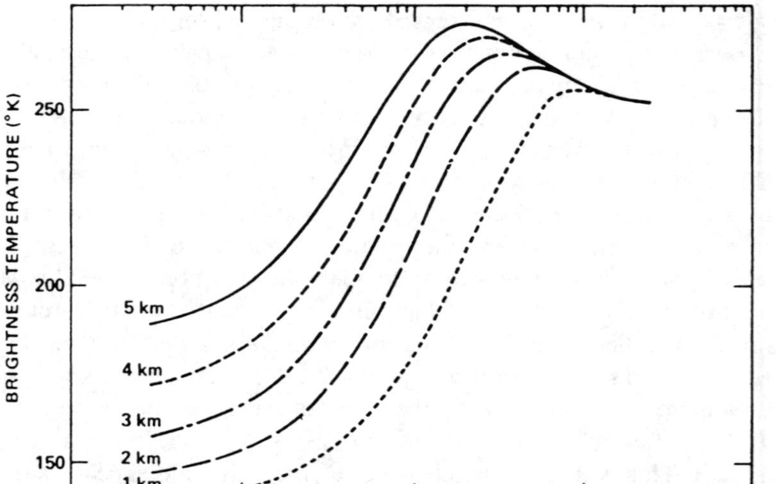


**Figure 2.32** Calculated brightness temperature at 1.55 cm (19.35 GHz) as a function of rain rate for freezing levels of 1–5 km (from [Wilheit *et al.*, 1977]).

sections) are very small, whereas the effective scattering cross sections remain large (cf. the values of the refractive index terms in Appendix 4). Under these conditions, the estimation of the precipitation is based on the relationship between the rainfall rate on the ground and the quantity of frozen hydrometeors that, at the top of the precipitation column, determine (by scattering) the brightness temperature. Figure 2.33 shows the value of $T_B$ calculated by Wilheit *et al.* [1982] for a frequency of 92 GHz as a function of $R$ for various values of the thickness of the frozen hydrometeor layer. We can see that $T_B$ decreases with $R$ and reaches very low values. Because of this, there is a possibility of confusion between the low values of $T_B$ due to precipitation and those measured when the radiometer sees the sea. This ambiguity can be resolved by measurements at two polarizations, because the sea emission is strongly polarized whereas the scattering by hydrometeors is hardly polarized. The value of the scattering method is that it can be used over land and for estimating precipitations of strong intensity; therefore, it complements the emission method.

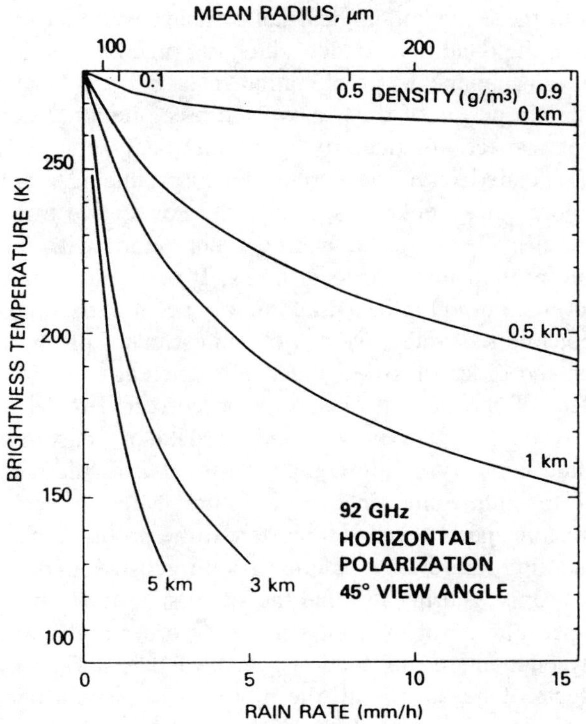


**Figure 2.33** Calculated brightness temperature at 0.33 cm (92 GHz) as a function of rainfall rate. The parameter is the thickness of the ice layer. The assumed freezing level is 5 km (from [Wilheit *et al.*, 1982]).

However, it is an indirect method, and the interpretation of the brightness data comes up against the problem of inadequate knowledge of the relationship between the intensity of ground rainfall and the granulometric properties of the population of frozen hydrometeors at the top. Now, the scattering is strongly dependent on the numerical concentration, the size, the shape, and the density of these hydrometeors. For these reasons, it seems unrealistic to expect a high degree of precision from the scattering method [Wilheit, 1986].

Between the high and low microwave frequencies lies the propagation window situated around 35 GHz (precisely, the frequency used is 37 GHz). At this frequency $T_B$ may be dominated by either the emission mode or the scattering mode, depending on the intensity of the observed precipitation. The opacity threshold of the emission mode is reached at approximately 4 mm $h^{-1}$. Spencer [1986] suggested using the 37 GHz frequency to estimate the rain above oceans by scattering, by combining the $H$ and $V$ polarization measurements. The 37 GHz frequency may also be used for measuring rainfall of low intensity in emission mode above oceans; that is, to complete the 19.35 GHz frequency measurement.

In addition to these methodological deficiencies are other sources of potential error. One involves the conditions under which the precipitation is sampled: the satellite estimates the instantaneous precipitation rates to be used to obtain values integrated over a certain period of time. Now, the precipitation fields are highly variable over time and space, particularly in the tropical zone, where most of the precipitation is associated with convective clouds producing intense and localized rainfall. Furthermore, these fields, especially the convective tropical ones, have a very marked diurnal cycle. Now, a satellite synchronous with the sun always observes a given site at the same two local times. If one ignores the diurnal cycle, it can be shown that, on a monthly basis, taking into account the statistical distribution of the instantaneous rainfall rates, we can obtain estimates of the mean with a standard deviation on the order of 10%, which is sufficient to validate models of the climate. The effect of the diurnal cycle can be corrected for if another means of measurement is available, such as VIS and IR data of geostationary satellites to obtain its phase and intensity. Another solution involves using a satellite that is not synchronous with the sun. Some time in the future, when the techniques of assembling large antennas in space have been mastered, the problem of time sampling can be tackled by installing microwave radiometers on geostationary satellites.

Microwave beams, bearing in mind the small size of the antennas used on the platforms, lead to resolution of the footprint on the order of 10 km. Problems arise, like these already encountered in Section 2.5.7, of partial filling of the beam and homogeneity defects of the target, all the more problematical because the transfer function between $T_B$ and $R$ is strongly nonlinear. Solutions allowing this problem to be partly overcome rely on suitable parameterization and measurements at several frequencies and at two polarizations [Spencer, 1986]. The precision of the estimates made by these various methods is not uniform and depends on the observation conditions, yet we cannot always identify these conditions and so define the accuracy.

### 2.9.4 Orbital Radars

The installation of a radar on board a satellite platform to observe precipitation is a very attractive proposition. The value of such an arrangement lies not only in the direct measurement of the precipitation field but also in the complementarity with other satellite techniques for estimating rain, because it can provide the necessary information for removing ambiguities and refining the interpretation of radiometric data: presence of precipitation, measurement of the freezing level, height of the precipitation column. Space radar to some extent presents an ideal solution to the problem of optimizing the installation site of the radar: extent of the observed area, no attenuation between the radar and the target, no masking, almost constant resolution cell, and so forth. No space radar has yet been used to measure precipitation, but there are a large number of studies on instrumental design and algorithms for deriving precipitation rates from measurements.

The project has some serious technical difficulties that, however, are not insurmountable. Contrary to radiometers, radar consumes large amounts of power (the radiation has to be emitted and the performance of MW transmitters is mediocre). Now, resources are limited on board a satellite. The efficiency of the system is based on two conditions: the level of the backscattered signal must be sufficient, bearing in mind the attenuation in $r^{-2}$ to allow the detection of low intensity precipitation, and the resolution cell (size of the footprint of the beam on the ground) must be small enough to allow a correct measurement of the rainfall, bearing in mind the spatial heterogeneity of the rain fields. These conditions depend mainly on three parameters: the radar-target distance, the wavelength, and the size of the antenna. These dictate low orbits ($\sim$ 500 km), wavelengths less than 3 cm and sensitive to attenuation by clouds and precipitation, and antenna sizes of several meters.

Several methods may be used to estimate precipitation from radar in orbit. They depend on the number of parameters measured by the radar. These methods derive from those discussed in relation to the estimation of precipitation by ground-based radar (cf. Section 2.6): measurement of attenuation across the rain for a single or several wavelengths and measurement of reflectivity. They must take into account that generally there is no method of ground calibration (oceans, inhospitable regions). The proposed algorithms [Atlas *et al.* 1982; Meneghini, Eckerman, and Atlas, 1983; Goldhirsch, 1988] combine these various methods. The decisive advantage for satellite measurement is the possibility of measuring the attenuation of the energy reflected at the surface. The attraction of this method is due to the quasi-linearity of the relationship between the attenuation and rain rate and the independence of the size distribution of the drops at the frequencies used [Atlas and Ulbrich 1977]. It does not require absolute calibration by measurement on the ground; however, it does require a higher accuracy of measurement to reduce statistical errors and provides only an average rate. In the algorithms that use a single wavelength, the power returned by the surface seen across the precipitation is compared with that obtained for the closest observation without rain. The difference is proportional to the quantity of rain traversed. The shortcomings of this method are due to the fact that we assume that the radar cross section of the ocean in the absence of precipitation is representative of that in the presence of precipitation and that the rain rate is constant with height. Furthermore, at values of $R$ below approximately 0.5 mm $h^{-1}$, the attenuation is insufficient to allow $R$ to be estimated and we have to return to using the $Z$-$R$ relationship. The retrieval of the vertical profile of $R$ makes it necessary to deal with the profile of $Z$.

After receiving the echo of the surface in the presence of rain, the radar receives a mirror image, due to a double reflection at the surface and a backscattering at the hydrometeors. Using the relationship between the mirror signal and the direct signal to obtain the vertical attenuation profile and the rain rate has been suggested [Meneghini and Atlas, 1986]. The value of this method is that it does not depend on alterations of the surface by the rain; but it cannot be used at low rain levels (the mirror signal cannot be detected).

The other drawbacks of the single-wavelength method are because at the frequencies considered, the liquid water clouds (nonprecipitating hydrometeors), although not detected, are responsible for significant attenuation and it is not possible to cover the entire range of precipitating rates. In fact, if the rain has a low attenuation coefficient, the low rates cannot be measured; and if the rain has a high attenuation coefficient, the surface signal disappears. Using a second wavelength may appreciably increase the extent of the range of measurable rates. The algorithm for two wavelengths uses the differential attenuation of the surface and the vertical profile of the signal backscattered by the rain.

The antenna is a pointed spike. If the necessary radiating surface is too large to be incorporated directly on the platform, a folded structure may be used underneath the cap of the launcher and unfolded after the satellite has been placed in orbit. The beam can be steered by electronic scanning with the help of an array of sources directed at the deployable reflector. The beams must have low level sidelobes because these hit the surface at the useful range.

The precipitating field may be covered by scanning in the plane perpendicular to the movement of the platform. However, the available time is a critical factor, taking into account the speed of the platform, the area to be covered, and the minimum time the beam has to stay at a point to obtain an accurate mean. This results in taking samples following nonadjacent footprints and using a code that makes it possible to increase the number of independent samples without increasing the beam's residence time.

Of the projects for orbital radar for measuring rainfall, two have undergone far-reaching feasibility studies and have been specified in detail. These are the *Bilan Energètique du Système Tropical* (BEST), Energy Balance of the Tropical System, [CNES, 1988] and *Tropical Rainfall Measuring Mission* (TRMM) [Simpson, Adler, and North, 1988]. The objective of these projects is to estimate precipitation fields on a climatic scale.

The main objectives of TRMM are to determine the distribution and the variability of the precipitation and the liberation of latent heat in monthly averages of approximately $10^5$ $km^2$. The scientific goals are to improve short-term models of the climate and global circulation models and provide a better understanding of the hydrological cycle; in particular, how it is affected by tropical oceanic precipitation and its variability. The satellite is planned for an orbit of 320 km inclined from 30 to 35° to the equator. Taking into account the precision of the orbit, the time necessary to cover a diurnal cycle is approximately three weeks. It is equipped with VIS and IR (10 $\mu$m) radiometers and uses three microwave frequencies 19, 37, and 90 GHz with double polarization, of 10 km resolution with swaths of 600 km. The radar is a pulse system using two wavelengths, 2 cm (14 GHz) and 1.25 cm (24 GHz) with a footprint of 4 km, a radial resolution of 250 m, and a swath of 220 km. The radar is intended to measure precipitations for rates lying between 0.5 and 50 mm $h^{-1}$. These different resources are complementary and partly redundant.

In conclusion, we can say that the solution obtained with a radar in orbit, even with two wavelengths, in itself does not make it possible to solve the problem of estimating rainfall completely. Clearly, by exploiting the complementarity of the passive and active techniques and direct and indirect methods (i.e., by combining radar and VIS, IR and MW radiometry) it is possible to obtain the optimum solution.

Different approaches have been suggested for overcoming the problem of time sampling for microwave radiometric and radar measurements; for example, the use of geostationary platforms when the techniques of assembling large antennas in space have been mastered. The use of several nonstationary satellites spread around the planet makes it possible to repeat an observation on a site in less than a few hours, which thus meets most needs and in particular those of short-term forecasting. It is plausible that such systems, in view of their efficiency, would rapidly become economically more attractive than earth-based networks for measuring precipitation for the meteorological monitoring of continents and operational meteorology. The correlation and assimilation of the information from the different nonstationary and geostationary satellites should improve the quality of the results. High-resolution satellites can validate the low-resolution but high-coverage satellites.

# Chapter 3
# Velocity Measurements

The observation of the velocity field is complex because of the large scale movements involved in atmospheric circulation, their three-dimensional nature, and the range of values to be measured. Radar provides efficient solutions to overcome these problems. We saw in Section 1.2.2 that the target moving with radial velocity $V_r$ ($> 0$ when the target is moving away) with respect to the radar, receives an electromagnetic wave whose frequency is shifted by an amount $\Delta f = -V_r/\lambda$ with respect to the transmitted wave. The field radiated by the target to the radar has again a frequency shift $\Delta f = -V_r/\lambda$; that is, a total shift equal to $-2V_r/\lambda$ for the backscattered wave (the total shift being 0 for the wave scattered forward). In this chapter we will drop the subscript $r$ and write simply $V$ for $V_r$. The technical conditions allowing access to the measurement of the radial component of the velocity of a target with the help of a coherent radar are described in Chapter 1. They can be summarized in the following manner: in the phase demodulator, the received signal, which is a real function of time, is mixed with two sinusoidal functions in quadrature, forming the phase reference, so that the two components in quadrature of the received signal are obtained at the output. These two components are those of the coherent bipolar video, an exact representation of which in the complex plane is given by the expression

$$y(t) = A(t)\,\mathrm{e}^{-\mathrm{i}\phi(t)} \tag{3.1}$$

$A$ and $\phi$, the amplitude and phase of the signal, define the resultant vector of the addition of vectors corresponding to individual scatterers included in the resolution volume of the radar, whose amplitudes ($a_i$) and phases ($\varphi_i$) are distributed randomly if the effects of secondary scattering are negligible and if the movements of the individual targets are independent; these conditions generally are satisfied in clouds and precipitation with the possible exceptions of snow and intense precipitation. Because meteorological targets contain many elements, application of the central limit theorem indicates that $\phi$ has a uniform statistical distribution in the interval $[0, 2\pi]$

and that the probability density of the amplitude $A$ obeys a Rayleigh distribution; the components of the coherent bipolar signal have a Gaussian distribution with a zero mean.

## 3.1 THE DOPPLER SPECTRUM

### 3.1.1 Spectral Parameters

The number of cloud or precipitation particles contained in the pulse volume is very large so that the distribution of their velocity can be approximated by a continuous distribution. Therefore, the received signal, the sum of the contributions of each particle present in the pulse volume (each being the origin of a spectral line determined by its velocity and weighted by its effective cross section and its position in the beam), is not monochromatic. The distribution of the received power as a function of the velocity or the frequency has a certain broadening determined by the radar characteristics (pulse volume, bandwidth, *et cetera*), the target structure, and by dynamic phenomena acting on the scattering elements; this distribution is the power spectrum.

The signal arriving from a specific resolution volume situated at a given range can be considered the sampling of a Gaussian process; its amplitude, given by expression (3.1) is a complex random function of time described by statistical properties. The spectral density function $Y(f)$ representing the process, defines the amplitude spectrum; it is the Fourier transform of the complex signal. For a continuous signal, assuming that the process is a steady-state process, it is given by [see, for example, Bendat and Piersol, 1971]

$$Y(f) = \int_{-\infty}^{+\infty} y(t)\, e^{-i2\pi ft}\, dt \tag{3.2}$$

The corresponding power spectral density (or energy spectrum) is

$$S'(f) = |Y(f)|^2$$

The power spectral density can also be obtained from the complex autocorrelation function $R(\tau)$, which is defined by

$$R(\tau) = \lim_{T\to\infty} \frac{4}{T} \int_{-T/2}^{T/2} y(t)y^*(t + \tau)\, dt \tag{3.3}$$

where $\tau$ is the autocorrelation interval and $y^*$ is the complex conjugate quantity.

$S'(f)$ is the Fourier transform of $R(\tau)$:

$$S'(f) = \int_{-\infty}^{+\infty} R(\tau)\, e^{-i2\pi f\tau}\, d\tau \tag{3.4a}$$

The transform (3.4a) has the associated reciprocal transform:

$$R(\tau) = \int_{-\infty}^{+\infty} S'(f)\, e^{i2\pi f\tau}\, df \tag{3.4b}$$

The distribution of $S'(f)$ as a function of the frequency forms the doppler spectrum. The power in a frequency interval between $f$ and $f + df$ is $p(f) = S'(f)\, df$. It is proportional (cf. equation (1.81)) to the sum of the average powers $\Sigma_i^{A^2}$ backscattered by the particles whose doppler frequencies lie between $f$ and $f + df$. The summation of $S'(f)$ over the entire spectrum gives the average received power:

$$\bar{P}_r = \int_{-\infty}^{+\infty} S'(f)\, df \tag{3.5}$$

The doppler velocity spectrum corresponds to the frequency spectrum via the relationship $f = -2V/\lambda$.

It is convenient to use a power spectral density normalized by the average received power, $S(f) = S'(f)/P_r$, which is also the probability density of the distribution. We can associate statistical moments of order $n$ to $S(f)$, given by the general expression:

$$m_n = \int_{-\infty}^{+\infty} f^n S(f)\, df \tag{3.6}$$

The main characteristics of the doppler spectrum (Figure 3.1) are the average frequency $f$ and the variance $\sigma_f^2$ which are written as:

$$\overline{f} = \int_{-\infty}^{+\infty} f S(f)\, df = m_1 \tag{3.7}$$

$$\sigma_f^2 = \int_{-\infty}^{+\infty} (f - \overline{f})^2 S(f)\, df = \overline{f^2} - (\overline{f})^2 = m_2 - m_2^2 \tag{3.8}$$

Therefore, the average frequency is the moment of order 1 of the spectrum and the variance is the centered moment (or moment with respect to the mean) of order 2. By analogy, we say that $P_r$ is the moment of order 0 of the nonnormalized spectrum. The standard deviation $\sigma_f$ (i.e., the square root of the variance $\sigma_f^2$) can be interpreted as a measurement of the width of the spectrum.

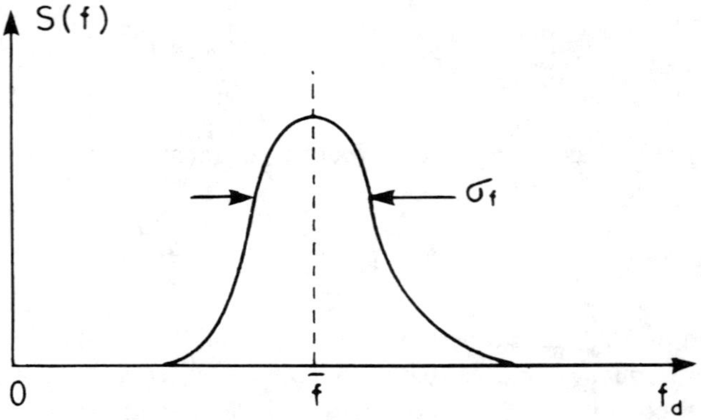


**Figure 3.1** The doppler spectrum. $S(f)$ is the power spectral density normalized by the average received power as a function of $f_d$; $\bar{f}$ is the average doppler frequency and $\sigma_f$ is the standard deviation of the spectrum.

Note that the derivative of order $n$ of the autocorrelation function of the spectral power density can be written as

$$\frac{d^n R(\tau)}{d\tau^n} = (2\pi i)^n \int_{-\infty}^{+\infty} f^n S'(f)\, e^{i2\pi f\tau}\, df \tag{3.9}$$

We thus have the following relationship between the derivatives of order $n$ of $R(\tau)$ evaluated at $\tau = 0$ and the moments of order $n$ of the spectrum

$$\overline{m_n} = \left\{ \frac{d^n R(\tau)}{d\tau^n} \Big/ [(2\pi i)^n R(\tau)] \right\}_{\tau=0} \tag{3.10}$$

and, in particular,

$$\bar{P}_r = R(0) \tag{3.11}$$

$$\bar{f} = \frac{R'(0)}{2\pi i R(0)} \tag{3.12}$$

$$\overline{f^2} = \frac{R''(0)}{(2\pi i)^2 R(0)} \tag{3.13}$$

The velocity spectrum is the image of the frequency spectrum, and we can write

$$S(f)\, df = S(V)\, dV$$

The preceding equations therefore can be transposed in terms of velocity. We have the following relationships between the characteristic parameters of velocity and frequency:

$$\bar{V} = -\frac{\lambda}{2}\bar{f} \tag{3.14}$$

$$\sigma_V^2 = \frac{\lambda^2}{4}\sigma_f^2 \tag{3.15}$$

### 3.1.2 Discrete-Fourier Transform

The analysis of the doppler velocities from the signal returned by a scattering volume starts just as the estimation of powers of the noncoherent radar, by a "demultiplexing" as described in Section 1.5.7. Figure 3.2 shows an example of a time series and the spectrum obtained in a gate with a doppler radar, eq. (1.8) and (1.80). The raw data for the time series occupy a considerable volume. Consider, for example, a sampling using 256 range gates with a repetition frequency of 3000 Hz; the direct recording of the complex time series in the form of 9-bit words represents $256 \times 3000 \times 2 \times 9 \approx 1.3 \times 10^6$ bits per second.

The selection of processing procedures involves a compromise between the needs for accuracy, speed for making results available, data compression to keep the volume at an acceptable level, and cost. Moreover, certain quantities may have to

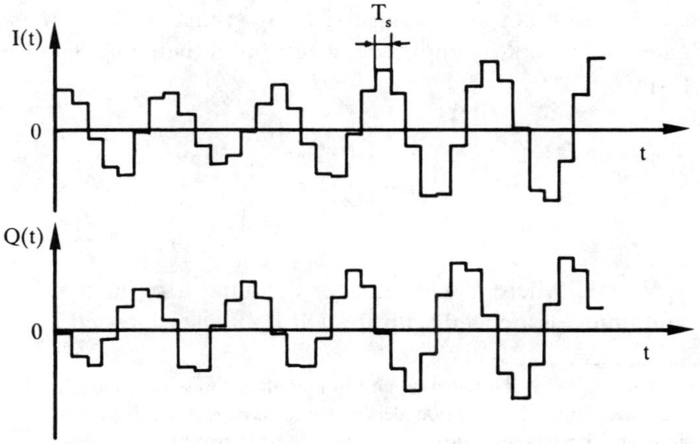


**Figure 3.2** Time series (36 pulses) of the two components of the complex signal obtained by sampling of the coherent video in a range gate, Re($y_i$) and Im($y_i$), eq. (3.1), or $I(t)$ and $Q(t)$, eq. (1.8) or (1.75). Note that during the interval $T_s$, separating two consecutive samples, the signal is maintained at the level of the last reading to produce a video synthesis.

be available in real time, for example, to subsequently adjust the observation conditions (reflectivity and velocity fields), whereas others can be processed after an elapsed time (study of spectra, variances).

*Spectrum*

The techniques for processing a doppler signal can either be analog or digital. The simplest analog technique for obtaining the doppler spectrum consists of using a spectrum analyzer; the apparatus either contains a single filter, in which the central frequency is shifted, or a battery of filters, in which the number of elements fixes the number of lines in the spectrum [Skolnik, 1962]. This type of procedure is unsuitable for handling the considerable amount of data necessary for analyzing movement fields in the atmosphere.

Digital techniques offer an attractive solution because they use especially fast and efficient algorithms for calculating the *discrete-Fourier transforms* (DFT). These algorithms, called *fast Fourier transforms* or (FFT)[1] [Cooley and Tukey, 1965], allow tens of spectra to be obtained in 1 s, which is compatible with real-time processing of the data. Digital methods for signal processing are now common in all domains of science and technology. Many publications give a detailed account of time series spectral analysis and discuss the precautions to be taken to arrive at the correct results [Bendat and Piersol, 1971; Papoulis, 1977; Priestly, 1981; among others]. We restrict ourselves to a summary of the subject.

The first operation consists in storing over a duration $N/f_r$, the complex time series of $N$ samples each corresponding to $M$ gates; that is, $N \times M$ pairs of values $y_{nm}$. In each gate, the complex amplitude spectrum is then calculated by a discrete-Fourier transform:

$$Y(k) = \sum_{n=0}^{N-1} y_n \, e^{-i2\pi nk/N} \tag{3.16}$$

with $0 \leqslant k \leqslant N - 1$, where $k$ is an integer denoting a frequency interval. $Y(k)$ is the complex amplitude associated with the $k$th frequency interval.

[1]*Fast Fourier transform* is the name given to rapid algorithms for calculating Fourier transforms used in computers. These algorithms, introduced in 1965, have had a considerable impact on the procedures and equipment for analyzing time series because of their convenience for calculating spectral density. Previously, the necessary calculation time (and cost) was such that the spectra had to be evaluated via previously calculated truncated correlation functions. In this way, the calculation of Fourier transforms relating to all the values of the samples considered was avoided. With the FFT procedures, the spectra could be obtained by direct Fourier transforms of the time series with a gain in time by a factor of 10 over previous methods, even using considerably simplified machines.

The reciprocal transform is

$$y_n = \frac{1}{N} \sum_{k=0}^{N-1} Y(k)\, e^{i2\pi nk/N} \tag{3.17}$$

The frequency resolution of the spectrum depends on the number of samples of the series; a series of $N$ samples let us obtain a spectrum containing in the Nyquist $f_r$ interval, at most $N$ frequency intervals (or lines) with widths $f_r/N$ corresponding to a minimum velocity resolution:

$$\Delta V = \frac{\lambda}{2} \frac{f_r}{N} \tag{3.18}$$

The maximum frequency that can be measured without ambiguity is $f_r/2$. The unambiguous measurement domain is presented on both sides of $f = 0$, between $-f_r/2$ and $+f_r/2$, with $N/2$ lines on each side; that is, $N + 1$ lines with the line $f = 0$.

The complex amplitude spectrum leads to the power spectrum $S'(k)$:

$$S'(k) = Y(k)Y^*(k) = [R_e Y(k)]^2 + [I_m Y(k)]^2 = |Y(k)|^2 \tag{3.19}$$

Sometimes $S'(k)$ is called the *periodogram* (this term was introduced in the past in connection with the use of $S'(f)$ to find "hidden" periodicities).

### *Window Effect*

The discrete-Fourier transform of necessity is applied to a signal of limited duration, here the duration $NT_S$ of the time series. The Fourier transform $Y(f)$ of the signal $y(k)$ is a periodic function. The samples of this signal in fact are the coefficients of the expansion of $Y(f)$ in a Fourier series. Now, the representation of a periodic function in a Fourier series requires an infinite number of coefficients, particularly when there are discontinuities. The limitation of the duration of the series therefore affects the spectrum. Simple limitation by truncation of the series at its ends is equivalent to multiplying the signal by a rectangular function consisting of $N$ samples of unity associated with each of the $N$ coefficients of the time series and equal to zero everywhere else.

If we denote the terms of the series of infinite duration $y(k)$, we have

$$y_N(k) = y(k)w_N(k) \tag{3.20}$$

with

$$w_N(k) \begin{cases} = 1 \text{ for } 0 \le k \le N-1 \\ = 0 \text{ everywhere else} \end{cases}$$

Functions other than rectangular may be used with the aim of attenuating the adverse effects on the spectra of limiting the duration. Such functions are called *time windows* or *data windows* because they limit (and modify) the data sequence we see in a series, which in principle is infinite. The consequence of limitation on the spectrum is called the *window effect*.

The Fourier transform of a time window is called the *spectral window*. In the frequency domain, the convolution of the Fourier transform of the signal of infinite duration with the spectral window corresponds to the product of the time series with the data window in the time domain. In other words, the coefficients $Y(f)$ of the DFT of the time series seen through the time window are approximations to the coefficients of the series of infinite duration, and the quality of the approximation depends on the form of the spectral window. Generally speaking, the spectral window consists of a main lobe and parasite sidelobes due to the truncation, in which the amplitude decreases with distance from the main lobe. When $N$ increases, the ratio of the amplitude of the central peak and the first sidelobes hardly varies, but there is a larger number of oscillations and they are narrower. The data windows aim to smooth the transition at the ends of the time series to reduce the sidelobes. The sidelobes therefore are significantly reduced in this way but the width of the principal lobe is increased.

Diverse types of data windows have been proposed. By way of example we can give the following simple relationships:

$$w_N(k) \begin{cases} = \alpha + (1-\alpha)\cos(2\pi k/N) & \text{for} \quad |k| \le N/2 \\ = 0 \text{ everywhere else} \end{cases} \tag{3.21}$$

This relationship is the generalized Hamming window:

for $\alpha = 0.54$ we obtain the Hamming window

for $\alpha = 0.5$ we obtain the Von Hann window

for $\alpha = 1$ we return to the rectangular window

$$w_n(k) = \left\{ \begin{array}{ll} \dfrac{\lambda}{\pi}\left|\sin\dfrac{\pi k}{\lambda}\right| + \left[1 - \dfrac{|k|}{N}\right]\cos\dfrac{\pi k}{N} & \text{for} \quad |k| \le N/2 \\ 0 \text{ everywhere else} & \end{array} \right\} \tag{3.22}$$

This relationship is the Papoulis [1973] window.

Figure 3.3 illustrates the window effect. Starting from a noisy random Gaussian spectrum with 2048 points, standard deviation $\sigma = 0.6$ m s$^{-1}$ and signal-to-noise ratio of 35 dB, the coefficients of a time series of 2048 points are calculated by reverse Fourier transform. A series of 64 samples is used to calculate a periodogram through a rectangular window (i.e., without window) and with a Papoulis window. The periodogram with uniform weighting has a significant level of sidelobes with a slow fall off. The periodogram using the Papoulis weighting shows a less concentrated principal lobe but well free of the sidelobes. The use of the window therefore proves to be very efficient.

*Variance of the Periodogram*

The power observed in a line of the periodogram has the distribution of an exponential variable (Section 1.5.2) and therefore fluctuates from one periodogram to the other. To reduce the variance of the periodogram, we calculate a mean over several independent periodograms. The observed time series of $M$ samples is divided into $L$ segments of $N$ samples $y_1(k)$ with $L = M/N$. The $L$ single periodograms are then evaluated. The averaged periodogram calculated given by

$$\overline{S'(k)} = \frac{1}{L}\sum_{l=1}^{L} S_l(k) \tag{3.23}$$

and we have

$$\mathrm{Var}[\overline{S'(k)}] = \frac{1}{L}\,\mathrm{var}[S_l(k)] \tag{3.24}$$

The variance of the averaged periodogram decreases in proportion to the duration of the observation.

The spectral moments are calculated from the averaged periodogram. When the spectra have a Gaussian characteristic, which is often the case in meteorological spectra (there are obviously exceptions, particularly those of spectra obtained with vertical angle of elevation, Section 3.2.1), the first two moments completely define them.

*Noise Elimination*

The doppler spectrum is always contaminated by the radar system noise. To calculate the moments, in particular the variance or the limits of the spectrum, it is useful to know the threshold value of the spectral density below which the instrument noise dominates in such a way as to eliminate the noise components.

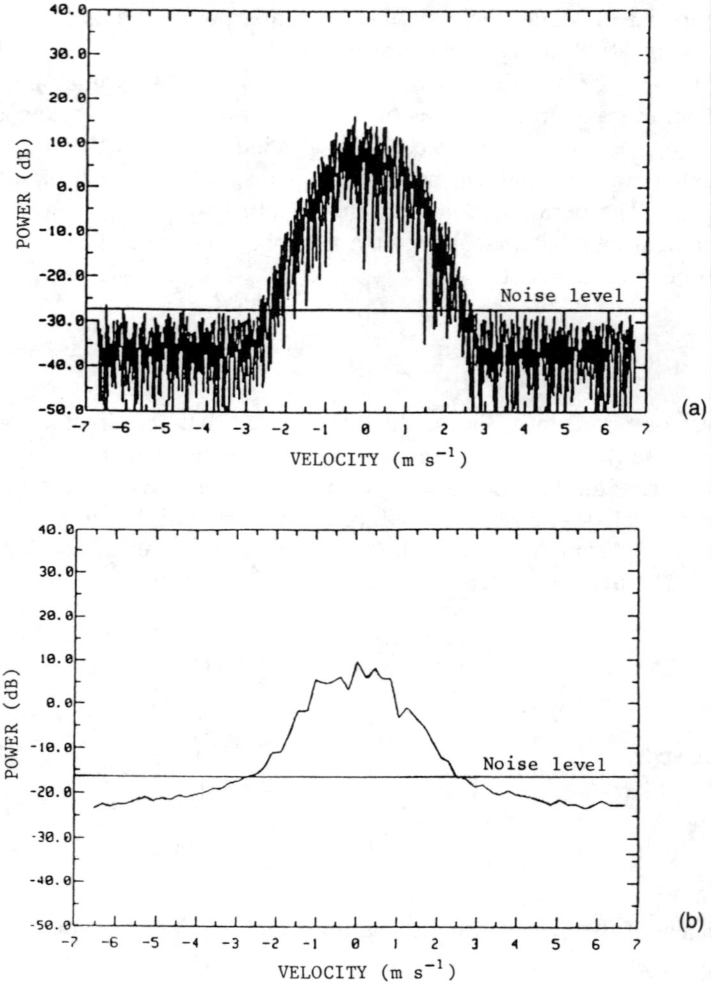


**Figure 3.3** Illustration of window effect: (a) initial Gaussian power spectrum used to generate the time series; (b) power spectrum calculated without window (that is, rectangular window; (c) power spectrum calculated through a Papoulis window, the number of points is 64 (courtesy B. Campistron).

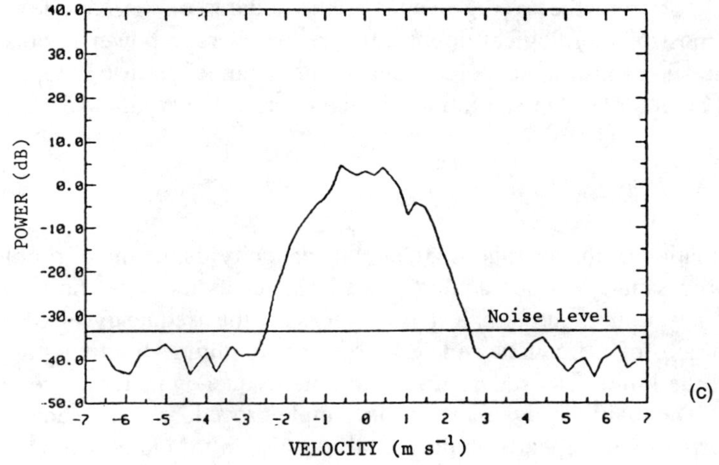


**Figure 3.3** (continued)

The elimination can be effected in a brutal manner by suppressing all the normalized spectral components lying below a level, arbitrarily defined with respect to the spectral peak. Thresholds of $-10$ to $-15$ dB have been used. An objective method of determining the noise threshold, proposed by Hildebrand and Sekhon [1974], states that the radar noise is white and Gaussian and that for such a noise the standard deviation of the spectral densities is equal to the mean spectral density. The threshold is the level below which this property is verified. This method is in current use.

### 3.1.3 Estimators of Spectral Moments

Knowledge of the spectra is useful only in particular applications, such as those that concern the microphysics of the precipitation (Section 3.2) or some aspects of the turbulence (Section 3.4). Most often moments of order 0 (intensity of the signal) and 1 (average speed) and, to a lesser degree, those of order 2 (variance) contain the required information. Under these conditions, it therefore is worth preprocessing the time series to limit storage and handling to the spectral moments of order $< 3$ because then the volume of the data is appreciably reduced (by a factor close to $3/N$ for the periodograms or $3/2N$ for complex time series). Furthermore, preliminary knowledge of the spectra is not always necessary for calculating spectral moments; some estimators work directly from the time series.

*Average Power*

We have seen that in doppler radars the average power of the signal can be estimated by calculating the sum of the average power in each of the lines of the spectrum, eq. (3.5), or from the average value of the squares of the complex amplitude. In fact, in most meteorological doppler radar, the average power is obtained by averaging the signal in a noncoherent logarithmic channel as in nondoppler radar. This solution avoids possible saturation of the coherent linear channel.

*Average Velocity*

The estimation of the average frequency (or velocity) is the major problem in doppler signal processing. Various analog procedures let us measure the signal frequency. Without any doubt, the simplest is to measure the frequency of the bipolar video with the help of a frequency meter across an amplitude-limiting stage. The output signal of the limiter is a square wave, and the frequency meter is used as a crossover detector. The main disadvantage of this simple procedure is its inadequate precision, which furthermore depends on the signal-to-noise ratio. Other analog estimators suggested, such as the phase comparator or the phase feedback loop, have the same disadvantages. The best way of estimating the average frequency is to use digital methods. The properties of the various algorithms have been compared by Sirmans and Bumgarner [1975a].

*Fast Fourier Transform.* The most obvious method for calculating the average frequency of the doppler spectrum is the estimate of the spectral density $S'(k)$ obtained from the time series (3.19):

$$\hat{f}_{\mathrm{FFT}} = \sum_i f_i S'(f_i) \Big/ \sum_i S'(f_i) \tag{3.25}$$

where $\hat{f}_{\mathrm{FFT}}$ is the estimate of $\bar{f}$ and $f_i$ is the frequency interval, denoted by $k$ in (3.16) and (3.17).

In fact, the estimate relates to the average frequency of the *combined* sum of the signal and noise, and we can show that this is not a good estimator of the average frequency of the signal except for high ratios of signal to noise. Therefore it is desirable to improve the estimator by limiting the influence of the noise. Two techniques have been proposed.

One technique simply subtracts the noise spectral density $N(f_i)$ from the spectral density of the combined sum and calculates the average frequency of the resulting signal:

$$\hat{f}_{\mathrm{FFT}/N} = \sum_i f_i[S'(f_i) - N(f_i)]/\Sigma[S'(f_i) - N(f_i)] \tag{3.26}$$

The estimator obtained this way is not "contaminated" by noise, even at low values of the signal-to-noise ratio; however, it is sensitive to aliasing and underestimates the average frequency in absolute value when the true value approaches the Nyquist limit.

Another technique for suppressing the influence of the noise when the modal value of the signal spectral density is large compared to the average noise eliminates from the signal plus noise sum values below a certain threshold, determined so that the noise components disappear from the spectrum but most of the signal components remain. The appropriate threshold level depends on the standard deviation of the signal spectral density and the signal-to-noise ratio. If the threshold is chosen at a level $A$ dB below the modal value, we have

$$\hat{f}_{\mathrm{FFT}/A} = \sum_i f_i S(f_i) \Big/ \Sigma S(f_i) \tag{3.27}$$

with

$$S(f_i) = \begin{cases} S(f_i) & \text{for } 10 \log[S(f_{\mathrm{mode}})/S(f_i)] \leq A \text{ dB} \\ 0 \text{ otherwise} \end{cases}$$

$A$ can be proportional to the signal-to-noise ratio or a constant; a current value is $-12$ to $-15$ dB.

This estimator is not contaminated by the noise so long as the signal-to-noise ratio is higher than the threshold and the standard deviation approaches that obtained with the direct FFT and the FFT/$N$ for high values of the signal-to-noise ratio. This technique is most used currently in Fourier transform machines because of its simplicity, that it requires no knowledge of the shape of the noise spectrum and it eliminates parasite signals due to the machine.

*Argument of the Autocorrelation Function.* An estimator of the average frequency using the autocorrelation function (or covariance) of the time series has been suggested by Rummler [1968] and applied to the doppler spectra of meteorological echoes by Lhermitte [1972] and Groginsky [1972]. This technique, called *pulse pair processing* (PPP), is based on estimating the complex autocorrelation function starting from $N$ successive complex samples of the signal in a range gate, assuming that the spectrum is symmetric. This is a consequence of the relationship given earlier in (3.10) between the derivatives of the autocorrelation function and the spectral moments.

The complex autocorrelation function is related to the power spectral density by expression (3.4b), which, with $\tau = T = f_r^{-1}$, can be written as

$$R(T) = e^{i2\pi\bar{f}T} \int_{-\infty}^{+\infty} S'(f)\, e^{i2\pi(f-\bar{f})T}\, df \tag{3.28}$$

where $\bar{f}$ is the first moment of this spectrum.

When the spectrum is even around $\bar{f}$, that is, if $S'(f - \bar{f}) = S'(\bar{f} - f)$, the integral on the right-hand side of (3.28) is real, because

$$\int_{-\infty}^{+\infty} S'(f) \sin[2\pi(f - \bar{f})T]\, df = 0 \tag{3.29}$$

In this case, the function $R(T)$ has the argument $2\pi\bar{f}T$, and we have

$$\bar{f} = \frac{1}{2\pi T} \arctan\left[\frac{\text{Im}[R(T)]}{\text{Re}[R(T)]}\right] \tag{3.30}$$

The estimate of the complex autocorrelation function is obtained from successive pulse pairs by the expression (analogous to (3.3))

$$\hat{R}_m(T) = \frac{1}{N} \sum_{n=0}^{N-1} y_{n,m} y^*_{n+1,m} \tag{3.31}$$

where $y_{n,m}$ and $y_{n+1,m}$ are samples in the $m$ gate. The successive pairs are assumed to give independent estimates of $R(T)$.

For a time series of $N$ complex samples of the coherent radar signal, $y_n = a_n + ib_n$, we have

$$\hat{f}_{\text{PPP}} = \frac{f_r}{2\pi} \arctan\left[\frac{\sum_n \text{Im}(y_n y^*_{n+1})}{\sum_n \text{Re}(y_n y^*_{n+1})}\right] \tag{3.32a}$$

That is, by denoting the vector angle in the complex plane by $\phi_n$:

$$\hat{f}_{\text{PPP}} = \frac{f_r}{2\pi} \arctan\left[\frac{\sum_n |y_n y^*_{n+1}| \sin(\phi_{n+1} - \phi_n)}{\sum_n |y_n y^*_{n+1}| \cos(\phi_{n+1} - \phi_n)}\right] \tag{3.32b}$$

or

$$\hat{f}_{\text{PPP}} = \frac{f_r}{2\pi} \arctan\left[\frac{\Sigma\, a_n b_{n+1} - \Sigma\, b_n a_{n+1}}{\Sigma\, a_n a_{n+1} + \Sigma\, b_n b_{n+1}}\right] \tag{3.32c}$$

The complex product $y_{n,m}y^*_{n+1,m}$ is a vector with a magnitude proportional to that of $y_{n,m}$ and $y_{n+1,m}$ and a phase proportional to the doppler phase shift over the interval $T$. The $N$ samples of $R_m(T)$ are added vectorially, and the average doppler frequency is estimated from the average of the phase shifts weighted by the power of the sampling interval $T$. It is simple to calculate $f_{\text{PPP}}$ by using expressions (3.32), and it can be implemented using inexpensive digital circuits; this is an advantage of the method.

When the spectrum is asymmetric, the estimator is wrong; however, the error usually is small and can be tolerated in most cases. In particular, for a narrow spectrum and if $T$ is small, eq. (3.29) can be written as

$$\int_{-\infty}^{+\infty} S'(f)2\pi(f-\bar{f})T\,\mathrm{d}f = 0$$

or

$$\bar{f} = \int_{-\infty}^{+\infty} fS'(f)\,\mathrm{d}f \bigg/ \int_{-\infty}^{+\infty} S'(f)\,\mathrm{d}f \tag{3.33}$$

which shows that for a narrow spectrum, the symmetry disappears.

Various authors have tested this estimator and defined its statistical properties [Miller and Rochwarger, 1972; Sirmans and Bumgarner, 1975b; Zrnic', 1977]. The conclusions of these studies demonstrate the advantages this estimator provides. The estimate is not changed by the presence of noise and does not depend on the signal-to-noise ratio. In fact, when the signal-to-noise ratio is small and the noise contaminates the spectrum, the classical estimation leads to a weighted sum that can be different from the true average frequency. With the autocorrelation function estimator, the phase of the noise samples is taken into account. Now, because successive samples are statistically independent, the error due to the noise is zero on average. The effect of the noise on the standard deviation of the estimate is negligible for values of the signal-to-noise ratio higher than approximately 15 dB.

Furthermore, the covariance argument estimator minimizes the problem of velocity ambiguity; in fact, when the spectrum extends beyond the Nyquist limits (i.e., $f_r/2$), the estimate of $\bar{f}$ is not altered as long as its value stays within the unambiguous interval, which is not the case for the classical estimator deriving from eq. (3.7), as this is a weighted sum extending over the entire observed spectrum. When $\bar{f}$ crosses the Nyquist limits, a clear and therefore detectable discontinuity of amplitude $f_r$ is observed.

Finally, analysis of average frequency (and also variance) estimates by the covariance argument estimator, in the case of a Gaussian signal and Gaussian white noise, shows that it retains its properties for independent pulse pairs, whether separated or contiguous. Interlaced pairs of pulses can also be used to increase the velocity-range ambiguity region if the signal spectrum is not too broad [Zrnic', 1977].

*Variance*

*Measurement by Noncoherent Radar: R Meter.* We have seen, equations (1.81) and (1.82), that the fluctuations of the instantaneous power result from movements of the scatterers with respect to each other. The frequency component $F$ of the fluctuation spectrum $S_F(F)$ results from the beats of the signals from all the scatterers that have a radial velocity difference $v_i - v_j$ and a doppler frequency difference $f_i - f_j = F$ (with $F = 2(v_i - v_j)/\lambda$. Its power is the sum of $a_i^2 a_j^2$ over all the scatterers for which $f_i - f_j = F$. Now, $a_i^2$ and $a_j^2$ are proportional to the average power in the doppler spectrum with frequency $f_i$ and with frequency $f_j = f_i + F$, respectively. We therefore have

$$S_F(F) = \int_{-\infty}^{+\infty} S(f)S(f + F)\,\mathrm{d}f \tag{3.34}$$

The fluctuation spectrum is the integral of the doppler spectrum multiplied by itself after having been shifted by the frequency fluctuation $F$. Because $F$ takes positive and negative values, we have

$$S_F(F) = S_F(-F) \quad \text{and} \quad \bar{F} = 0$$

In addition,

$$E^2 = \overline{F^2} = \int_{-\infty}^{+\infty} F^2 S_F(F)\,\mathrm{d}F = 2\sigma_f^2 \tag{3.35}$$

The variance of the fluctuation spectrum of the signal power ($E^2$) is equal to twice that of the doppler spectrum. The spectrum $S_F(F)$ can be measured with a noncoherent radar although it usually is limited to estimating its variance, which makes it possible to obtain the moment of order 2 of the doppler spectrum.

The data from a noncoherent radar therefore let us obtain information on the width of symmetric doppler spectra. This opportunity provided by the fluctuation spectrum has been little exploited. Limited information on the fluctuation spectrum is given by the "*R* meter," which measures the number of times the signal crosses a power threshold with given value during a sampling interval [Rutkowski and Fleisher, 1955]. This number is related to the fluctuation rate of the signal. Lhermitte [1963] shows and experimentally verifies that the number of times per second $L/2$ that a fluctuating signal, with a probability density given by (1.85) crosses a reference level $A = \bar{A}$ with positive slope, is

$$L/2 = (\overline{F^2})^{1/2} \tag{3.36a}$$

where $F$ is the fluctuation frequency of the signal.

The number of positive and negative crossings $L$ is proportional to the standard deviation of the doppler velocity spectrum $\sigma_V$ [Rogers, 1963]; we have

$$L = a\sigma_V/\lambda \tag{3.36b}$$

where $a$ is a constant equal to 5.74 for a receiver using a linear detector and 5.22 for a square-law detector. The transfer function of the receiver therefore has little effect on the value of the variance of the doppler spectrum given by the $R$ meter. On the other hand, the carrier frequency instabilities of noncoherent radars may lead to significant errors in the estimate of $\sigma_V$ by this technique [Bechtel and Prinsen, 1967].

The theory by Rutkowski and Fleisher is restricted to incoherent continuous wave radars in the absence of noise and with linear detection. It is called the *conventional R meter theory*. Equation (3.36b) applies in this case.

The $R$ meter concept has been extended to incoherent pulse radars taking the receiver noise into account [Lob, 1968]; we can show that eq. (3.36b) applies only when the signal-to-noise ratio is higher than approximately 15 dB. Walker and Cronenwett [1974] verify the Lob theory, but show that the results are affected by aliasing. Thus, the $R$ meter provides a possibility of estimating the spectrum width with the help of incoherent radars; but the standard deviation of the estimates is larger than that obtained by processing a coherent signal.

An application of noncoherent radar to turbulence measurements consists of exploiting the velocity differences between two pulse volumes separated by an arbitrary distance. This is based on the fact that signals from these two volumes are coherent with respect to each other [Atlas and Srivastava, 1971]. It uses measurements of power and fluctuation spectrum variance in the two volumes considered and the corresponding measurements of the sum of signal predetection in these two volumes. The latter is obtained by delaying the signal from $r_1$ ($< r_2$) so that it can be made to coincide with the signal from $r_2$. We can show that

$$(\bar{V}_1 - \bar{V}_2)^2 = \left(\frac{\lambda^2}{8}\right) \frac{(P_1 + P_2)}{P_1 P_2} [(P_1 + P_2)E_{1,2}^2 - P_1 E_1^2 - P_2 E_2^2] \tag{3.37}$$

In this expression, $P$ is the received power, $V$ is the radial velocity, and $E^2$ is the variance of the fluctuation spectrum; the subscripts 1 and 2 refer to the two range gates considered and the double subscript denotes the sum signal. Because all the quantities situated on the right-hand side of eq. (3.37) can be measured with a noncoherent radar, with such an instrument obtaining the term $(\bar{V}_1 - \bar{V}_2)^2$, and therefore the velocity structure function (cf. Section 3.4.2), is thus possible. The same information obviously can be obtained with a coherent radar from direct measurement of the average doppler velocities in each pulse volume considered. The estimates of turbulent fluctuations given by a coherent radar are better than those that can be obtained by a noncoherent radar [Serafin, Peach, and Atlas, 1972].

*Fast Fourier Transform*. The discrete Fourier transform of the coherent video that defines the doppler spectrum obviously makes it possible to estimate the width. However, the spectral density obtained by this technique is that of the superimposition of the signal as well as the noise and its image; therefore, the estimated spectral width is affected by the noise in a manner that depends on its level with respect to the signal. Consequently, an acceptable estimate is obtained only for large signal-to-noise ratios or after noise suppression. This method therefore places technical demands on the system.

The estimate of the doppler spectrum variance from the spectral density is given by

$$\sigma_f^2 = \frac{\sum_i f_i^2 S(f_i)}{\sum_i S(f_i)} - \left( \frac{\sum_i f_i S(f_i)}{\sum_i S(f_i)} \right)^2 \tag{3.38}$$

$S(f_i)$ is the spectral density above a threshold with given value; that is, by eliminating the base of the spectrum. This truncation of the spectrum produces a systematic error on its width because part of the spectrum is eliminated. Study of the quality of this estimator shows that the standard deviation of the estimate is weakly affected by noise for $S/N > 15$ dB and increases rapidly at low values.

*Autocorrelation Function Argument*. The estimate of the variance by the autocorrelation function argument technique (PPP) uses the calculation of the signal correlaton. The estimator can be written [Miller and Rochwarger, 1972]

$$\sigma_f^2 = (2\pi^2)^{-1} f_r^2 \left( 1 - \frac{|R(T)|}{R(0)} \right) \tag{3.39a}$$

or

$$\sigma_f^2 = (2\pi^2)^{-1} f_r^2 \left\{ 1 - \frac{\left| \sum_n y_{n+1} y_n^* \right|}{\sum_n y_n y_n^*} \left[ \frac{1 + 10^{(S/N)/10}}{10^{(S/N)/10}} \right] \right\} \tag{3.39b}$$

where S/N denotes the signal-to-noise ratio.

The second expression includes a corrective term for the noise (term in brackets). This estimator gives good results when the standard deviation of the spectrum ($\sigma_f$) lies between 0.05 and 0.3 of the Nyquist interval. The same does not apply to narrow spectra and low signal-to-noise ratios because the estimator tends to produce a negative value of $\sigma_f^2$ due to the numerical approximation of the derivatives. The standard deviation of the estimator also varies with the spectral width and noise. For a Gaussian signal, the use of independent pulses, whether spaced or contiguous, does not alter the properties of the estimator [Zrnic', 1977]. As far as the effect of symmetry defects of the spectrum on the estimate of the variance by PPP is concerned, the conclusions are the same as for the estimate of the mean; these defects do not have any serious effect on the estimate.

An estimate of the variance derived from the preceding considerations has been proposed, making possible a better correction for the effects of noise [Srivastava, Jameson and Hildebrand, 1979]. Without explicitly involving the S/N ratio, the method uses the complex autocorrelation function for two intervals $T$ and $2T$. We can show that, when the doppler spectrum of the meteorological signal can be approximated by a Gaussian function, the velocity variance is given by the expression

$$\sigma_v^2 = \frac{\lambda^2}{2\pi^2 T^2} \ln\left(\frac{R(T)}{R(2T)}\right) \tag{3.40}$$

and that the signal-to-noise ratio can be expressed in the form

$$S/N = |R(T)|^{4/3} / [R(0)|R(2T)|^{1/3} - |R(T)|^{4/3}] \tag{3.41}$$

The estimate of $\sigma^2$ by this method is in good agreement with that given by the method (4.39).

### 3.1.4 Factors Affecting the Width of the Doppler Spectrum

Many factors cause the doppler spectrum to broaden: the properties of the observed meteorological target and technical factors specific to the radar, the observation procedure, and signal processing [Sloss and Atlas, 1968; Atlas, Srivastava and Sloss, 1969; Waldteufel, 1976]. A distinction is made between the following main causes.

*General and Permanent Causes*

1. Finite width of the distribution of the terminal fall velocities of the scatterers.
2. Movements of the atmosphere on a scale smaller than the sampling volume or turbulence.
3. Movements of the atmosphere on a scale larger than the sampling volume or shear.
4. Beamwidth.
5. Rotation of the antenna (or relative movement of the sampling volume and the target).

Factors 1–3 are meteorological in nature whereas factors 4 and 5 are specific to the radar.

*Specific Causes*

6. Receiver noise.
7. Statistical uncertainty.
8. Nonmeteorological factors: ground echoes, nonmeteorological targets in the atmosphere (airplanes, birds), external radioelectric signals.
9. Meteorological factors: secondary lobes, second trip echoes.
10. Factors linked to the receiver: instability, nonlinearity.
11. Factors linked to signal processing: aliasing, broadening due to the limited duration of the sample, errors associated with the processing technique.

We assume that the various spectrum broadening factors are independent of each other. Under these conditions, the total variance is the sum of the partial variances due to each one:

$$\sigma_v^2 = \sum_i \sigma_i^2 \tag{3.42}$$

If it is possible to isolate one of the terms in this sum by eliminating or sufficiently minimizing the others, the variance can be interpreted as a measurement characterizing the corresponding cause. Only the first two terms (factors 1 and 2) are of practical value because they are linked to the granulometric distribution of the precipitation and to the turbulent kinetic energy of the air with smaller scales than the dimensions of the resolution volume.

*Contribution of the Different Terms*

The dispersion of the fall velocities of the particles is the cause of a variance $\sigma_P^2$ in which the value depends on the width of the granulometric distribution. This factor appears in the doppler spectrum (proportional to the sine of the elevation angle); the effect is maximal at vertical elevation, whereas it is negligible for elevations less than approximately 12°. We have

$$\sigma_p^2 = (\sigma_{p_0} \sin\alpha)^2 \tag{3.43}$$

where $\sigma_{p_0}$ is the spectral width due to the distribution of the fall velocities of the precipitation in calm air and $\alpha$ is the elevation angle.

The variance therefore depends on the nature of the precipitation. At vertical elevation, the observed variances are less than 1 $m^2\ s^{-2}$ in snow and rain, outside the generating convective zones, and are independent of the size distribution [Lhermitte, 1963]; in hail, the variance may reach and exceed 20 $m^2\ s^{-2}$.

Small-scale air movement carrying cloud and precipitation particles (or any other tracer) are responsible for $\sigma_t^2$, the variance due to turbulence. Small-scale turbulence is understood as turbulence whose components have dimensions equal to or less than the resolution volume of the pulse; the size of the movements concerned therefore depends on the equipment characteristics and range as far as filtering due to the resolution volume is concerned [Srivastava and Atlas, 1974]. The inertia of the particles determines their response to the fluctuations in the velocity of the fluid that carry them. Small drops and snowflakes faithfully reproduce turbulent movements; on the other hand, hail and large drops tend to filter out small-scale movement. The orders of magnitude of the contribution of the turbulence to the standard deviation of the spectrum $\sigma_t$ are 0.5 m $s^{-1}$ for small turbulence, to 5 m $s^{-1}$ for moderate or strong turbulence, and to 10 m $s^{-1}$ for extreme turbulence.

A wind shear carrying scattering particles is reflected by a variation in velocity across a beam of finite width that gives a contribution $\sigma_S^2$ of the shear to the variance; a vertical shear of the horizontal wind results in a maximum broadening of the spectrum for a horizontal elevation in the direction of the wind. The effect increases with the beamwidth and so with the range.

We can assume that $\sigma_s^2$ consists of the sum of three independent terms corresponding to the perpendicular shear $\sigma_{sp}^2$, the vertical shear $\sigma_{sv}^2$, and radial shear $\sigma_{sr}^2$:

$$\sigma_s^2 = \sigma_{sp}^2 + \sigma_{sv}^2 + \sigma_{sr}^2 \tag{3.44}$$

where $\sigma_{sp}^2$ is the variance due to the gradients of radial velocity $K_p$ measured perpendicularly to the vertical plane passing through the beam axis; it is proportional to $(K_p r\theta)^2$, where $r$ is the distance between the radar and the center of the pulse volume considered and $\theta$ is the 3 dB beamwidth. The term $\sigma_{sv}^2$ is the variance due to the gradients of radial velocity $K_v$ measured vertically in the vertical plane passing through the beam axis; it is proportional to $(K_v r\theta \cos\alpha)^2$, where $\alpha$ is the elevation angle. The term $\sigma_{sr}^2$ is the variance due to the gradients of radial velocity $K_r$ measured radially or radial shear; it is proportional to $K_r^2 h^2$, where $h$ is the spatial length of the pulse.

The total variance due to shear may be written [Nathanson, 1969; Sirmans and Doviak, 1973]

$$\sigma_s^2 = 0.09\theta^2 r^2 (K_p^2 + K_v^2 \cos^2\alpha) + 0.021 K_r^2 h^2 \tag{3.45}$$

with $r$ and $h$ in $m$, $\theta$ in rad, and $K$ in $s^{-1}$.

For a maximum shear of $K_p = K_v = 2 \times 10^{-2}$ s$^{-1}$, a beamwidth of 1°, and at a range of 50 km with $\alpha = 0$, we have $\sigma_s^2 \approx 55$ m$^2$ s$^{-2}$, neglecting the term in $K_r$ for the radial shear. This term is always less than 0.5 m$^2$ s$^{-2}$.

The differences in the value of the projection of the velocity on the direction of propagation due to the divergence of the beam cause a variance $\sigma_b^2$ that increases with the increasing width of the beam and increasing velocity. We can write [Nathanson, 1969]

$$\sigma_b^2 = 0.09\theta^2 V^2 \sin^2\beta \tag{3.46}$$

where $V$ is the average velocity of the wind (in m s$^{-1}$). The term $\beta$ is the azimuthal angle of the beam with respect to the direction of the wind, and $\theta$ is in rad.

Thus, for $\beta = \pi/2$; $V = 50$ m s$^{-1}$, and $\theta = 1°$, we have

$$\sigma_b^2 = 0.07 \text{ m}^2 \text{ s}^{-2}$$

The rotation of the antenna broadens the spectrum because the movement of the beam within the scattering volume modifies the properties in the sampled volume, which partly affects the coherence of the received signal during the acquisition of the sample. This decorrelation is proportional to the square of the velocity of rotation of the antenna $\alpha'^2$. Nathanson [1969] writes the following equation:

$$\sigma_r^2 = 0.017\left(\frac{\alpha'\lambda}{\theta}\right)^2 \tag{3.47}$$

where $\lambda$ is the wavelength (in m). For a velocity of rotation of 10° $s^{-1}$ a beamwidth of 1°, with $\lambda = 0.1$ m, this term is 0.017 $m^2$ $s^{-2}$. It therefore is small.

*Other Causes*

Broadening also can result from inadequate instrumental precautions or just be due to specific circumstances. These include two internal causes: the receiver noise, which if it is not suitably subtracted can add appreciably to the variance, and the uncertainty relating to the statistics of the signal. There are also external causes, in particular, ground echoes, which have to be taken into consideration for low elevations when the absolute values of velocity seem to be abnormally small. The presence of airplanes or birds or insects in the beam is also a cause of temporary error, just like the reception of an external radioelectric transmisson.

The detection by secondary lobes of meteorological echoes does not usually produce any significant perturbation because the relative power is small; but there are exceptional cases (ground reflection, intense precipitation cell) where the contribution of secondary lobes will be predominant. Similarly, in the presence of double-trip echoes (from targets situated beyond the ambiguous range), the measurement of the variance is falsified. The abnormal broadening of the spectrum created by these disturbing factors often is perceptible. The ambiguity concerning the presence of double-trip echoes can be removed by intermittent reduction of the repetition frequency of the radar.

Factors linked to the equipment can produce "offset voltages," superimposed on the complex signal (for example, at the level of the digital-analog converters), instability or phase shift between the components of the complex signal, defects in the linearity of the receiver (for example, by saturation). Saturation has little effect in the spectral domain except when it is reflected by clipping in the baseband (or doppler band). This clipping can be responsible for parasite harmonics the level of which can be quantitatively significant (clipping alters the estimation of variance).

One of the most important factors associated with data processing is aliasing, which affects the variance by sharing the spectral mode between the two ends of the Nyquist cointerval; this can introduce significant error. The limited duration of the sample used for calculating the variance in the spectral domain has a corresponding broadening that cannot be negligible. We can correct the spectrum and totally eliminate such effects by deconvolution, assuming that the doppler spectrum is Gaussian. It is also possible to eliminate the effects of the finite duration of the sample by truncating the spectrum and limiting the spectral band. Finally, the techniques used to correct the estimation of the spectral width themselves can cause error in certain cases, which can be manifested in an underestimation of the width on the order of $1\ \mathrm{m \cdot s^{-1}}$.

Most of these various factors affect the fast Fourier transform or autocorrelation function argument estimators with the exception of those related to the receivers or data processing. This analysis therefore shows that the effects of a particular factor from the spectral width can be measured only if from the start we are sure that other nonnegligible terms can be either suitably eliminated or evaluated and corrected. The problem must be examined in detail, considering the characteristics of the receiver, the choice of the estimator, the observation procedure, the nature of the target, and the order of magnitude of the measurement. Finally, under suitable technical observation conditions, taking into account the properties of the radar and meteorological conditions, it is possible to make the contribution of all terms negligible, with the exception of the terms associated with the transverse shear (coefficients $K_p$ and $K_v$), the turbulence, and the dispersion of the velocity of fall of the drops. These last two in fact are the only ones of practical value for observing the atmosphere.

### 3.1.5 Ground Clutter Suppression

The presence of ground echoes mixed with meteorological echoes, presenting a minor disturbance for certain qualitative observations on a screen, becomes a major problem when automatic procedures are used for quantitative evaluation. This is the case particularly for hydrological measurements (Section 2.6) at low elevation, above a hilly terrain, or containing strongly scattering obstacles (such as metal pylons). The problem is especially acute in panoramic radar installed at points of high elevation and used for observation over a large range. Therefore, efforts are made to eliminate fixed echoes.

The simplest procedure involves recording ground echoes in the absence of atmospheric echo for a given elevation and then eliminating data corresponding to ranges and directions for which ground echoes have been observed when under meteorological observation. The implementation of this technique involves procedures of varying complexities, from simple manual elimination to automatic elimination in real time or elapsed time. However, this procedure is inefficient for ground echoes associated with abnormal propagation. In addition ground echoes can modify with time (season, and so on).

With a coherent radar, the ground echoes can be eliminated by filtering the doppler component with zero frequency. In fact, the ground echoes either do not fluctuate (building, bare soil) or barely fluctuate (crops, forest), so that the spectra are narrow with an average speed of zero (with possible exceptions, for example, rotary anemometers when the wind is not zero). When the doppler spectrum occupies a domain including this frequency, the suppression of the peak corresponding to the fixed echoes involves the meteorological component superimposed on it; however, this does not have a significant effect on the properties of the actual spectrum in most practical cases, because of the small width of the eliminated spectral band. Furthermore, the spectrum can be retrieved by interpolation using values lying on both sides of the suppressed line.

In noncoherent radar, the elimination of ground echoes can be made in real time from the examination of the statistical properties of the echoes. The operation can be carried out in several different ways. The fluctuations in the signal can be measured and then compared to a threshold determining the rejection of signals with low fluctuation. The threshold is chosen to reject most of the ground echoes and as small as possible a proportion of meteorological echoes. The "measurement" of the fluctuations is expressed by the difference between the maximum value and the average value of the signal or by the average of the difference between successive pulses over the $N$ pulses of an integration cycle. For example, when the number of integrated pulses is greater than or equal to 32, and elimination rate of ground echoes greater than 70% without loss of meteorological echoes can be obtained, the remaining 30% consists largely of weak echoes contaminated by the receiver noise and fluctuates as meteorological echoes; the strongest ground echoes are practically all eliminated. However, to increase the rate of elimination, we must accept the rejection of precipitation echoes superimposed on fixed echoes situated in the same volume [Geotis and Silver, 1976].

An alternative to the observation of precipitation begins with the fact that when the echo of the target fluctuates as a Rayleigh variable, eq. (1.85), the average power $\overline{A^2}$ is proportional to the variance of the amplitude of the signal at the output of a linear receiver.

$$\overline{A^2}/\sigma_A^2 = \text{constant} \tag{3.48}$$

When the ground echoes are superimposed on the precipitation echoes, the proportionality between the power of the meteorological component $\overline{A^2}$ and the variance of the total signal (ground echo + meteorological echo) is approximately maintained. Under these conditions, Tatehira and Shimizu [1978] suggested estimating the average power of the meteorological echo from the measurement of the variance of the total signal.

## 3.2 DOPPLER SPECTRA AT VERTICAL INCIDENCE

With a vertically pointed radar, only the fall of precipitation and vertical air movements affect the doppler velocity. These are important physical variables in meteorology.

If we denote the doppler velocity, the vertical component of air velocity, and the terminal velocity of the hydrometeors in still air by $w_d$, $w_a$, and $w_t$, respectively, all taken positively moving upward, we have

$$w_d = w_a + w_t \tag{3.49a}$$

For example, for an updraft of $w_a = 10$ m s$^{-1}$ and precipitation with $w_t = -5$ m s$^{-1}$, the doppler velocity is $w_d = +5$ m s$^{-1}$ (sometimes in radar meteorology doppler velocity is taken as positive when the target is moving toward the radar).

If $w_a$ is known, zero, or negligible, the measurement of $w_d$ leads to the velocity of fall of the precipitation and to its size spectrum. If, on the other hand, $w_t$ is known, or negligible with respect to $w_a$, $w_d$ is a measure of the vertical velocity of the air.

### 3.2.1 Size Distribution of Precipitation

Outside regions occupied by convective movements or dynamic perturbations of various origins, the vertical velocity of air is low, on the order of a few cm s$^{-1}$, and the turbulence is of little importance. If the divergence of the radar beam is small, the doppler spectrum reflects mainly the dispersion of the velocity of fall of the precipitation and, consequently, that of their dimensions or their mass $m$. To pass from the spectrum of $w_t$ to the size distribution, we must know the relationship between $w_t$ and $D$ (Section 2.2.5). For ice crystals and snowflakes, this is no simple relationship, because $w_t$ depends on a shape factor; that is, hydrometeors of the same mass can have different values of $V_t$.

If $w_a$ is not negligible, the velocity spectrum may also be interpreted in terms of particle size distribution, after subjecting it to a translation equal to $\bar{w}_a$ parallel to the velocity axis. We have

$$S(w_t)\,dw_t = S(w_d - \bar{w}_a)\,dw_t = N(D)\sigma(D)\,dD/\eta \tag{3.50}$$

where $S$ denotes the normalized spectrum: the relationship between $w_t$ and $D$ is introduced by differentiating $dD/dw_t$.

For rain and with Rayleigh scattering, we have

$$S(w_d - \bar{w}_a)\,dw_t = N(D)D^6\,dD/Z \tag{3.51}$$

that is,

$$N(D) = D^{-6}ZS(w_d - \bar{w}_a)\,dw_t/dD$$

Therefore, measurement of the doppler spectrum and the reflectivity factor lets us calculate the size distribution, provided that $w_a$ is known or negligible.

These relationships can be applied mainly to stratiform precipitation below the generating convective layer; that is, where the turbulence does not modify the velocity spectra too much. At low fall velocity as for snow or precipitation of small particles, the method suffers from a lack of precision.

### 3.2.2 Vertical Air Velocity

With vertically pointed radar, the trajectory of the cloud that drifts horizontally at the speed of the average winds present at this level brings it above the radar, which thus observes it in height-time coordinates. The observation therefore suffers from fixed vertical sighting, because the observed volume is not chosen and the observation is not instantaneous (during the time necessary to observe a structure of given scale, it may evolve in a nonnegligible manner). On the other hand, it also has the advantages of precision linked to the slowness of the observation and a maximum resolution because the observation range is minimal.

To obtain the air velocity with vertically pointed radar it is necessary to correct the doppler speed with the velocity of fall of the hydrometeors. Several methods of correction have been proposed.

The simplest [Battan, 1964] is based on the limiting velocity of fall of the smallest particles that can be detected by the radar. The particle size is known approximately, so their limiting velocity of fall in calm air is also known. Under these conditions, the air velocity corresponds to the upper limit of the spectrum (if the ascending velocities are taken $> 0$; the lower limit applies in the opposite case) corrected by this velocity of fall. For example, for a radar with wavelength of 3 cm and standard characteristics, these particles will have a size of approximately 300 $\mu$m and a terminal fall velocity of 1 m s$^{-1}$. The procedure assumes that these small particles are always present in the observed spectra, which is not always true (in particular, if the wind shear carries out a sort on the drop spectrum; cf. Section 5.5.2). Furthermore, the minimum detectable dimension depends on the range; but above all, where the vertical velocity of the air is not zero, in general a significant turbulence is found that modifies the spectral limits (broadening).

A more elaborate procedure consists of making an assumption on the form of the size distribution of the particles and deriving from this a relationship between $w_t$ and $Z$; measuring $w_d$ and $Z$ by doppler radar lets us calculate $w_a$. Equation (3.49a) is first rewritten in terms of the mean velocity:

$$\bar{w}_a = \bar{w}_d - \bar{w}_t \tag{3.49b}$$

where $\bar{w}_d$ is the mean velocity given by the first moment of the normalized spectrum and $\bar{w}_t$ is the mean terminal velocity derived from the reflectivity.

A first approximation [Rogers, 1964] considers a population of spherical, liquid Rayleigh scatterers, obeying the Marshall and Palmer exponential distribution with limiting velocity of fall given by $v = 1420D^{0.5}$ with $v$ in m s$^{-1}$ and $D$ in cm. For rain we obtain

$$\bar{w}_t = 3.8Z^{0.072} \tag{3.52}$$

with $w_t$ in m s$^{-1}$ and $Z$ in mm$^6$ m$^{-3}$.

This calculation can be carried out for other types of precipitations and distributions. A similar expression established from granulometric measurements of ground rainfall [Joss and Waldvogel, 1970] is

$$\bar{w}_t = 2.6Z^{0.107} \tag{3.53}$$

For snow, relationships (2.25) lead to this expression:

$$\bar{w}_t = 0.817Z^{0.063} \tag{3.54}$$

with the same units in these last two equations as in (3.52).. In the case of snow, an average velocity of fall of 1 m s$^{-1}$ can often be considered as a reasonable approximation.

Relationships (3.52)–(3.54) apply close to sea level; with altitude, the right-hand side has to be multiplied by the approximate correction factor $(\rho/\rho_0)^{0.4}$ (Section 2.2.5). The equations can be used for reflectivities lying between 10 and $2 \times 10^5$ mm$^6$ m$^{-3}$. The accuracy is at best $\pm 1$ m s$^{-1}$. The main drawback of this method comes from the fact that rain distribution changes from one measurement to another (the MP distribution is valid for an average) so that the coefficients of the $w_t - Z$ are not constant in nature.

In an attempt to correct this, Hauser and Amayenc [1981] have suggested a so-called three-parameter method that makes it possible to determine at the same time $\bar{w}_a$ and the two parameters $N_0$ and $\lambda$ of the distribution, assumed to be exponential, of the hydrometeors by a least squares fit. The results show a significant dispersion of the three estimated parameters (and the values of precipitation intensity and water content deduced from them) due to the influence of the spectral broadening created by the turbulence. The méthod is suitable only when the assumption of exponentiality is satisfied and when the turbulence has negligible effects on the spectrum.

To measure the detailed distribution of the hydrometeors from doppler spectra in all cases, therefore we must be able to correct for the effects of turbulence. Clear air radar opens up one avenue in this direction, because it lets us obtain information on the structure of the air movement and the velocity spectrum of the hydrometeors (cf. Section 4.3). A method has been suggested by Wakasugi *et al.* [1986] for VHF doppler radar spectra with vertical sighting. The hydrometeors are assumed to be raindrops obeying an exponential distribution of the parameters $N_0$ and $\lambda$ (equation 2.15). In calm air, the relationship between the normalized doppler spectrum $S_1(w)$ and distribution of drops $N(D)$ is given by equation (3.51) with $w_a = 0$ and $w_t$ related to $D$ by equation (2.28). The turbulent velocities also are assumed to have a Gaussian distribution whose normalized spectrum is

$$S_2(w) = (2\pi\sigma^2)^{-1/2} \exp(-w^2/2\sigma^2) \tag{3.55}$$

where $\sigma$ is the spectral broadening.

The doppler spectrum $S_0(w)$ measured by the VHF radar is the sum of the contribution of the rain and that of the turbulent air as both are detected by the radar. The contribution of the rain is the convolution of $S_1$ and $S_2$ so that $S_0$ can be written

$$S_0(w) = P_1S_1(w)*S_2(w) + P_2S_2(w) \tag{3.56a}$$

where $P_1$ and $P_2$ are powers associated with the precipitation and the fluctuations in the refractive index, respectively.

In the presence of an average velocity $\bar{w}$, we have

$$S_0(w) = P_1S_1(w - \bar{w})*S_2(w) + P_2W_2(w - \bar{w}) \tag{3.56b}$$

With the help of these expressions, a least squares fit lets us obtain the six parameters ($P_1N_0$, $\lambda$, $\bar{w}$, $\sigma$, $P_2$, and $N$) where $N$ is the noise level. Comparison of the results obtained by ignoring and correcting for the effects of turbulence by this method shows that the error on the values of $N(D)$ and water content of the precipitation may reach an order of magnitude (Gossard, 1988).

Recently, Lhermitte (1988) has proposed to differentiate air vertical motion and raindrop terminal velocity, in a spectrum observed at vertical incidence from the Mie backscattering oscillations at the 3.2 mm radar wavelength. At this wavelength, $\sigma(D)$ (cf. Figure 2.9) exhibits maxima at $D = 1.0$, 2.26, 3.44, and 4.62 mm and minima at $D = 1.67$, 2.86, 4.04, and 5.22 mm. These oscillations modulate the doppler spectrum at vertical incidence and thus can serve as a means to identify particle size. The occurrence of air velocity $\bar{w}_a$ shifts these oscillations in the doppler scale without destroying their integrity. Therefore, comparing an observed doppler spectrum with a predicted spectrum evaluated from the Mie function leads to a measurement of $w$ and then, correcting for $w$, a measurement of $N(D)$.

## 3.3 MEASUREMENT OF THE VELOCITY FIELDS WITH A SINGLE DOPPLER RADAR

A doppler radar allows the measurement of only one component of the velocity of the scatterers, the radial component. In the general case, air movement is three-dimensional and variable over time and space. Therefore, in principle, to describe it completely, we must make simultaneous measurements with three doppler radars (provided that a homogenous tracer is available). Most of the numerous studies on the methodology of the measurement of velocity fields by doppler radar discuss the information that can be obtained with an incomplete instrumental system; that is, two or most often a single doppler radar. Obviously, we are forced to make simplifying assumptions on the structure of the field under study or, what amounts to the same thing, to accept an incomplete description of the local air circulation. In the first instance, we are interested in measurements with a single Doppler radar.

### 3.3.1 Analysis of the Mean Field by the VAD Method

The general principle of these methods, because the radar "sees" only the radial velocity, involves using the variation of the terms required for measurement as a function of the beam direction (i.e., a trigonometric dependence).

*Uniform Wind Field*

The simplest case is that of a uniform wind field observed on the basis of the movements of a precipitation whose velocity of fall is also uniform. This case has been studied by Lhermitte and Atlas [1961], who show how the measurement of the velocity along circles with center at the radar by azimuthal scanning at a constant angle of elevation makes it possible to retrieve the vertical profile of the wind in terms of speed and direction.

Consider (Figure 3.4(a)) a Cartesian coordinate system $0xyq$, whose origin 0 is the radar, with vertical axis $0z$ directed upward and axes $0x$ and $0y$ directed to the east and the north, respectively. The spherical coordinates $r$, $\alpha$, and $\beta$ of the measurement point are associated with the Cartesian system, where $r$ is the range, $\alpha$ is the elevation angle, and $\beta$ is the azimuthal angle starting from the north. The components of the velocity of scatterers (doppler velocity) along $0x$, $0y$, and $0z$ are

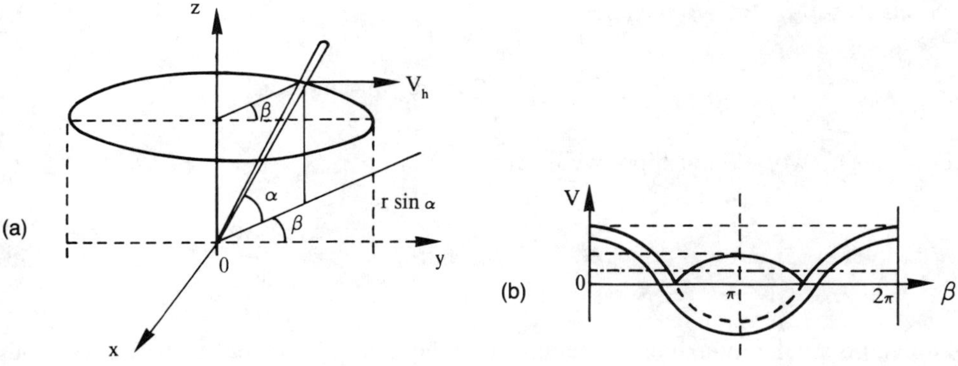


**Figure 3.4** (a) Geometric representation of VAD type scanning. (b) Diagrammatic representation of the average velocity as a function of the azimuth for a VAD scan.

denoted $u$, $v$, and $w$, respectively, with $w = w_d = w_a + w_t$, as defined in eq. (3.49) (for simplicity we drop the $d$ subscript for the vertical doppler component). We assume that cloud and precipitation particles follow the horizontal air movement. Thus, the components of the air velocity are $u$, $v$, and $w_a$. We also assume in what follows that the observed air volume around the radar is occupied by an extended precipitation allowing observation in all directions and that the effect of refractivity on ray curvature is negligible. Under these conditions the radial velocity of the scatterers measured by the radar at range $r$ can be written

$$V_r = \mu \cos\alpha \sin\beta + v \cos\alpha \cos\beta + w \sin\alpha \tag{3.57a}$$

or

$$V_r = r^{-1}(ux + vy + wz) \tag{3.57b}$$

with

$$r = (x^2 + y^2 + z^2)^{1/2}$$

When the scatterers move in a horizontally uniform wind field with constant vertical velocities, a scanning of the antenna along the azimuth at constant elevation angle ($\alpha$) makes $V_r$ vary following a sinusoidal function (Figure 3.4(b)) the maximum ($V_1$) and minimum ($V_2$) of which determine the direction and the sense of the wind. If the horizontal velocity of the wind is denoted by $V_h$ and its direction by $\beta_0$, eq. (3.57) can be simply written as

$$V_r = V_h \cos\alpha \cos(\beta - \beta_0) + w \sin\alpha \tag{3.58}$$

When $\beta = \beta_0$, we have

$$V_1 = V_h \cos\alpha + w \sin\alpha$$

When $\beta = \beta_0 + \pi$, we have

$$V_2 = -V_h \cos\alpha + w \sin\alpha$$

From these two relationships, we derive

$$\bar{V}_h = \frac{V_1 - V_2}{2 \cos\alpha} \quad \text{and} \quad \bar{w} = \frac{V_1 + V_2}{2 \sin\alpha} \tag{3.59}$$

Plotted on a velocity-azimuth diagram, the function $V_r(\beta)$ therefore has a continuous component equal to $w \sin\alpha$, a phase shift $\beta_0$ indicating the direction of the wind, and a maximum and minimum that can be used to deduce $\bar{V}_h$ and $\bar{w}$. This type of scanning is called *velocity-azimuth-display* (VAD) because in 1960 the sinusoidal signal was displayed on an oscilloscope screen. The term has been retained for all methods using scanning with constant angle of elevation. The representation of $V_r(\beta)$ shows deformations due to uniformity defects of $V_h$ and $w$. Small-scale fluctuations are reflected by a broadening of the trace (Section 3.4.3).

At each level ($r \sin\alpha$) where there is a sufficient quantity of scatterers, $\bar{V}_h$ and $\bar{w}$ can be estimated by a least square fit and vertical profiles of wind and vertical doppler velocity result. The interpretation of $\bar{w}$ in term of $\bar{w}_a$ or $\bar{w}_t$ can be done as discussed in Section 3.2.2. It is not necessary to make a complete revolution or have a continuous circle to obtain the required parameters; a few points suffice.

### *Linear Wind Field*

The technique of analyzing velocity fields by VAD has been perfected by Browning and Wexler [1968], who studied a local wind field varying almost linearly so that the velocity components could be approximated by a Taylor series expansion limited to first derivatives. In this way, we express that the velocity field inside the observed domain is described by the sum of the value at the center and the gradient term. Because the measurement circle is horizontal, it is sufficient to consider the expansions in terms of $x$ and $y$. We have

$$\begin{aligned} \bar{u} &= \bar{u}_0 + \frac{\partial \bar{u}}{\partial x} x + \frac{\partial \bar{u}}{\partial y} y \\ \bar{v} &= \bar{v}_0 + \frac{\partial \bar{v}}{\partial x} x + \frac{\partial \bar{v}}{\partial y} y \end{aligned} \tag{3.60}$$

where the subscript 0 denotes the value at the center of the circle being scanned.

These expressions are introduced into (3.57) with

$$x = r\cos\alpha\sin\beta,\ y = r\cos\alpha\cos\beta,\text{ and } z = r\sin\alpha \tag{3.61}$$

and we obtain

$$\begin{aligned}\bar{V}_r(\beta) = {} & \frac{1}{2} r\cos^2\alpha\left[\frac{\partial\bar{u}}{\partial x} + \frac{\partial\bar{v}}{\partial y}\right] + \bar{w}\sin\alpha + \left[r\sin\alpha\frac{\partial\bar{w}}{\partial x} + \bar{u}_0\right]\cos\alpha\sin\beta \\ & + \left[r\sin\alpha\frac{\partial\bar{w}}{\partial y} + \bar{v}_0\right]\cos\alpha\cos\beta + \frac{1}{2} r\cos^2\alpha\sin 2\beta\left[\frac{\partial\bar{u}}{\partial y} + \frac{\partial\bar{v}}{\partial x}\right] \\ & + \frac{1}{2} r\cos^2\alpha\cos 2\beta\left[\frac{\partial\bar{v}}{\partial y} - \frac{\partial\bar{u}}{\partial x}\right]\end{aligned} \tag{3.62}$$

$\bar{V}_r(\beta)$ is a periodic function with base period $2\pi$ that can be written in the form of a Fourier series expansion:

$$\bar{V}_r(\beta) = a_0 + \sum_{n=1}^{\infty}(a_n\cos n\beta + b_n\sin n\beta) \tag{3.63}$$

Equation (3.62) has the form of such an expansion limited to the first three components. The Fourier coefficients are given by

$$a_0 = \frac{1}{2} r\cos^2\alpha\left(\frac{\partial\bar{u}}{\partial x} + \frac{\partial\bar{v}}{\partial y}\right) + \bar{w}\sin\alpha \tag{3.64}$$

$$a_1 = \bar{v}_0\cos\alpha \tag{3.65}$$

$$b_1 = \bar{u}_0\cos\alpha \tag{3.66}$$

$$a_2 = -\frac{1}{2} r\cos^2\alpha\left(\frac{\partial\bar{u}}{\partial x} - \frac{\partial\bar{v}}{\partial y}\right) \tag{3.67}$$

$$b_2 = \frac{1}{2} r\cos^2\alpha\left(\frac{\partial\bar{u}}{\partial y} + \frac{\partial\bar{v}}{\partial x}\right) \tag{3.68}$$

The kinematic properties of the local field are derived directly from the Fourier coefficient. The horizontal divergence:

$$\frac{\partial\bar{u}}{\partial x} + \frac{\partial\bar{v}}{\partial y} = \frac{2}{r\cos^2\alpha}(a_0 - \bar{w}\sin\alpha) \tag{3.69}$$

The components of the mean wind:

$$\bar{u}_0 = b_1/\cos\alpha \qquad \text{and} \qquad \bar{v}_0 = a_1 \cos\alpha \tag{3.70}$$

The stretching deformation:

$$\frac{\partial \bar{u}}{\partial x} - \frac{\partial \bar{v}}{\partial y} = -\frac{2a_2}{r\cos^2\alpha} \tag{3.71}$$

The shearing deformation:

$$\frac{\partial \bar{u}}{\partial y} + \frac{\partial \bar{v}}{\partial x} = \frac{2b_2}{r\cos^2\alpha} \tag{3.72}$$

The average vertical velocity can also be calculated from the continuity equation:[2]

$$\frac{\partial \bar{u}}{\partial x} + \frac{\partial \bar{v}}{\partial y} + \frac{\partial \bar{w}_a}{\partial z} = 0 \tag{3.73}$$

Integrating between levels $z$ and $z_0$, eq. (3.73) yields

$$\bar{w}_a(z) = \bar{w}_a(z_0) - \int_{z_0}^{z} \left(\frac{\partial \bar{u}}{\partial x} + \frac{\partial \bar{v}}{\partial y}\right) \mathrm{d}z \tag{3.74}$$

or, with eq. (3.69),

$$\bar{w}_a(z) = \bar{w}_a(z_0) - \int_{z_0}^{z} \frac{2}{r\cos^2\alpha} (a_0 - \bar{w}\sin\alpha)\, \mathrm{d}z \tag{3.75}$$

The maximum value of the elevation angle that can be used for such observations depends on the nature of the precipitation; the more regularly spaced out is the precipitation and the lower is the terminal velocity of fall, the higher is this value. Snow

[2]The continuity equation that is an expression of the conservation (or continuity) principle of mass establishes that the net mass flow per unit volume and per unit time is equal to the local density variation rate. In its general form, the continuity equation (3.73) therefore has a right-hand side equal to $(-1/\rho)(D\rho/Dt)$, where $\rho$ is the density of air and $t$ the time. However, in the preceding equations, local variations of the air density and horizontal density advection are assumed to be negligible, and the right-hand side is zero. Under these conditions, knowing the horizontal velocities (horizontal divergence) lets us deduce the vertical velocities with the help of (3.73).

precipitation corresponds to optimum conditions. The analysis of the various causes of error leads to the conclusion that in the presence of extensive precipitation, the method lets us measure the divergence and the deformation over a radius of 20 km for observations at low elevation; we can assume that the velocities of fall are negligible for $\alpha < 9^\circ$ in snow and $\alpha < 7^\circ$ in rain. However, if corrections are made, measurements at higher elevations are possible. This analysis technique is particularly useful for the observation of extratropical depression systems and also in the stratiform part of convective systems.

### 3.3.2 The VVP Method

In the preceding methods, derived from the VAD method, the antenna scanning is at a constant angle of elevation. A more complete discussion of the analysis of a linear wind field based on data from a single radar may be developed if we consider scanning at several values of the angle of elevation, because then the vertical gradients of the wind field parameters can be accessed. This is done by the method called *volume velocity processing* (VVP) proposed by Waldteufel and Corbin [1979].

VVP is an extension and generalization of the preceding considerations. As with VAD, the velocity components in the general expression for $V_r$, eq. (3.57), are replaced by the expansion in a Taylor series limited to first derivatives, but this time the terms relating to the vertical $z$ are not excluded, because volume data are being considered. A sum of trigonometric functions is obtained that can be discriminated and reconstituted from their dependence on the angle of elevation and azimuth. A brief description of this method follows.

The motion components of the scatterers $u$, $v$, and $w$ that are assumed to vary linearly around its value ($u_0$, $v_0$, $w_0$) at the point ($x_0$, $y_0$, $z_0$) are written as

$$u \approx u_0 + u'_x(x - x_0) + u'_y(y - y_0) + u'_z(z - z_0), \tag{3.76}$$

and similarly for $v$ and $w$. The overbars have been omitted for simplicity. The primed quantities are the partial derivatives with respect to the coordinate of the subscript.

By substituting in (3.57) and rearranging, we obtain

$$\begin{aligned} V_r = {} & \cos\beta \cos\alpha(v_0 - v'_x x_0 - v'_y y_0 - v'_z z_0) \\ & + \sin\beta \cos\alpha(u_0 - u'_x x_0 - u'_y y_0 - u'_z z_0) \\ & + \sin\alpha(w_0 - w'_x x_0 - w'_y y_0 - w'_z z_0) + r \cos^2\beta \cos^2\alpha v'_y \\ & + r \sin^2\beta \cos^2\alpha u'_x + r \sin\beta \cos\beta \cos^2\alpha(u'_y + v'_x) \\ & + r \sin^2\alpha w'_z + r \cos\beta \sin\alpha \cos\alpha(v'_z + w'_y) \\ & + r \sin\beta \sin\alpha \cos\alpha(u'_z + w'_x) = \sum_{j=1,9} Q_j \end{aligned} \tag{3.77}$$

A linear field is described by 12 parameters: $u_0$, $u'_x$, $u'_y$, $u'_z$, $v_0$, $v'_x$, $v'_y$, $v'_z$, $w_0$, $w'_x$, $w'_y$, $w'_z$. Equation (3.77) shows that it is possible to obtain at most nine of them. Therefore, we must accept restrictive assumptions to be able to use it. Waldteufel and Corbin point out (1) that, as the horizontal vorticity $v'_x - u'_y$ cannot be discriminated from $u_0$, $v_0$, because they share the same trigonometric dependence, it is preferable to choose $x_0 = y_0 = 0$ to retrieve $u_0$ and $v_0$; and (2) that, in situations where the linear wind approximation can be made, we can neglect $w'_x$ and $w'_y$ compared to $u'_z$ and $v'_z$. Under these conditions, (3.77) may be written

$$\begin{aligned} V_r &= u_0 \sin\beta \cos\alpha + v_0 \cos\beta \cos\alpha + w_0 \sin\alpha + r u'_x \sin^2\beta \cos^2\alpha \\ &\quad + r v'_y \cos^2\beta \cos^2\alpha + r(u'_y + v'_x) \sin\beta \cos\beta \cos^2\alpha + w'_z(r \sin\alpha - z_0) \sin\alpha \\ &\quad + u'_z(r \sin\alpha - z_0) \sin\beta \cos\alpha + v'_z(r \sin\alpha - z_0) \cos\beta \cos\alpha = \sum_{j=1,9} V_j \end{aligned} \tag{3.78}$$

Equation (3.78) consists of nine terms that in principle can be retrieved from radar data. When applied to a circle and neglecting vertical gradients, eq. (3.78) reduces to that for VAD.

Koscielny, Doviack, and Rabin [1982] have discussed the conditions of applying the VVP method and evaluated the errors that result from the possible choices of the number $m$ of parameters that can be discriminated. The method can lead to considerable errors if there is a linearity defect in the wind field. They propose an extension of VVP, called *modified VVP* (MVVP), that involves considering a limited angular sector; that is, a reduced volume in which the linear approximation is more easily satisfied. Applied to the measurement of the divergence in a planetary boundary layer in a prestorm situation over a reduced sector (30 km, 30° azimuth, 1° elevation above flat ground), they obtained an accuracy of $4 \times 10^{-5}\ s^{-1}$. However, when the assumption of linearity is not satisfied, the accuracy is impaired without it being possible to quantify this.

### *Other Methods: DVAD, EVAD*

The methodological extension obtained by implementing two VADs at the same time, using two closely situated doppler radars so that the scanned regions overlap, called *dual VAD* (DVAD), has been discussed by Scialom and Testud [1986]. The method lets us determine all the first derivatives of the horizontal wind field and a representation of the three-dimensional flow in stratiform situations for which the assumptions of steady-flow and linearity of the field are satisfied. It should be compared with the methods using several doppler radars (Section 3.5) for the study of nonlinear flow (convective part of the perturbations) that requires almost simultaneous measurements and with which the effective zone is reduced to regions for which the two radars see the measurement point from different angles. Provided that the assumptions of linearity and steady state are satisfied, the DVAD method is not subject to this latter restriction and so allows a more extensive region to be covered.

The *extended VAD* (EVAD) method [Srivastava, Matejka, and Lorello, 1986] takes as its point of departure eq. (3.69), which is rewritten

$$\frac{2a_0}{r' \cos\alpha} = \text{DIV} + 2\bar{w}_t \frac{h}{r'^2} \tag{3.79}$$

where $r'$ is the radius of the circle VAD (or horizontal range $r' = r \cos\alpha$) and DIV is the horizontal divergence. We assume that in the observed region, the vertical component of the air velocity is negligible compared to the terminal velocity of the hydrometeors; that is, $w \approx w_t$. We also assume that DIV and $w_t$ are constant in the circular zone of low altitude considered by the method. Under these conditions, the values of $2a_0/r' \cos\alpha$ as a function of $h/r'^2$ are distributed over a straight line. The slope and intercept of the line determined by least squares fit yield $w_t$ and DIV. These quantities are determined in successive thin-altitude sections, in which it is assumed that the assumptions are satisfied. In this way we obtain the vertical profiles of DIV and $w_t$.

The common problem with all these methods is that they are subject to highly restrictive assumptions. Certainly, there are circumstances in nature when these assumptions are satisfied but it is not clear where and when. We have to conclude, therefore, that these methods are essentially qualitative [see, in particular, Smith and Rabin, 1989]. In nonlinear flow (i.e., in the general case of actual atmospheric situations), single doppler radar lets us obtain information or signatures that, although semiquantitative or purely qualitative, are still very useful (an example is the detection of wind shear) for the description of local weather and for short-term forecasting.

### 3.3.3 Display of the Radial Velocity

The representation of a scalar field such as reflectivity by variations in shades of gray or in colors results in images that can be directly interpreted because of their intuitive nature. The same is not true for a vector field such as velocity. With a single doppler radar, the quantity to be represented is limited to the radial component. This is why most often just the algebraic value of this is represented. Various techniques have been proposed to display the radial velocity shear, such as the *plan shear indicator* (PSI) [Armstrong and Donaldson, 1969].

In general, the amplitude of the velocity is represented simply by a color scale on the conventional image in circular coordinates. The isodops (i.e., curves with constant doppler velocity) correspond to the lines where the colors change. In what follows, we briefly discuss the interpretation of the PPI image. The amplitude of the velocity is written (3.58)

$$V_r = V_h \cos\alpha \cos(\beta - \beta_0) + w \sin\alpha$$

For panoramic scanning at $\alpha$ = constant > 0, if a laminar flow is involved with $w$ = constant, $w \sin\alpha$ is constant and only the first term of the right-hand side varies: the shape of the isodops describes the variation of $V_h$ and $\beta_0$ as a function of $z$, with $z = r \sin\alpha$. The isodops always have a pole symmetry of 0. Any symmetry defect reveals a homogeneity defect of the field observed by the radar (Figure 3.5).

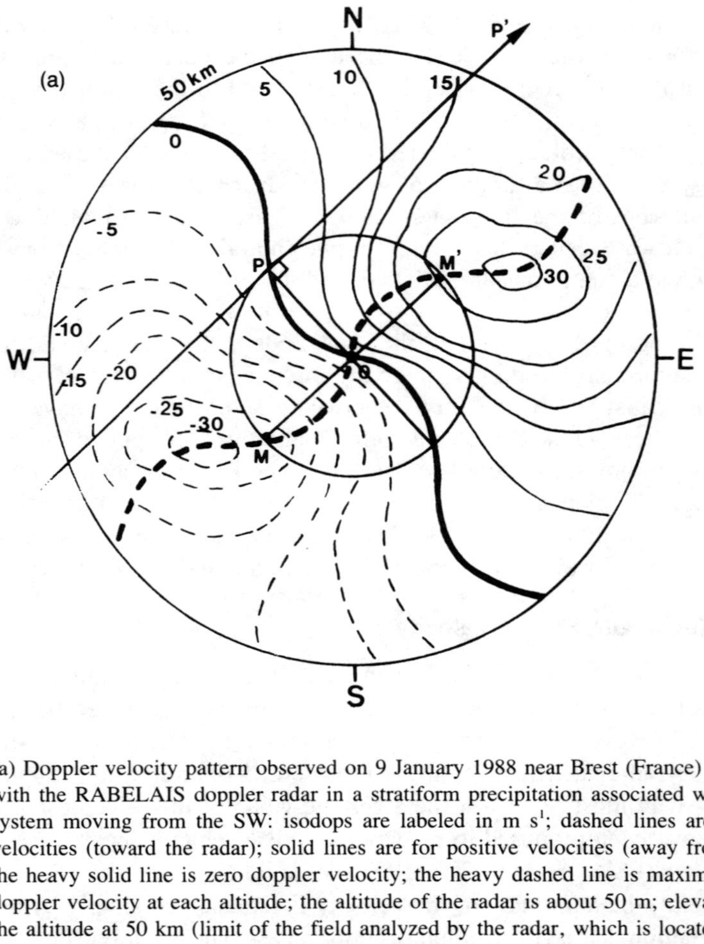


**Figure 3.5** (a) Doppler velocity pattern observed on 9 January 1988 near Brest (France) at 1543 GMT with the RABELAIS doppler radar in a stratiform precipitation associated with a cyclonic system moving from the SW: isodops are labeled in m s$^1$; dashed lines are for negative velocities (toward the radar); solid lines are for positive velocities (away from the radar); the heavy solid line is zero doppler velocity; the heavy dashed line is maximum (absolute) doppler velocity at each altitude; the altitude of the radar is about 50 m; elevation is 5°; so, the altitude at 50 km (limit of the field analyzed by the radar, which is located at the point 0) is 4.36 km; closed lines correspond to wind maximum ($|V| > 30$ m s$^{-1}$). For example, on the circle of radius $0P$ (range of 19 km, altitude of 1.68 km), the wind direction is given by the perpendicular to $0P$: $PP'$, wind velocity can be read in $M$ (negative value) or in $M'$ (positive value). (b) The wind profile corresponding to (a).

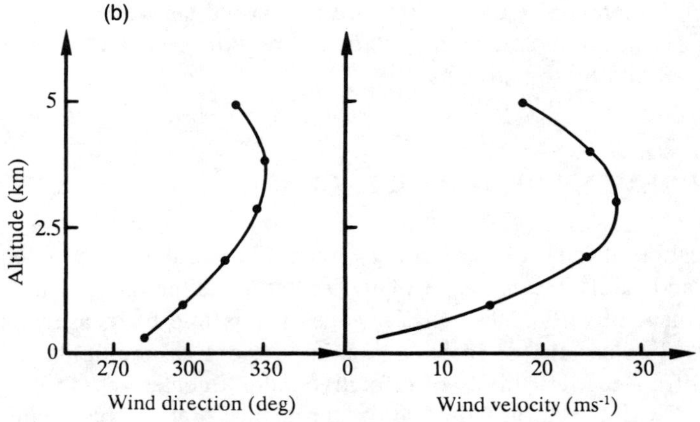


**Figure 3.5** (continued)

The line $v_{dop} = 0$ reproduces $\beta_0(z)$ exactly with a shift of $\pm\pi/2$ depending on whether the velocities to the radar are to the right or the left of the curve for increasing $r$. The $V_{rmax}$ line is derived from the $V_r = 0$ line by a 0 pole rotation of $\pi/2$ in the convenient direction. Along this line, we have $V_r = V_h \cos\alpha + w \sin\alpha$; that is, $V_h = (V_r - w \sin\alpha)/\cos\alpha$. If $w \sin\alpha$ is negligible, for example, in a snow precipitation or stratiform rain at low elevation, we have $V_h = V_r/\cos\alpha$. Therefore, an examination of the shape of the isodops on the panoramic screen lets us deduce approximately the wind profile in terms of speed and direction (Figure 3.5).

Let us assume that $w \sin\alpha$ is negligible; we then have

$$V_r \propto V_h(z) \cos[\beta - \beta_0(z)]$$

In this case, the following conditions hold:

- If $V_h(z)$ = constant, $V_r(z)$ depends only on $\beta_0$, and all the isodops reproduce the profile of $\beta_0$; that is, they have their origin at point 0 and are superposed by rotation about 0: in particular, if $\beta_0(z)$ = constant, the isodops are radial. This also is the case when $\alpha$ is zero or very small, as for almost horizontal angles of elevation.

- If $V_h(z)$ is monotonic, the isodops are not closed curves.
- If $V_h(z)$ is not monotonic, that is, there are extremes, the isodops are closed curves around these extremes.

## 3.4 MEASUREMENT OF TURBULENCE

In the application of radar to meteorology, reflectivity and average velocity are used most of all and rarely the variance of the spectrum, although this has potential in the measurement of turbulence. Radar makes it possible to measure turbulence in several ways. In clear air (Chapter 4), making various assumptions, turbulence can be measured from the reflectivity of refractive index fluctuations. In this section, we are concerned with the information on the turbulence that can be obtained just from measurement of the doppler velocity. The tracer of the movements can be hydrometeors (the most frequent case), fluctuations in the refractive index, insects, or artificial tracers. This information is more general than that which can be derived from the reflectivity. If Kolmogorov's hypotheses are satisfied, an estimate of the rate of dissipation of turbulent kinetic energy can be derived from the measurement of the velocity.

### *Pulse Volume Filtering*

The mean doppler velocity is a weighted average of the radial velocity of the scatterers in the pulse volume, $V_p$. The effect of the pulse volume averaging is to smooth out the fluctuation of scales below $V_p$; this is called *pulse volume filtering*. Such filtering and the relationship between the velocity at a point and the mean doppler velocity have been analyzed by Rogers and Tripp [1964], Srivastava and Atlas [1974], and Frisch and Clifford [1974].

A distinction must be established therefore between the fluctuations of the average velocity providing access to movements of scale above $V_p$ and the width of the doppler spectrum that takes into account movements of scale below $V_p$, weighted to show the distribution of the energy in $V_p$ (illumination, pulse shape). For a locally homogeneous turbulence, the variance of the velocity at a point $\sigma_p^2$ can be written as a sum of two terms [Rogers and Tripp, 1964]:

$$\sigma_p^2 = \sigma_v^2 = \sigma_{\bar{v}}^2 + \langle \sigma_v^2 \rangle \tag{3.80}$$

where $\sigma_{\bar{v}}^2$ is the variance of the mean velocity and $\langle \sigma_v^2 \rangle$ is the variance of the doppler spectrum averaged over an ensemble. The measurement of each of these two terms lets us determine the distribution of the turbulent kinetic energy between the two scales of movement.

### 3.4.1 The Inertial Domain

For a locally isotropic turbulence in an incompressible medium characterized by a sufficiently high Reynolds number, Kolmogorov's hypotheses [see, for example, Monin and Yaglom, 1975] consider the factors determining the transfer and dissipation of the kinetic energy for high values of the wave number at the end of the turbulence spectrum. They predict the existence of an equilibrium domain, consisting of an energy transfer zone and a viscous dissipation zone. The energy transfer zone, called the *inertial domain,* is bounded by the wave numbers $k_0$ and $k_m$ (Figure 3.6). At scales above $L_0 = 2\pi/k_0$ (typically on the order of 10 to the order of 100 m), the air movement no longer is isotropic; large anisotropic vortex are found, whch are considered sources of energy transferred to the small scales. At scales below $l_m = 2\pi/k_m$ (typically, from 1 to a few tens of centimeters), the vortexes rapidly lose energy by thermal dissipation due to the forces of molecular viscosity. In the inertial domain, between $L_0$ and $l_m$, the vortexes lose all structure; they are homogeneous and isotropic; the viscosity is little involved; and there is a continuous transfer of energy from the vortexes with dimensions greater than $L_0$ to smaller ones. In this part of the spectrum, the energy as a function of the size of the vortexes is distributed in proportion to $k^{-5/3}$; and the local average properties of the turbulent movement,

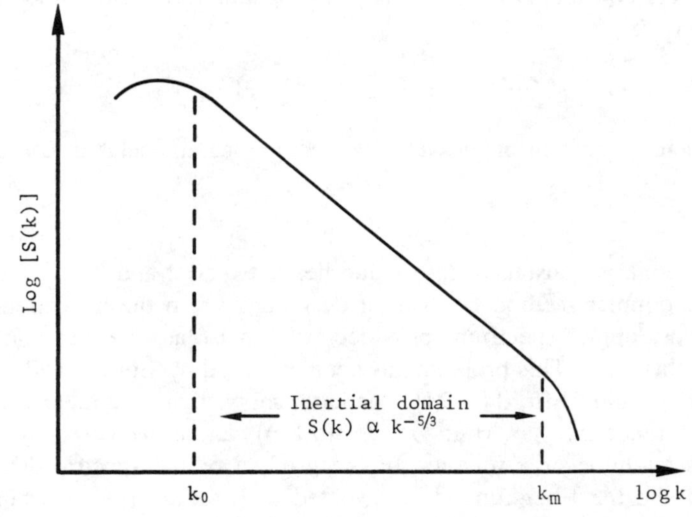


**Figure 3.6** Schematic representation of kinetic energy spectrum $S(k)$ of turbulence in the inertial domain.

homogeneous and isotropic, are determined solely by $\varepsilon$, the rate of dissipation of turbulent kinetic energy per unit of mass (see eq. (4.4)). The boundaries of the inertial domain vary as a function of the dynamic and thermodynamic parameters of the atmosphere.

The lower limit $l_m$ increases with the kinematic viscosity $\nu$ and decreases with $\varepsilon$; that is, the stronger the turbulence, the larger is $\varepsilon$ and the more the lower limit of the inertial domain $\lambda_t$ is pushed down toward smaller scales. We have [Hill, 1978]

$$\lambda_t \propto (\nu^3/\varepsilon)^{1/4} \tag{3.81}$$

Figure 4.1 shows the variations between extremes of the limits of the inertial domain with altitude.

### 3.4.2 Measurement of the Rate of Dissipation of Turbulent Kinetic Energy

The term $\varepsilon$ is frequently used to parameterize the small-scale atmospheric turbulence, when the field is locally homogeneous, isotropic, and obeys the Kolmogorov theory. We shall see in Chapter 4 that, in clear air, the radar reflectivity is a measure of $C_n^2$, the structure constant of the fluctuations in the refractive index, based on which we can estimate $\varepsilon$.

With doppler radar, a method of estimating $\varepsilon$ consists of analyzing the time series of the average velocity. When the observed field satisfies the Kolmogorov assumptions, Tatarski [1961] shows that we can describe the fluctuations in velocity by their structure constant $D_v(r)$, where $r$ is the spatial scale of the fluctuation. This can be written

$$D_V(r) = \langle [V(x + r) - V(x)]^2 \rangle = C_V^2 r^{2/3} \tag{3.82}$$

$C_V^2$ is the structure constant of the velocity fluctuations and related to $\varepsilon$ by

$$C_V^2 = a_V^2 \varepsilon^{2/3} \tag{3.83}$$

$a_V^2$ is a dimensionless constant whose value lies between 2 and 2.2.

Another doppler method consists of deriving $\varepsilon$ from the measurement of the variance of the doppler spectrum, provided we can isolate the contribution of the turbulence to this term. This problem has been analyzed by Gorelik and Mel'Nichuk [1966] and Frisch and Clifford [1974]. The latter shows that for a radar with a narrow beam, the difference $\sigma_p^2 - \sigma_{\bar{v}}^2$ (i.e., $\sigma_v^2$, eq. (3.80)) can be expressed by the integral over the pulse volume of a function of the product of the spectral density of the velocity field and the filter function associated with the energy distribution in the beam. They make the terms of this product explicit in the case of a Kolmogorov turbulence and a three-dimensional Gaussian distribution of the energy in the pulse volume and, after integration, arrive at an expression for the variance of the doppler spectrum as a function of the rate of dissipation and the dimensions of the beam. After inversion, this expression, generalized by Labitt in 1979 [see in Gossard and Strauch, 1983] can be written as

$$\varepsilon = \delta^{-1}\left[\frac{\sigma_v^2}{1.35\alpha\left[1 - \frac{\gamma^2}{15}\right]}\right]^{3/2} \tag{3.84}$$

with

$$\left.\begin{aligned}\delta &= a\\ \gamma^2 &\approx 1 - (b/a)^2\end{aligned}\right\} \text{ for } \quad b \leq a$$

$$\left.\begin{aligned}\delta &= b\\ \gamma^2 &\approx 4[1 - (a/b)^2]\end{aligned}\right\} \text{ for } \quad a < b$$

where $a = r\sigma_\theta$ is the width of the beam ($\sigma_\theta$ is the standard deviation of the energy distribution in the beam) and $b$ is the length of the pulse.

With somewhat different approximations, this method has been used to study the relationships between the kinematic structure and the turbulent dissipation in a tornadic storm [Istok and Doviak, 1986]. Brewster and Zrnic' [1986] have compared the estimates of $\varepsilon$ derived, on the basis of the same sample of data collected in a storm, from the width of the doppler spectrum and spatial spectra of average velocity; at low range the estimates agree to within 10%.

### 3.4.3 The Turbulence Field

The factors involved in the determination of the variance of the doppler signal have been described in Section 3.1.4. If all the terms contribute to the total variance at the same time, no interpretation can be made. However, we have seen that, provided adequate experimental precautions are taken and possibly after elimination of the shear term, only the terms due to turbulence and precipitation are significant; and we can write

$$\sigma_v^2 = \sigma_t^2 + \sigma_p^2 \tag{3.85}$$

Under these conditions, if we know $\sigma_p^2$ or have an acceptable estimate of it, the measurement of $\sigma_v^2$ lets us obtain $\sigma_t^2$, which gives the turbulent energy. For example, the case of a vertical elevation in snow favors such an observation because the shear effects are eliminated, and in addition, the velocity spectrum of the snowfall is narrow. From measurements of $\sigma_v^2$ carried out under such circumstances, Rogers and Tripp [1964] found, by assuming that $\sigma_p^2$ is less than 0.25 $m^2 s^{-2}$, that three-quarters of the turbulent energy corresponds to wavelengths less than 100 m, and that, in addition, the root mean square of the turbulent fluctuations of the velocity lies between 0.4 and 2 m $s^{-1}$ [Borresen, 1971].

The opportunities provided by the measurement of the variance of the fluctuation spectrum of the radar signal and the doppler spectrum have been exploited in the study of the rate of dissipation of the turbulent kinetic energy in clouds and precipitation [Gorelik and Mel'Nichuk, 1966; Gorelik, 1968, among others]. Mel'Nichuk, Smirnova, and Chernikov [1968] show that, when the spectral distribution of the turbulent energy obeys the $-5/3$ law predicted by the Kolmogorov theory for the inertial domain, the contribution to the variance due to the beamwidth and precipitation can be eliminated by an observation made with pulses of different lengths. Turbulence in clear air with the help of artificial tracers has also been studied [Lhermitte, 1969a and 1969b; Chernikov *et al.*, 1969].

At scales greater than the size of the resolution volume of the radar, the turbulence is manifested by fluctuations around the average radial velocity. VAD-type conical scanning makes it possible to scan the turbulence under quite favorable conditions. This procedure has been used to describe the turbulent field in snow, finding that the turbulent zones sometimes are strongly orientated and associated with the presence of a pronounced wind shear [Borresen, 1971; Lhermitte, 1968b].

Lhermitte [1968b] showed that, by neglecting the variance of the terminal fall velocity, the variance of the mean radial velocity as a function of the angular position of the beam can be written, from (3.57a), in the form

$$\sigma_{\bar{v}_r}^2 = \sigma_{\bar{u}}^2 \cos^2\alpha \sin^2\beta + \sigma_{\bar{v}}^2 \cos^2\alpha \cos^2\beta + \sigma_{\bar{w}}^2 \sin^2\alpha$$
$$+ \operatorname{cov}(\bar{u}\bar{v}) \cos^2\alpha \sin 2\beta + \operatorname{cov}(\bar{u}\bar{w}) \sin 2\alpha \sin\beta$$
$$+ \operatorname{cov}(\bar{v}\bar{w}) \sin 2\alpha \cos\beta \qquad (3.86)$$

where cov denotes the covariance. Equation (3.86) has a special feature. In the expression for the variance of $\bar{V}_r$, it introduces the covariances of these components, which are related to the moment transfers due to the turbulent fluctuations.

This analysis has been extended to the extraction of the intensity of turbulence and the moment flux for the scales smaller than $V_p$ with the help of measurements made over the entire VAD circle [Wilson, 1970]. By assuming that the turbulent field is horizontally homogeneous and steady state, Wilson writes the mean square of the fluctuations of the radial velocity in the form

$$\sigma_v^2 = \overline{V_r'^2(\beta)} = (\overline{u'^2} + \overline{v'^2})\frac{\cos^2\alpha}{2} + \overline{w'^2}\sin^2\alpha$$
$$+ (\overline{v'^2} - \overline{u'^2})\frac{\cos^2\alpha}{2}\cos 2\beta + \overline{u'v'}\cos^2\alpha \sin 2\beta$$
$$+ \overline{u'w'}\sin 2\alpha \sin\beta + \overline{v'w'}\sin 2\alpha \cos\beta \qquad (3.87)$$

where the primed quantities are fluctuations with respect to the mean with

$$\overline{u'^2} = \overline{(u - \bar{u})^2} = \sigma_u^2$$
$$\overline{u'v'} = \overline{(u - \bar{u})(v - \bar{v})} = \mathrm{cov}\,u'v'$$

and the same for $v$ and $w$.

This equation is expanded as a Fourier series as for the wind field, eq. (3.62), and we obtain Fourier coefficients from which the properties of the average turbulence field can be derived. We have

$$a_0 = \left(\frac{\overline{u'^2} + \overline{v'^2}}{2}\right)\cos^2\alpha + \overline{w'^2}\sin^2\alpha \qquad (3.88)$$

$$a_1 = \overline{v'w'}\sin 2\alpha \qquad (3.89)$$

$$b_1 = \overline{u'w'}\sin 2\alpha \qquad (3.90)$$

$$a_2 = \left(\frac{\overline{v'^2} - \overline{u'^2}}{2}\right)\cos^2\alpha \qquad (3.91)$$

$$b_2 = \overline{u'v'}\cos^2\alpha \qquad (3.92)$$

For low elevation angles, $a_0$ is an estimate of the horizontal turbulent kinetic energy; $a_2$ is an estimate of the isotropy of the turbulent field; and the coefficients $a_1$, $b_1$, and $b_2$ are estimates of the moment fluxes involved in the determination of the rate of mechanical generation of the turbulent kinetic energy.

The results obtained for the determination of the turbulent fields by the method of conical VAD scanning are encouraging. The turbulent kinetic energy profiles maintain a consistent relationship with each other for different values of the elevation angle and reveal an organized structure. This method has also been used for the study of turbulence in frontal zones [Bryant and Browning, 1975] (Section 5.5), in stratiform precipitation [Harris, 1977], and in the convective boundary layer [Kropfli, 1986].

Frisch, Martner, and Gibson [1989] suggest extending the method to the treatment of third-order moments. Still, making the assumption of horizontal homogeneity of the turbulence and integrating over the VAD circle, we obtain

$$\int_0^{2\pi} \overline{(V_r - \bar{V}_r)^3}\, d\beta = 3\pi\overline{(u'^2w' + v'^2w')}\sin\alpha\cos^2\alpha + 2\pi\overline{w'^3}\sin^3\theta \quad (3.93)$$

For the particular value $\alpha = 50.8°$, the coefficients of the terms on the right-hand side of (3.93) are equal; and we have

$$\int_0^{2\pi} \overline{(V_r - \bar{V}_r)^3}\, d\beta = (1.86\pi)\frac{\overline{w'(u'^2 + v'^2 + w'^2)}}{2} \quad (3.94)$$

## 3.5 MEASUREMENT OF THE VELOCITY FIELDS WITH SEVERAL DOPPLER RADARS

The different methods for measuring the velocity fields with a single doppler radar leave unsolved the problem of nonlinear wind fields. This is the case of the tridimensional structure of movements in deep convective fields. Various attempts have been made to determine velocities in convective clouds with the help of a single radar from successive analyses of the same cell, which because of its movement with the surrounding winds can be seen at different angles from the same point. In this case, we must assume that the average circulation in the cloud is almost stationary over periods of 5–10 min [Peace, Brown and Camnitz, 1969]. This method has an operational interest for an approximate observation of the intensity of storms. However, air circulation associated with convective clouds only rarely satisfies the stationary assumption even over durations of a few minutes. Under such conditions, simultaneous measurement with the help of several doppler radars is necessary.

The methodology of retrieving the three-dimensional wind field from multiple doppler radars has been discussed by a large number of authors, among others Lhermitte [1968a], Armijo [1969], Doviak *et al.* [1976], Rabin and Zrnic' [1980], Ray *et al.* [1980]. For the methodology with two doppler radars, see Testud and Chong [1983], Chong, Testud, and Roux [1983], Chong and Testud [1983].

Let us consider in a Cartesian coordinate system $0xyz$, three radars situated at the coordinates $(x_i, y_i, 0)$, with $i = 1, 2, 3$, simultaneously aimed at a point with coordinates $(x, y, z)$; the components $u$, $v$, $w$ of the air velocity are defined as earlier, $w$ and $w_t$ are related to $w_a$ by equation (3.49). Under these conditions, the three radars measure the radial velocity components, eq. (3.57b) given by

$$V_i(x, y, z) = r_i^{-1}[(x - x_i)u + (y - y_i)v + z(w_a + w_t)], \qquad (3.95)$$

$$r_i = [(x - x_i)^2 + (y - y_i)^2 + z^2]^{1/2} \qquad (3.96)$$

$$i = 1, 2, 3.$$

By denoting the elevation angles by $\alpha_i$ and the azimuthal angles $\beta_i$ (relative to the north) for the three radars, the horizontal component of the air velocity which points in the direction $\beta_0$ with respect to the north by $V_h$, we can also write

$$V_i = V_h \cos(\beta_i - \beta_0) \cos\alpha_i + (w_a + w_t) \sin\alpha \qquad (3.97)$$

with $i = 1, 2$, and 3.

The three equations (3.95)–(3.97) contain four unknowns $u$, $v$, $w$, and $w_t$. The solution can be obtained using the data of three doppler radars and a relation giving the terminal velocity $w_t$ as a function of the reflectivity factor (see eqs. (3.52) to (3.54)). Usually, this method is applied with only two doppler radars. Two velocity components and the reflectivity factor are measured and the third component is calculated by using the equation of continuity (3.73). In this case, the commonly used scanning procedure called COPLAN [Lhermitte and Miller, 1970] involves two radars simultaneously scanning the common planes passing through their baseline (Figure 3.7). The inclination of the planes is varied between scans to obtain the volume analysis. During a scan, the direction of the beams is controlled automatically to satisfy the condition

$$\frac{\tan\alpha_1}{\cos\delta_1} = \frac{\tan\alpha_2}{\cos\delta_2} = \tan\alpha_p = \text{constant} \qquad (3.97)$$

where $\alpha_p$ is the angle of the scanning plane with the horizontal, $\delta_1$ and $\delta_2$ are the azimuthal angles of the elevation of radars 1 and 2 from the baseline.

Coplane data are processed by using a cylindrical coordinate system. The two-dimensional velocity field in the common plane is computed from radar data. The component normal to the plane is obtained by solving the equation of continuity. The condition at the limits is that $w$ is zero at the ground or cloud top. Examples of results are shown in Chapter 5. The method assumes that the circulation is stationary throughout an analysis of the velocity fields within the cloud volume. This assumption represents the thorniest point of the multidoppler methods. There is also a difficulty due to the difference in the resolution volume of radars at different points of the grid; after interpolation, the estimates are made on large scales compared to the largest resolution cell.

The precision of the measurements of velocity with two doppler radars depends on the angle formed by the direction of the two beams, which must not be too small. As a result, the domain of measurements common to the two radars is restricted to two approximately circular areas situated on each side of the baseline. To extend

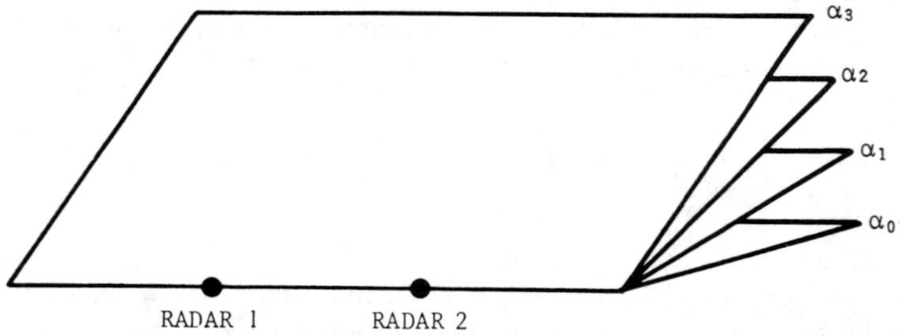


**Figure 3.7** COPLAN scanning procedure used for tridimensional doppler measurements.

the measurement domain, we must increase the number of radars, that is, create an array.

The configuration of an array of more than two radars depends on the type of observation planned. For three-dimensional movement fields in convective fields and three radars, the optimum configuration consists of placing the radars at the vertex of an equilateral triangle; a coordinate scanning procedure must then be used [Carbone, 1977]. For an array of more than three radars, the position of the radars must be determined by an objective procedure [Ray *et al.*, 1978].

## 3.6 RETRIEVAL OF THE THERMODYNAMIC AND MICROPHYSICAL FIELDS

As with the wind and reflectivity fields the distribution of the thermodynamic and microphysical quantities in convective systems are complex and change rapidly over time. Now, it is useful to determine them to understand and model convective processes. This is why methods have been found with a view of deriving them from measurements accessible to doppler radar. These methods are based on equations that relate the movement, the thermodynamic variables, and the changes in the state of water.

Gal-Chen [1978] and Hane, Wilhelmson, and Gal-Chan [1981] were the first to discuss the retrieval of the perturbations in the pressure and temperature fields from three-dimensional wind fields measured by multiple doppler radars, by treating the equations of motion. However, the movement makes possible access to only relative values (the perturbation). To obtain absolute values, we must have a reference or calibration, for example, by a few measurements in situ, but this then poses the problem of the simultaneaneity of the data. Roux [1985] has suggested obtaining this reference by using a thermodynamic equation giving the constant part of the vertical gradient. Attempts also have been made to retrieve the fields of liquid water content and water vapor in a similar manner from radar data with the help of models by using the equation of continuity for the total water and various microphysical parameterizations. These methods have been applied, for example, by Rutledge and Hobbs [1984] to the study of the development of precipitation in the precipitation bands of frontal systems of middle latitudes, by Ziegler [1985] for diagnosing the temperature and the mixing ratio in a storm, by Hauser and Amayenc [1986] for analyzing the distribution of water in tropical squall lines. The results offer reasonable approximations and are a valuable resource for meteorological measurement.

In conclusion, the simultaneous operation of several doppler radars is a solution to the three-dimensional measurement of the movements of the atmosphere. However, applied to the intense convection that is its main area of interest, the method has severe limitations. First, the small range of the domain common to a pair of radars can capture only a sample of the phenomenon. Then, we come up against the conflicting requirements on the duration of acquisition: this must be as brief as possible to remain compatible with the steady-state assumption for the flow to avoid unacceptable errors; yet, the speed of antenna scanning must not exceed the limit compatible with the preservation of the quality of the reflectivity and velocity measurements, and the density of the measurement points (i.e., the angular spacing of successive scans) that controls the final resolution of the observation (spatial filtering) must be high.

The impossibility of satisfying these different conditions makes it necessary to accept a degradation of the resolution of the observations. After interpolation, the minimum scale observed (sometimes called the *cutting wavelength*), which is 5 to 10 times the value of the grid step [see Carbone, Carpenter, and Burghart, 1985] is several kilometers. Under these conditions, certain major flow characteristics may escape analysis (for example, the ascending current of the convection lines associated with the surface cold front of cyclones at middle latitudes is only a kilometer wide) and certain important processes cannot be studied. This obviously points in the direction of developing radars with rapid scanning and finding other scanning procedures.

## 3.7 AIRBORNE RADAR

Most modern transport airplanes are equipped with noncoherent meteorological radar used mainly to detect storms. Thanks to the radar information, the plane can navigate in perturbed zones by avoiding regions presenting safety risks. Such radar is not designed for high performance, because it is used at short range and partly qualitatively. The radar explores in the direction the airplane is going.

The various limitations of ground-based radar, in particular the smallness of the domain of observation common to a pair of doppler radars, make airborne doppler radar very attractive for research. Such an instrument can go to where the "objects" of study are without limitation (oceans, tropical zones), accompany them while they move and so analyze complete cycles, approach zones where increased resolution is useful, and so forth.

Measurement of air movement with an airborne doppler radar goes back a long way [Lhermitte, 1971] but only at the beginning of the 1980s were tests made, first with a radar with fixed vertical beam and then with a radar with moving beam, exploring the space perpendicular to the plane's path. These tests involved mainly convective perturbations either observed by the airborne radar by itself or by combining the airborne radar with ground radars. The results as well as the measurement methodology have been discussed by Hildebrand and Mueller [1985], Mueller and Hildebrand [1985], Ray, Jorgensen, and Wans [1985], Ray and Jorgensen [1988], Testud and Amayenc [1989].

As with ground-based doppler multiradar systems, the crucial points concern the sampling conditions and the consequences for the minimum observable spatial resolution as well as for the accuracy of the measurements. To retrieve the three-dimensional circulation, we must measure at least two components of the movement field with a single radar in motion. The two components at the same point in space can be obtained only for two different positions of the plane. If the radar is scanning in only one plane, the plane's attitude has to be altered periodically to observe the useful volume twice. By using an antenna with two beams (i.e., two antennas) scanning, about an axis colinear with the trajectory, two conical surfaces of constant aperture, symmetric with respect to the plane ($P$) perpendicular to the axis of rotation passing through the radar, these two components are obtained following the movement of the airplane. The third component is calculated by integrating the equation of continuity in cylindrical coordinates. However, the two components are not measured simultaneously. For example, for beams inclined at 20° with respect to the plane $P$ and an airplane speed of 100 m $s^{-1}$, the time interval between the sampling at a point with front and rear aiming is 3 min at 25 km of distance and double this at 50 km. By way of comparison, in a tropical squall line (cf. Section 5.3.3), the convective part has typically a width of 50 km and progresses at approximately 60 km $h^{-1}$.

Present radar techniques are inadequate with regard to the time available for data acquisition; we must use rapid scanning (one to two rotations per second) and reduce the acquisition time, for example, by a multifrequency system (cf. Section 1.5.5). The performance that results from these various technical constraints may be expressed in terms of the measurement accuracy (of the reflectivity and the average velocity) and scale (or wavelength) of trimming due to spatial filtering of the sampling. These two parameters are related, varying inversely to each other and degrading with the range.

It remains to choose the wavelength and the antenna. Constraints of space make it necessary to choose flat antennas (i.e., array antennas) rather than parabolic ones and make short wavelengths more attractive. However, the problems of attenuation limit the possibilities in this direction. In fact, although we can correct the effects of attenuation by rain, attenuation by clouds that the radar does not detect is not as easy to correct for (cf. Section 2.5.7). At present, the technical and methodological concepts relating to airborne doppler radar are in the full throes of development.

# Chapter 4
# Observation of Clear Air

## 4.1 INTRODUCTION

Quantitative observation of scattering by turbulent media, which is a relatively recent conquest of radar, has its historical roots in the studies on long-range propagation undertaken around 1940. At this time, it was acknowledged that the atmospheric mechanisms responsible for transhorizon radioelectric propagation should involve scattering by the turbulent irregularities of the refractive index of air as well as partial specular reflection phenomena on the structures in horizontal layers. The need to explain long-range tropospheric propagation led Booker and Gordon [1950] and Megaw [1950] to develop the basic theoretical principles for the scattering of radioelectric waves by turbulence; the validity of this theory was subsequently established experimentally [see du Castel, 1966; Bean and Dutton, 1968].

With the advent of radar, it seemed possible to make more detailed investigations on the atmospheric structure in clear air [Friend, 1940–1949]. In fact, it very soon appeared that radar echoes were being detected from atmospheric regions apparently devoid of any target or any visible meteorological phenomenon capable of explaining their origin. The difficulties encountered at the beginning in providing a completely satisfactory explanation of this and the mysterious nature of the phenomenon gave rise to the ironical description "angel echoes" or "phantom echoes." However, it was fairly quickly realized that two main causes could be responsible for angel echoes: very strong gradients of the refractive index associated with turbulence or discontinuities of the atmosphere; and birds, and especially insects, whose distribution in air, as will be seen, is strongly influenced by the thermodynamic structure of the lower atmospheric layers. A certain confusion in the interpretation of the observations lasted until around 1966. The considerable technological advances made as the result of high-power radars with large and directional antennas made it possible to clarify the problem [Hardy, Atlas, and Glover, 1966] on the basis of appreciably enhanced observations. Although the subject is far from exhausted at the research level, there no longer is any mystery today and it has been possible to give a correct interpretation of the main types of echo and their properties [Atlas *et al.*, 1966; Glover *et al.*, 1966; Browning and Watkins, 1970; Atlas *et al.*, 1970; among others].

The possibility of detecting fluctuation in the refractive index of air has opened up new horizons in radar meteorology. Even in the absence of significant turbulence, the atmosphere, which is always in motion, presents such fluctuations. These heterogeneities move with the wind and act as a natural tracer, making it possible to measure velocity under all circumstances. The search for the most favorable technical conditions for observation has led to the concept of *stratosphere troposphere* (ST) radar. This type of radar finally made possible obtaining a signal in the absence of hydrometeors; that is, in the clear atmosphere. The term *radar observation in clear air,* which is too restrictive, is often used in this connection although the fluctuations in the refractive index of air nonetheless are detected in clouds. The principal application of this technique is wind measurement.

Although in themselves insects are of little interest for meteorological measurement, under certain circumstances they have been observed in the lower layers in such quantities that they constitute a cause of scattering that cannot be neglected and even are efficient tracers of the structure and movement of air. Because of their low concentration, the signals they send back often resemble the echo of a point target moving in the beam: they fluctuate little. The behavior of birds depends less on atmospheric conditions but sometimes reveals certain relationships with local meteorological events. Finally, in the absence of local targets, the use of artificial radar tracers, the "resonant dipoles," offers interesting possibilities (Section 4.6).

In this chapter, we first present a summary of the theoretical relationships between radar reflectivity, the microstructure of the refractive index, and the turbulence intensity. We then consider the application of the detection of fluctuations in the refractive index of air to the observation of the atmosphere (i.e., the ST radar) and then we discuss the radar properties of insects and artificial tracers. Some examples of observation are given in Chapter 5.

## 4.2 SCATTERING OF ELECTROMAGNETIC WAVES BY A TURBULENT MEDIUM

### 4.2.1 General Relations

At frequencies for which the scattering effects are negligible, which is the case for the metric, decimetric, and centimetric frequencies used in radar, we have seen that the radioelectric refractive index of the atmosphere $n$, depends on the pressure $p$, the temperature $T$, and the partial pressure of water vapor $e$ (Section 1.2.7 refraction). We can write

$$N = (n - 1)10^6 = (77.6/T)(p + 4810\,e/T) \tag{4.1}$$

where $N$ is the refractivity.

Small-scale dynamic perturbations produce local modifications to the field of the refractive index of air. In fact, during a turbulent mixing process, the "air particles"[1] are rapidly displaced from each other and do not immediately mix in a complete and homogeneous manner; they temporarily preserve their identity: the pressure is equal to that of the environment but the potential temperature[2] and the specific humidity are maintained. The higher the average initial gradients and the greater the displacements (i.e., the turbulence is strong) the more marked are the local variations in these variables at the boundary of the air particles.

The study of the scattering of electromagnetic waves by turbulent media has been approached by many authors. Among these, Tatarski (1961) has a general account of the theoretical results that served as a basis for interpreting a variety of experimental studies (Hardy *et al.*, 1966; Atlas *et al.*, 1966; Ottersten, 1969; among others]. This has made it possible to define the relationships between the concepts used to describe the phenomena and the measurable quantities characterizing atmospheric turbulence. Ottersten (1969-a) produced a summary on this subject to which frequent reference will be made in the following discussion.

Tatarski (1961) defines the spectral power density of the refractive index in space by the function $\phi_n(\mathbf{k})$, where $\mathbf{k}$ is the wave number vector. The function $\phi_n$ is obtained by the Fourier transform of the autocorrelation-function of the refractive index. Tatarski showed that the reflectivity is related to the random fluctuations of the refractive index by a general expression of the form

$$\eta(\mathbf{k}) = \frac{\pi^2}{2} k^4 \phi_n(\mathbf{k}) \tag{4.2}$$

where $\mathbf{k}$ is the wave number vector whose direction is that of the radar propagation with $|\mathbf{k}| = 4\pi/\lambda$. Although $\phi_n(\mathbf{k})$ is defined for the entire space of the wave numbers, only spatial fluctuations whose scale, following the direction of propagation, is close to $\lambda/2$ produce additive phases and are perceived by the radar; however, detection takes place only if there is sufficient spectral energy on the $\lambda/2$ scale, which is selectively observed by the radar.

---

[1]Air particle is an approximately uniform volume of air to which various dynamic and thermodynamic properties can be attributed (synonym, turbulent element).

[2]The potential temperature, usually denoted by $\theta$ and expressed in K is that which an air particle would have following an adiabatic transformation (i.e., with no exchange of heat and matter between the particle and the surrounding air), bringing it from its current pressure level by expansion or compression to a standard pressure conventionally taken as equal to 1000 hPa. Therefore, adiabatic transformations of atmospheric air do not modify $\theta$. The potential temperature is said to be dry or humid depending on whether the partial pressure of the water vapor in the air particle considered is below or equal to the saturation pressure.

The integration of $\phi_n(\mathbf{k})$ along two dimensions of the space leads to a one-dimensional fluctuation spectrum that runs over positive and negative values of $k$. The actual one-dimensional spectrum $S_n^{\alpha}(k_\alpha)$ measured by the displacement of a refractometer in the direction $\alpha$ exists only for $k_\alpha > 0$. For a given value of $k_\alpha$, in addition to the corresponding scale in the direction considered, the spectrum contains an infinite number of higher scales relating to other directions but whose projection onto the direction $\alpha$ is equal to $2\pi/k_\alpha$. Thus, the result of the measurement cannot be directly interpreted in terms of radar backscattering following the direction $\alpha$ [Bolgiano, 1963].

For a field with isotropic turbulence, the spectrum is independent of the direction and the spectral density $\phi_n(k)$ can be expressed as a function of the slope of the one-dimensional spectrum. Equation (4.2) can be written

$$\eta(k) = -(\pi/8)k^3 \frac{dS_n(k)}{dk} \tag{4.3}$$

where $S_n(k)$ is the one-dimensional spectrum with $k > 0$. This expression makes it possible to compare the measurements of radar reflectivity with the spectrum of the refractive index for an isotropic field with the scales considered.

### 4.2.2 Reflectivity in the Inertial Domain

The velocity fields inside the inertial domain are described by reference to the one-dimensional spatial spectrum $S(k)$. The spatial spectrum for the velocity fluctuations following any direction is given by

$$S(k) = A_1 \varepsilon^{2/3} k^{-5/3} \tag{4.4}$$

where $A_1$ is a dimensionless constant that depends on the orientation of the velocity component with respect to the direction defining the spectrum. It is determined experimentally and is equal to 0.50 to 0.55 for a longitudinal spectrum, that is, parallel to the velocity component, and 4/3 of this value for a transverse spectrum. Equation (4.4) shows that the dissipation rate $\varepsilon$ determines the intensity of the turbulence in the inertial domain; $\varepsilon$ is usually expressed in $cm^2\ s^{-3}$ or in $m^2\ s^{-3}$.

By treating the potential refractive index[3] as a conservative, passive, and additive variable, Tatarski [1961] showed on the basis of similarity theory that, in the inertial domain, the spectrum of the variations in the refractive index is given by

$$S_n(k) = A_{1n} \varepsilon_n \varepsilon^{-1/3} k^{-5/3} \tag{4.5}$$

[3]Refractive index of a particle of air brought adiabatically to a standard pressure of 1000 hPa.

In this expression, $\varepsilon_n$, which is analogous to an energy dissipation rate, represents the rate of reduction in the irregularities of the refractive index field due to molecular diffusion. The term $A_{1n}$ is a dimensionless constant with a value between 0.1 and 1. In the inertial domain, the fluctuations of any conservative, passive, and additive scalar, such as temperature or water vapor, is represented by equations analogous to (4.5).

For radar using metric, decimetric, or centimetric wavelengths, the spatial scale $\lambda/2$ corresponding to the observed signals is usually included in the inertial domain when this exists; that is, in the presence of a well-developed turbulent field. By substituting the derivative of (4.5) into (4.3), we obtain the expression

$$\eta(k) = \frac{5}{3}\frac{\pi}{8} k^2 S_n(k) \tag{4.6}$$

which is a particular form of (4.3), allowing the radar reflectivity to be related to the fluctuations in the refractive index and in particular to refractometer measurements.

Under the same conditions, a useful analogous relationship may be written down by using the structure constant of the fluctuations in the refractive index $C_n^2$, which is related to the structure function $D_n(r)$, where $r$ is the spatial interval [Tatarski, 1961], by

$$D_n(r) = \langle [n(x + r) - n(x)]^2 \rangle = C_n^2 r^{2/3} \tag{4.7}$$

with

$$C_n^2 = a^2 \varepsilon_n \varepsilon^{-1/3} \tag{4.8}$$

where the symbol $\langle\ \rangle$ denotes an ensemble average; $C_n^2$ is a measure of the mean square of the fluctuations in the refractive index in the inertial domain (in $\text{cm}^{-2/3}$). It can also be interpreted as a coefficient of the refractive index fluctuation spectrum. The term $a^2$ is a dimensionless constant approximately equal to $4A_{1n}$; that is, 3.2 to 4.0.

By substituting (4.8) into (4.5), we obtain

$$S_n(k) \approx \frac{1}{4} C_n^2 k^{-5/3} \tag{4.9}$$

and by substituting (4.9) into (4.6) with $k = 4\pi/\lambda$, we have

$$\eta(\lambda) \approx 0.38\, C_n^2 \lambda^{-1/3} \tag{4.10}$$

Thus, provided the scale determined by the half-wavelength of the radar is included in the inertial domain, the measurement of radar reflectivity makes it possible to estimate $C_n^2$. Atlas *et al.* [1966] discussed the limits to detection of turbulence by radar: at short wavelengths, corresponding to scales for which dissipation is involved in the shape of the spectrum, (4.10) is no longer exact; eq. (4.3), which is more general, remains valid because the field is isotropic in the dissipation domain.

Formula (4.10) indicates that in the inertial domain, the reflectivity is proportional to $\lambda^{-1/3}$; therefore, simultaneous radar measurements at several wavelengths make it possible in principle to test the physical reality of the inertial domain concept and, on the basis of wavelength dependency, to distinguish echoes due to fluctuations in the refractive index from those produced by clouds or precipitation proportional to $\lambda^{-4}$ in the Rayleigh region (if the two types of scatterer are not mixed in space).

### 4.2.3 Relationship Between Radar Reflectivity and the Average Atmospheric Field

The preceding equations establish relationships between the reflectivity and the fluctuations in the refractive index of the air. The latter depend on the intensity of the turbulence and the kinetic energy of the average flux. The formulation of quantitative relationships between these different terms, which is of obvious operational interest for the detection and prediction of turbulence in clear air, poses problems that are not completely solved. In the convective field, in particular, the distribution of small-scale fluctuations depends strongly on the three-dimensional structure of the air movement and the average fluxes, which are constantly varying [Hardy and Ottersten, 1969].

The stable atmospheric layers, where significant turbulence can develop, lend themselves to simplifications facilitating the study of the relationships between the reflectivity and the average field. Such layers are observed at all altitudes (Sections 4.3 and 5.7).

Considering a horizontally homogeneous average turbulent flux, in presence of wind shear, we assume [Ottersten, 1969] that, in the regions where the transfer of the wind kinetic energy to turbulent movements is sufficiently significant, the ratio $\varepsilon_n \varepsilon^{-1}$ is determined by the ratio of the mean production of quadratic fluctuations in the refractive index to the net production of turbulent kinetic energy. Based on this assumption and under sufficient steady-state conditions, eq. (4.8) takes the form

$$C_n^2 = a^2 \varepsilon^{2/3} (\mathrm{d}\psi/\mathrm{d}z)^2 (\mathrm{d}u/\mathrm{d}z)^{-2} (1 - R_f)^{-1} \tag{4.11}$$

where $\mathrm{d}u/\mathrm{d}z$ is the vertical shear of the average wind, $\mathrm{d}\psi/\mathrm{d}z$ is the average vertical gradient of the generalized potential refractive index[4] $\psi$ at the level considered, and

[4]Characterizing a particle of air brought adiabatically to an arbitrary reference pressure level $p_0$, usually chosen equal to the average value of the layer considered.

$R_f$ denotes the Richardson flux number.[5] Based on quite general assumptions, this equation shows that within layers where significant and well-established turbulent transfers take place, measurement of $C_n^2$ by radar, eq. (4.10), and the knowledge of the average profiles of the potential refractive index and the wind (measured by aircraft or sounding) make it possible in principle to determine $\varepsilon$, which measures the degree of turbulence in the inertial domain.

In the lower troposphere, $d\psi/dz$ depends mainly on the humidity gradient, which is a highly variable quantity in the lower layers, and $C_n^2$ is a poor predictor of the intensity of turbulence in clear air although the presence of echoes reflects the presence of turbulence. In fact, the most intense turbulence phenomena in clear air develop in the upper troposphere at altitudes where the temperature is low and the humidity therefore is low. Under these conditions, if we consider an atmospheric layer of small thickness, $C_n^2$ is proportional to the structure constant of the temperature fluctuations $C_T^2$. In other words, if we assume that the fluctuations in humidity and temperature are not correlated, the sum of the partial scattering due to each of these fluctuations is influenced only a little by the humidity. We thus can establish an expression similar to (4.11):

$$C_T^2 = a^2 \varepsilon^{2/3} \frac{\theta}{g} \frac{d\theta}{dz} \left( \frac{R_f}{1 - R_f} \right) \tag{4.12}$$

Provided that the conditions given for (4.11) are satisfied, this equation shows that the radar reflectivity of turbulent regions in which the humidity gradients are low, depend directly on the average vertical gradient of the generalized potential temperature $d\theta/dz$ and on $\varepsilon^{2/3}$; that is, on the intensity of the velocity fluctuations in the inertial domain.

### *Experimental Results*

To try to verify eq. (4.6), which relates the reflectivity to the spectral energy, assuming that the turbulent medium has an inertial domain, containing the half-wavelength

---

[5]The Richardson flux number,

$$R_f = \frac{g}{\theta} \frac{\overline{\theta' w'}}{\overline{u' w'} du/dz}$$

expresses the ratio of convective kinetic energy to kinetic energy produced by the shear. The term $w$ is the vertical component of the velocity, $\theta$ is the generalized potential temperature, $g$ is the acceleration due to gravity; the symbols with a prime represent the fluctuation of the various quantities with respect to the mean. The generalized potential temperature characterizes a particle of air brought adiabatically to an arbitrary reference pressure level $p_0$, usually chosen equal to the average value in the layer considered.

of the radar used, almost simultaneous measurements have been made by ground-based radar and airborne refractometer. The results are in agreement with the theory [Konrad and Robison, 1972].

One difficulty of measuring the reflectivity of turbulent regions by radar stems from possible uniformity defects in the scattering field of the beam, inside the pulse volume, which can underestimate the intensity of the phenomena. Thus the use of a radar with frequency modulation and continuous wave has made it possible to describe the measurements relating to layers with thickness on the order of 2 m in which, at a wavelength of 10 cm, reflectivity values higher than $10^{-12}$ $cm^{-1}$ have been observed [Atlas *et al.*, 1970].

The determination of radar reflectivity produced by turbulent regions at wavelengths lying between 3 cm and 7.1 m for altitudes between the surface and 22 km, lead to values of $\eta$ lying between $10^{-20}$ and $10^{-11}$ $cm^{-1}$ and values of $C_n^2$ between $10^{-21}$ and $10^{-11}$ $cm^{-2/3}$. As a whole, the values of $\eta$ measured close to the tropopause are 20–40 dB less than those observed in the lower layers [Hardy *et al.*, 1966; Kropfli *et al.*, 1968; Atlas *et al.*, 1970; Starr and Browning, 1972; Browning *et al.*, 1973; Metcalf, 1975a and 1975b]. Experiments have also made it possible to verify the relationship with the wavelength given by eq. (4.10) and confirm that it is possible to use radars with several wavelengths to obtain a signature from the scattering regions: refractive index turbulence in terms of $\lambda^{-1/3}$ or hydrometeors in terms of $\lambda^{-4}$.

## 4.3 ST RADAR

### 4.3.1 Influence of the Wavelength

The microwaves used in meteorological radar lie at the lower limit of the band of spatial scales occupied by the inertial domain, the shortest are excluded (cf. Section 3.4.1 and Figure 4.1). The threshold for detecting variations in current values of the refractive index in the lower layers of the atmosphere is just exceeded by highly sensitive high-performance radar operating in the *S* band. For such radar, the minimum detectable value of $C_n^2$ lies approximately between $10^{-14}$ and $10^{-15}$ $m^{-2/3}$ at 10 km; they make it possible to detect high values of $C_n^2$ that are found mainly in the convective boundary layer and turbulent stratifications (cf. Section 5.2). The longest wavelengths, band *L*, UHF, and VHF correspond to these spatial scales of turbulence, where there is considerable energy and it is easier to obtain a signal. The minimum detectable values of $C_n^2$ lie between $10^{-17}$ and $10^{-19}$ $m^{-2/3}$ at 10 km. A specific difficulty, at these frequencies, is obtaining beams that are sufficiently directive and orientable for a radar observation, taking into account the size of the antennas.

At the beginning of ST radars, installations built for radio astronomy and the study of ionospheric incoherent scattering (Thomson scattering) were used. The first

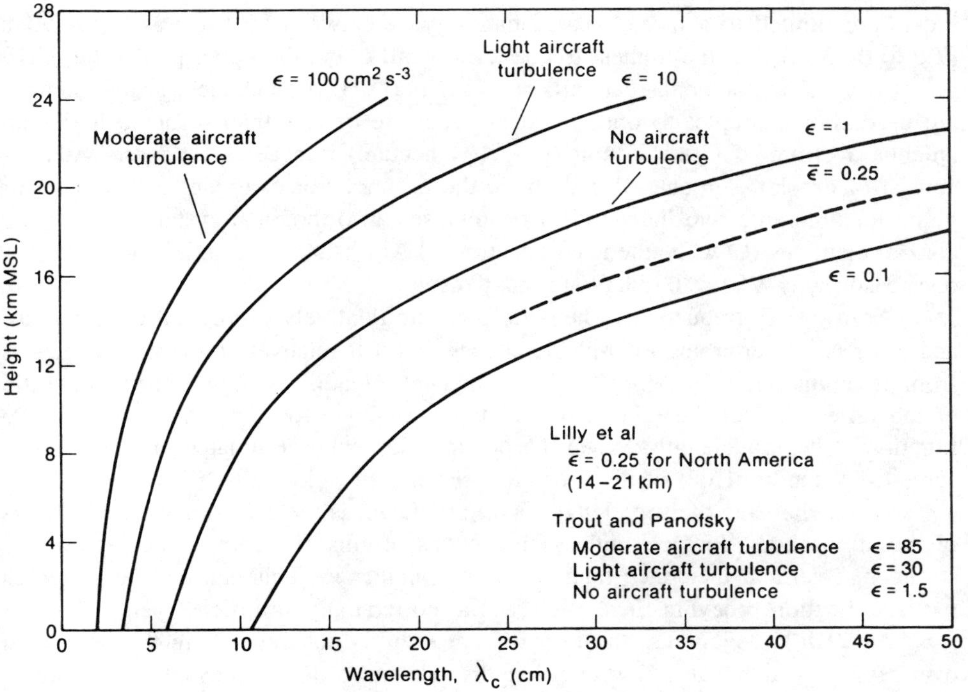


**Figure 4.1** Cutoff wavelength ($\lambda_c$) *versus* height for various levels of mechanical turbulence ($\varepsilon$). Routine clear-air wind profiling should be possible with radar wavelengths greater than $\lambda_c$, that is, to the right of the curves (from [Strauch *et al.*, 1984]).

observations were made with the help of a 50 MHz radar at Jicamarca, Peru [Woodman and Guillen, 1974], then at Chatanika, Alaska, with a 1290 MHz radar [Balsley and Farley, 1976], and at Arecibo, Puerto Rico, with a 430 MHz radar [Aso, Kato, and Harper, 1977]. The last two radars have sufficient sensitivity to detect fluctuations in the refractive index as far as 20 km altitude.

Several VHF ST radars were constructed specially for tropospheric and stratospheric observation, in particular a 40 MHz radar at Sunset, Colorado [Green *et al.*, 1975; Gage and Clark, 1978], in Germany (SOUSY 53.5 MHz radar; Czechowsky *et al.*, 1976), and in France [Crochet, 1985] as well as high-performance systems capable of observing as far as the mesosphere (≈100 km), called *mesosphere, stratosphere, troposphere* (MST) radar. Recent technical developments may be classified into two groups: systems intended for research and less powerful radars optimized for creating operational networks for making wind measurement.

Most ST radars are coherent pulse radars. Table A4.5, Appendix 4, gives the characteristics of some representative systems. The frequencies actually used for ST radars are limited to a few allocated bands lying between 40 MHz and 915 MHz (7.5 to 0.33 m). For the highest frequencies (band *L* and the upper part of the UHF region), parabolic antennas can still be used. For VHF frequencies, array antennas are used. For example, to obtain a beam aperture of less than 5° at 50 MHz, an antenna aperture of $(50 \times 50)\text{m}^2$ (i.e., 1/4 hectare) is necessary. These antennas consist of dipolar elements placed above the ground. For example, a current technique for obtaining two independent beams uses an orthogonal system of collinear coaxial elements (COCO antennas) constructed with standard coaxial cable. This is one reason why VHF ST radars are inexpensive.

At low VHF frequencies, the radar picks up relatively significant cosmic noise and can receive perturbing ionospheric echoes, but it is relatively insensitive to echoes from precipitations. The opposite is true for UHF radars. Figure 4.2 shows the ratio of the reflectivity of the rain to that of the turbulence for $C_n^2 = 10^{-16}\ \text{m}^{-2/3}$ as a function of the wavelength. When the precipitation echo dominates, the radar measures the velocity of fall of the drops and not the air velocity.

When choosing the wavelength, a critical factor is the detection of stable layers of the atmosphere. In fact, with VHF, echoes originating from stable regions are stronger at vertical incidence than at oblique incidence. This can be interpreted as partial reflection (obeying Fresnel's laws) at horizontally stratified layers.

At VHF wavelengths, these layers are seen as coherent along the horizontal over ranges greater than a few wavelengths. The reflections contribute to the signal only when the beam is vertical (see Section 4.3.3 and Figure 4.4). For an oblique elevation, only the isotropic turbulence returns the energy to the radar [see, for example, Röttger, 1980]. The value of detecting these layers is apparent for measurements at the level of the tropopause and the stratosphere.

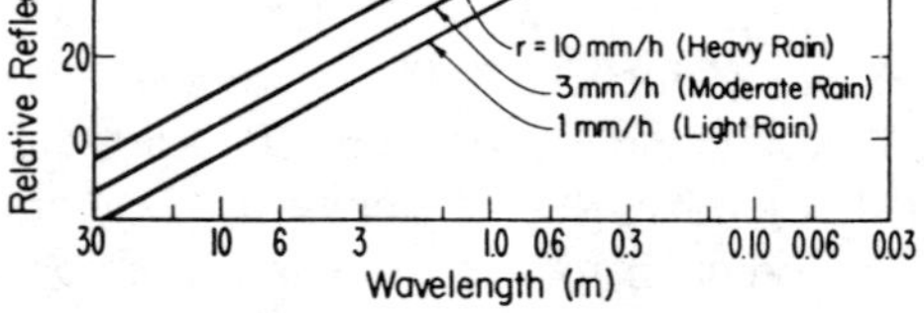


**Figure 4.2** Ratio of reflectivities of precipitation and turbulence as a function of wavelength ($C_n^2 = 10^{-16}\ \text{m}^{-2/3}$) (from [Balsley and Gage, 1982]).

In all, the VHF frequencies therefore are more favorable for ST radar. Their main limitation is the incapability of taking measurements in the lower layers; that is, at 1 to 2 km above the ground. This is due to several factors, the most important being the ground echoes and the internal reflections in the array, and the Fresnel distance, eq. (1.21), which extends over approximately 1 km. Now, for a large number of applications, it is important to have information on the wind in the boundary layer. A UHF frequency therefore is used for observations in the lower layers. Small, light 915 MHz radars are efficient at making wind measurements between approximately 100 m and 2–3 km [Eklund, Carter, and Balsley, 1988].

To obtain a complete system, a dual frequency system is often used, consisting of a frequency of around 900 MHz and a frequency of around 50 MHz. Another solution involves using a single intermediate frequency of around 400 MHz [see the comparative study in Frisch *et al.*, 1986]. The parameters of a 405 MHz ST radar are given in Table A4.5, Appendix 4. The methods that can be used for data processing and calculating the spectra and their moments are those described in Chapter 3. These are the same as for hydrometeor echoes. When calibrating ST radars, the subsystems are usually considered because it is not easy to sight a known target. Also, we can use precipitation echoes whose reflectivity has been measured by a conventional radar or by a disdrometer.

### 4.3.2 Wind Measurements

ST radars used for wind measurement are called *wind profilers*. Various techniques have been suggested for measuring profiles. When the beam can be easily orientated, as in the case of moving parabolic antennas, doppler methods of the VAD type can be used. With array antennas, successive sightings are made in fixed directions to produce a simplified VAD. In the minimum configuration, wind profilers observe two components of the wind velocity in two perpendicular planes intersecting at the vertical line of the radar. The beams are fixed and point at approximately 15° to the vertical. If the vertical velocity and small-scale movement are neglected (VAD assumption of local wind uniformity), an estimate of the wind profile in terms of speed and direction can be derived. Often a third beam points to the zenith for measuring the vertical component. We thus have a more complete description of the wind and a better detection of the tropopause. Systems with four beams or five beams (one vertical and four at 15° to North, South, East, and West) have also been tested. This method is sometimes called the *doppler method*. It is used by almost all wind profilers (Figure 4.3).

Another possibility, based on a method used for a large number of years to measure "ionospheric winds," has been demonstrated by Röttger [1980; Larsen and Röttger, 1989]. It uses a pulse system consisting of a transmission antenna with a vertically directed broad beam and two reception antennas arranged around the transmission antenna. The apparent horizontal movement of the heterogeneities of the

**Table 4.1**
Parameters of a 405 MHz ST Radar Used as a Wind Profiler (from [Frisch *et al.*, 1986])

| | | | |
|---|---|---|---|
| Radar | | | |
| Frequency | | 405.25 MHz | |
| Authorized bandwidth | | 1.00 MHz | |
| Peak power | | 30 kW | |
| Average power | | 0.4 kW maximum | |
| Pulse width | | 1, 3, 9 $\mu$s | |
| Pulse repetition period | | 100, 150, 300 $\mu$s | |
| Antenna aperture | | 9 m $\times$ 9 m | |
| Antenna pointing | | 15° off-zenith to north and east | |
| Antenna type | | phased array of Yagi-Uda elements | |
| Two-way beamwidth | | 4.3° | |
| Data processing | 1-$\mu$s pulse | 3-$\mu$s pulse | 9-$\mu$s pulse |
| Time domain averaging | 120 pulses | 75 pulses | 35 pulses |
| Spectral averages | 8 | 16 | 24 |
| Maximum radial velocity | $\pm$15.41 m s$^{-1}$ | $\pm$16.44 m s$^{-1}$ | $\pm$17.62 m s$^{-1}$ |
| Spectral resolution (64 points) | 0.48 m s$^{-1}$ | 0.51 m s$^{-1}$ | 0.55 m s$^{-1}$ |
| Height sampling | 1-$\mu$s pulse | 3-$\mu$s pulse | 9-$\mu$s pulse |
| First height | 0.4 km AGL | 2.4 km AGL | 4.0 km AGL |
| Height spacing | 100 m | 290 m | 870 m |
| Number of heights | 24 | 24 | 14 |

refractive index at a given level is derived from an analysis of the correlation between the signals received by the three antennas. The horizontal speeds are calculated from the time lags of the cross-correlation functions. The vertical speed also can be obtained by an analysis of the doppler spectrum. The main advantage of this so-called *spaced antenna* (SA) method is to prevent the loss of the doppler method signal at oblique elevations due to Fresnel reflection. The efficiency of the system therefore is better for measuring horizontal components at the level of the tropopause and in the stratosphere because the signal is stronger. This method can be used only at low VHF frequencies.

The accuracy of wind measurements by the doppler method is influenced by the radar's characteristics (beam aperture, pulse shape), the number of sightings used for the estimate, the structure of the observed flow, and the presence of rain. Compared to a four-beam profiler, a three-beam profiler gives a standard deviation of 1.3 m s$^{-1}$ for estimating the horizontal wind (for a month of measurements); the error becomes 1.7 m s$^{-1}$ for a two-beam profiler (i.e., when no correction is made for the vertical movement) [Strauch *et al.*, 1987; among other studies on the quality of measurements made with ST and MST radars].

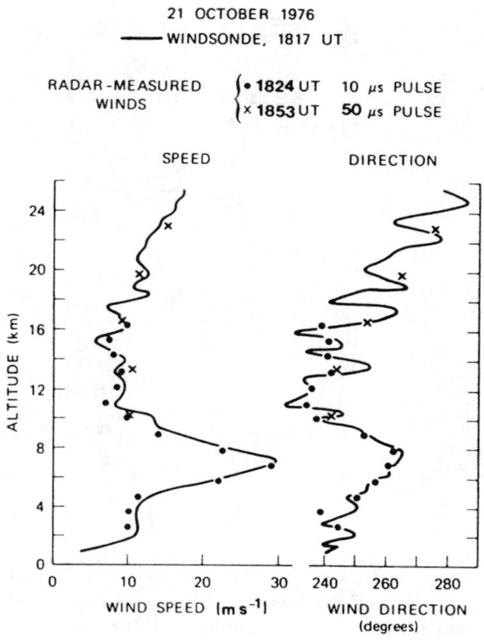


**Figure 4.3** Wind profile derived from ST radar (1290 MHz) by the doppler method and rawindsonde at the same time (from [Balsley *et al.*, 1977]).

### 4.3.3 Reflectivity

A large amount of semiquantitative information on the structure of the atmosphere can be extracted from the distribution of the reflectivity. The interpretation is made mainly in terms of atmospheric stability and turbulence.

*Stability*

We have seen that the contributions of the turbulence and the stable layers to the profiles of the reflectivity can be separated and measured. The difference between the vertical and oblique reflectivity gives information on the vertical profile of atmospheric stability. The relationship is shown in Figure 4.4, which contains the profiles of signal power (after the correction of the attenuation in $1/r^2$, where $r$ is the range) corresponding to an oblique elevation and a vertical elevation. We find that the profile at vertical incidence is correlated with the vertical profile of the potential temperature (at least up to the range of 14 km, beyond which the signal is

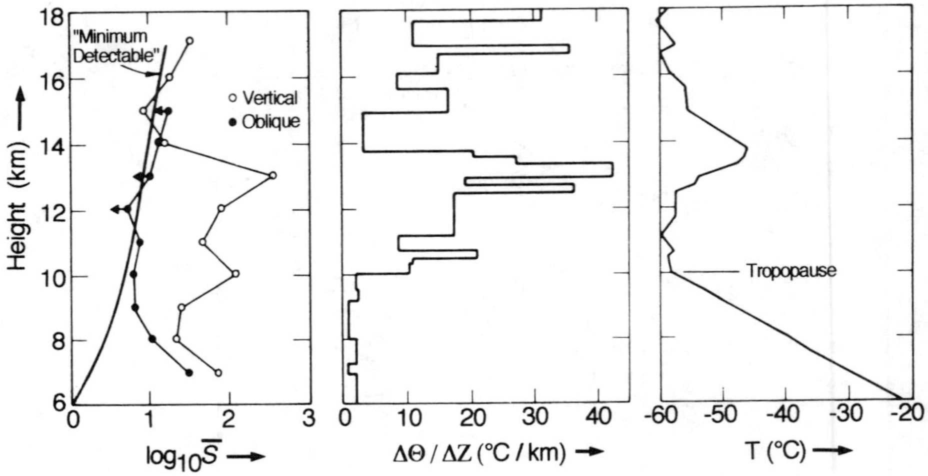


**Figure 4.4** Vertical profiles of the range-corrected ($r^{-2}$) received power on the vertical and oblique (30° off the zenith) antennas of the Sunset radar (40 MHz). Also shown are the potential temperature gradient ($\Delta\theta/\Delta Z$) and the temperature ($T$) from a simultaneous radiosounding (from [Gage and Green, 1979]).

weak). The signal increases above the tropopause as well as around 13 km. It therefore varies according to the static stability of the atmosphere. The level of the tropopause is determined very accurately. Because the coefficient of reflection is related to the vertical gradient of the temperature, we can derive a temperature profile from its profile above the tropopause [Gage and Green, 1982].

*Turbulence*

There is a considerable difference between the size of fluctuations creating radar backscattering ($\lambda/2$) and the size of large vortexes forming clear air turbulence, which, for example, affects the flight of aircraft in the upper troposphere and the stratosphere (Section 5.7). However, the inertial cascade that connects the small-scale turbulence to the large-scale vortexes responsible for injecting energy into the spectrum establishes a link between the two phenomena, and the measurement of the radar reflectivity of turbulent zones can be expected to provide information on the energy of large-scale movements. Many simultaneous observations by aircraft and radar have been studied to define this relationship. The results show that, practically, turbulence is always present in the place where the echoes are observed and that the more intense this turbulence the more probable is its detection. For so-called light or moderate turbulence the layers involved are characterized by a temperature inversion on the radio sounding profiles. In such layers, the value of $\varepsilon$ has been estimated as approximately 100 $cm^2/s^{-3}$ and the observed values of $d/\theta/dz$ are greater than $5 \times 10^{-5}$ $K/cm^{-1}$, which corresponds to reflectivities of approximately $3 \times 10^{-17}$ $cm^{-1}$ [in this connection see, for example, Barat and Bertin, 1984].

*Frontal Systems*

The continuity of the vertical profiles of reflectivity and the wind make it possible to construct height-time cross sections revealing the discontinuity zones or air mass transition zones [Larsen and Röttger, 1989]; in particular, frontal zones and tropopause folding and injections of stratospheric air. Figure 4.5 shows an example of a frontal zone in the reflectivity and velocity height-time surfaces.

### 4.3.4 Networks of ST Radar

The profiler has several advantages over the technique of wind measurement by balloon: increased temporal resolution, because it can operate continuously; appreciably improved accuracy of estimate; entirely automatic data acquisition and processing; high reliability (no moving parts); low operating cost; low initial outlay (ST radars are less expensive than tracking radars). Furthermore, because of the permanence of the measurements, the information from ST radars has excellent statistical representativity. For all these reasons, networks of wind profilers have been installed in many parts of the world: such as the United States, Europe, Japan, and Australia. The base data of the stations are wind data (possibly including the vertical component); turbulence profiles; and the altitude of the tropopause. These data provide an important resource for local short-term forecasting: detection of wind shear and various waves, convection monitoring, and so forth.

Processed as a whole, the data obtained from the network help describe the medium- and large-scale circulation with a high spatio-temporal coherency. It is possible, for example, to extract information on the temperature field from the movement field. Kuo and Anthes [1985] show that, on the basis of hourly profiles from a network of ST radars with grid size less than 200 km, we can obtain the divergence terms of the equation of motion from which the geopotential field and then the temperature field can be derived; the mean square errors obtained by simulation are less than 0.6 K. This accuracy can be improved by combining with temperature profiles determined by microwave radiometry from the ground or from space. The integration of the measured vertical speeds makes it possible to obtain average vertical movements over long periods with a resolution of a few cm $s^{-1}$. These few examples show that networks of ST radars have a significant potential for operational meteorology and forecasting not realized in current applications.

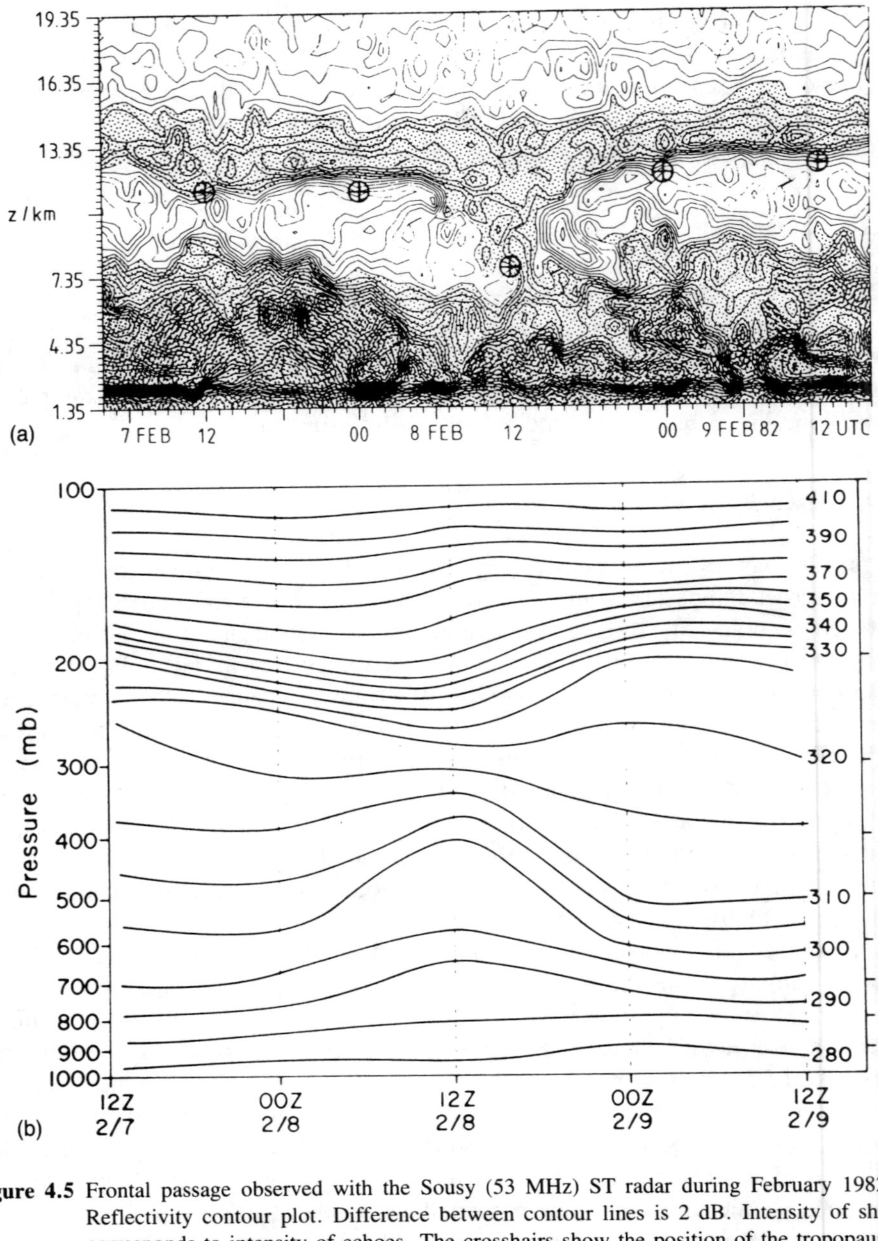


**Figure 4.5** Frontal passage observed with the Sousy (53 MHz) ST radar during February 1982: (a) Reflectivity contour plot. Difference between contour lines is 2 dB. Intensity of shading corresponds to intensity of echoes. The crosshairs show the position of the tropopause reported by radiosondes. (b) Pressure-time cross section of the potential temperature measured by radiosondes during the period corresponding to (a). The contour interval is 5 K in the troposphere and 10 K above 350 K isotherm. (c) Wind vectors measured by the radar as a function of time and height. The reference scale and directions are shown at the lower left-hand corner (from [Larsen and Röttger, 1982]).

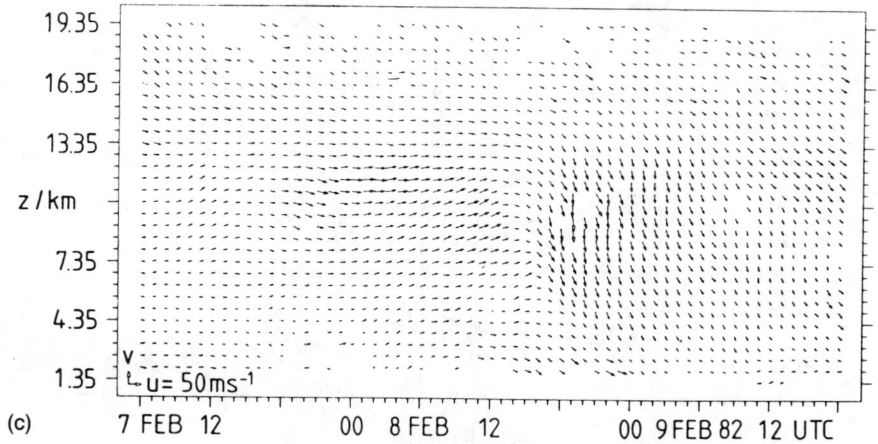


**Figure 4.5** (continued)

## 4.4 RASS

Doppler radar observes perturbations in the refractive index of air created by high-intensity acoustic pulses. The measurement of the velocity of acoustic propagation lets us calculate the temperature as a function of altitude. The acoustic wave is a local compression of the air followed by a depression that is propagated through the atmosphere. The acoustic and electromagnetic wavefronts are approximately spherical. When the acoustic source and the radar are situated at the same point, the electromagnetic energy returned by the gradient of the refractive index created by the acoustic wave is focused on the radar. Such a *radio acoustic sounding system* (RASS) avoids the attenuation in $r^{-2}$ undergone by the backscattering from the natural fluctuations in the refractive index and also focuses the scattered energy and the spatial coherence of the peaks of acoustic waves. The scattered fields will arrive in phase at the radar if the acoustic wave has a wavelength $\lambda_a$ equal to the half-wavelength of the radar $\lambda_e/2$. Under these conditions, we can show that the power of the scattered signal increases as the square of the number of wavelengths contained in the acoustic pulse. Figure 4.6 illustrates this point: for $\lambda_e/\lambda_a = 2$, the increase by a factor of 10 in the number of interacting waves produces a gain of 20 dB of the backscattered power.

The longer the duration of the acoustic pulse, the more sensitive is the value of the reflection coefficient to the tune between $\lambda_e$ and $\lambda_a$. Obviously, the performances of such a device are limited by the wind, turbulence, and the natural heterogeneities of the atmosphere, which affect the propagation of sound.

The doppler measurement of the velocity of propagation of sound $V_S$ is an approximate measure of the temperature $T$. In fact, in still air, $V_S = AT^{1/2}$, where

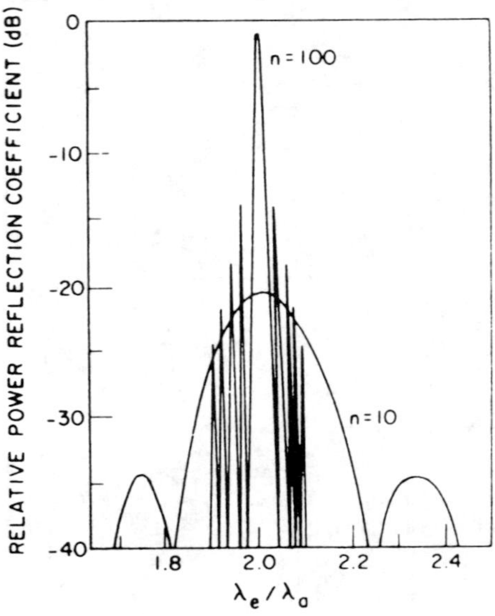


**Figure 4.6** Reflection coefficient *versus* wavelength ratio for acoustic pulse lengths of 10 and 100 wavelengths ($\lambda_e$, radar wavelength; $\lambda_a$ acoustic wavelength) (from [Barnes, 1972]).

$A$ is a term that depends only on the relative humidity. We therefore, with eq. (1.7), have

$$T = (f_d \lambda_e)^2/(4 A^2) \tag{4.13}$$

where $f_d$ is the doppler frequency (Hz), $\lambda_e$ is the radar wavelength ($m$); and $A = (\gamma R/M)^{1/2} = 20.05$, where $R$ is the gas constant, $\gamma \approx 1.4$ is the ratio of the specific heats of air, and $M$ is the average molecular weight of the constituents of the medium; $T$ is in K.

The drawback of such a system is the rapid dissipation of acoustic wave energy due to the damping during the propagation, which leads to a choice of the waves for which the ratio between the transmitted power and attenuation is the most favorable (metric and decimetric waves). The first significant measurements have been made with the help of an acoustic source with a wavelength of approximately 4 m and a radar wavelength of $\lambda_e = 8.15$ m [Barnes, 1972] and with $\lambda_a = 34$ cm and $\lambda_e = 68$ cm [Frankel, Chang, and Sanders, 1977].

The velocity of sound increases with the humidity of the air. However, the total contribution of the water vapor to the phase velocity of the sound is less than 1 m s$^{-1}$, which corresponds to a temperature correction of less than 1 K. In a calm atmosphere, a scattered signal usually is obtained up to a distance of 1.5 km and occasionally up to 5 km [Barnes, 1972]. The best signal-to-noise ratio is obtained for an acoustic pulse of approximately 50 $\lambda_a$. The results show that the obtained temperature profiles are in quite good agreement with the data of conventional radio sounding.

RASS has benefited from the technical advances of wind profilers. Table 4.2 gives the technical characteristics of two RASS systems, one developed around a VHF radar and the other around a UHF radar, as well as the altitudes covered. A comparison of the temperature measurements obtained with these two instruments and from radio sounding for varying meteorological conditions shows excellent agreement; the root mean square differences in temperature are about 1° C but can be reduced to 0.2° C with an optimized processor [May, Moran, and Strauch, 1989].

The height covered by the RASS may be increased considerably if we manage to adapt the direction in which the radar beam is pointed to maintain a normal incidence to the acoustic waves when their sphericity is altered by variation with the altitude of the temperature and the wind. A simulation study confirmed by tests with the Japanese *middle and upper* atmosphere (MU) radar shows that the RASS echoes can be obtained up to 8.4 km for a situation of strong wind and 22 km in calm wind [Masuda, 1988].

**Table 4.2**

System Parameters for the Denver and Platteville RASS (from [May *et al.*, 1989])

| | *Platteville* | *Denver* |
|---|---|---|
| | Radar | |
| Frequency (MHz) | 49.8 | 915 |
| Wavelength (m) | 6.0 | 0.33 |
| Antenna size ($m^2$) | 10 000 | 100 |
| Beamwidth (1 way) (deg) | 3 | 2 |
| Range resolution (m) | 300 | 150 |
| Mean power (W) | 200 | 100 |
| Sampling time (1 profile) (min) | 1.5 | 1 |
| | Acoustic | |
| Frequency (Hz) | ~100 | ~2 000 |
| Beamwidth (1 way) (deg) | 60 | 8 |
| Acoustic power (W) | 50 | 40 |
| | RASS Altitude Coverage | |
| Minimum altitude | 2.0 km | 200 m |
| Maximum altitude (km) | 5–10 | 1–2 |
| Observation period | Jun, Oct 1988 | Jul–Aug 1988<br>Nov 1988–Jan 1989 |

## 4.5 INSECTS

Most microwave radars used for meteorological observation ($\lambda \leq 10$ cm) have too short a wavelength ($\lambda < l_m$) or insufficient sensitivity to detect fluctuations in the refractive index of air ($C_n^2$). However, echoes frequently are observed in clear air with such radars, particularly during the hot season. As mentioned earlier, the interpretation of these microwave echoes in clear air by various authors up to 1966 was uncertain, due to inadequacies in the observation of the phenomena, in particular the dependence on the wavelength. Atlas [1964] has reviewed the first studies and results in this area; his article contains the arguments of the controversy that had arisen within the community of specialists regarding the various causes that could explain these echoes. Present knowledge makes it possible to validate the early observations in part and refine the interpretation.

At positive temperatures and except above deserts and oceans, there are always insects in the lower layers of the atmosphere. Chapter 5 will show with a few examples that their distribution is influenced by the dynamic and thermodynamic structure of the atmosphere. They are the main cause, and in many cases the sole cause, of the echoes in clear air at short radar wavelengths. However, at short range an additional contribution may be made by inert particulate scatters of millimetric size (vegetable debris, seeds) carried by the atmospheric currents. Little is known about this contribution. These scatterers provide an important resource for meteorological measurement in the low atmosphere: thickness of the boundary layer, visualization of the convection, localization of stable layers, and wind measurements.

### 4.5.1 General Characteristics

From the point of view of radar detection, insects possess a large number of similarities with hydrometeors. Their effective backscattering cross section $\sigma$ depends on their shape, size, and the value of the dielectric factor $|K|^2$. To simplify the discussion, they can be approximated by elongated ellipsoids with horizontal major axis consisting of a mixture of water and a dielectric material with low loss. Because the proportion of water is significant, on the order of 50 to 70% of the total weight, the insect has a significant effective backscattering cross section and so can be detected. The dry parts (e.g., the wings) scatter very little. This statement obviously applies to the inert particulate scatters; the smaller their water content, the more difficult it is to detect them. The value of $|K|^2$ for insects can be calculated by a relationship of the same form as (2.53) and (2.54) (Figure 4.7).

Insect size extends over a wide range, from a fraction of a millimeter to a few centimeters. The theories of Mie and Rayleigh-Gans apply to the scattering of equivalent ellipsoids and the discussion of this subject with regard to hydrometeors can

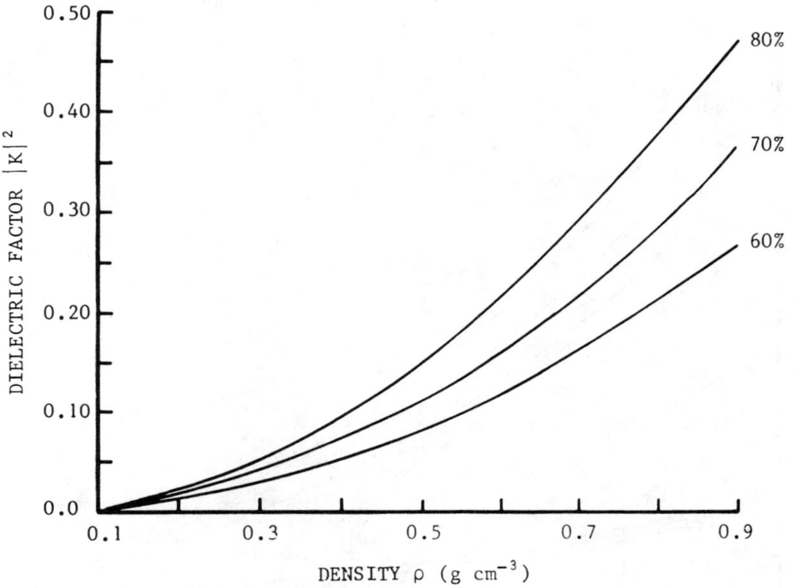


**Figure 4.7** Dielectric factor $|K|^2$ for insects *versus* density $\rho$ (g cm$^{-3}$) with liquid water content (in percentage) as a parameter (courtesy of C. A. Etty).

be carried over to insects. The number of insects in nature decreases with size. However, the comparison with hydrometeors stops there because the distribution of the sizes, at a given point in time and space, can be uniform if only one specie is present. The individual cross sections of insects reported by various investigators are between 1 and $10^6$ cm$^2$.

The insects of the atmosphere can be broadly divided into two categories: the large insects, called *dynamic insects*, with their own velocity independent of the wind (for example, 8 m s$^{-1}$ for a bee), which are few in number; and the small insects, which are the overwhelming majority. The small insects' own velocity is small; they are carried by the atmospheric currents and are good tracers of air movements for these reasons. The value of $\sigma$ for insects therefore depends on their attitude with respect to the direction of propagation of the waves and the variations in their shape. The ratio of the reflectivity for a polarization parallel to the principal axis to the reflectivity for orthogonal polarization ($Z_{DR}$) varies considerably as a function of the specie and may reach the order of 10 dB (Figure 4.8). Variations in shape result from the beating of wings and respiration. The frequency of wing beats decreases with increasing size and varies from a few Hz for the large insects to several hundreds of Hz for the small insects. It can be measured from a Fourier transform of the time series of the reflectivity. Because the humidity concentration is low in the wings

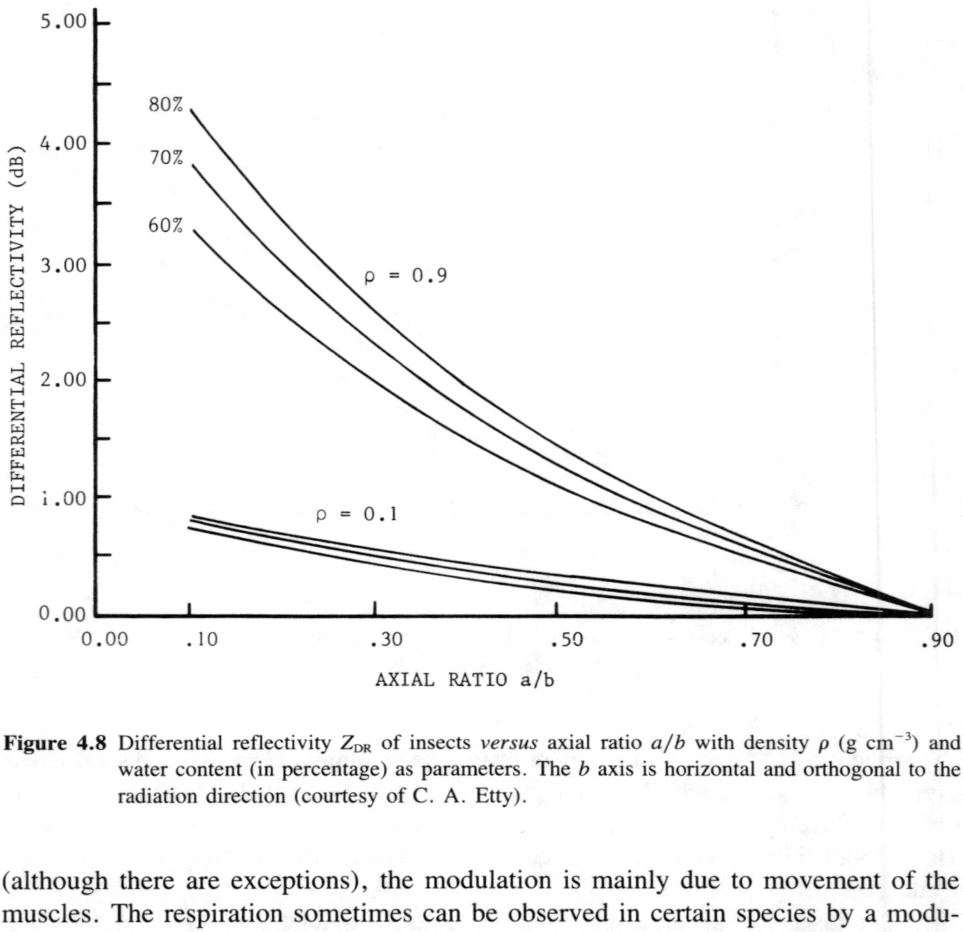


**Figure 4.8** Differential reflectivity $Z_{DR}$ of insects *versus* axial ratio $a/b$ with density $\rho$ (g cm$^{-3}$) and water content (in percentage) as parameters. The $b$ axis is horizontal and orthogonal to the radiation direction (courtesy of C. A. Etty).

(although there are exceptions), the modulation is mainly due to movement of the muscles. The respiration sometimes can be observed in certain species by a modulation of $\sigma$ of smaller amplitude (variation of the volume of the thorax). The spectral analysis of the temporal variation of $\sigma$ makes it possible to demonstrate these modulations [see, for example, Schaefer, 1976].

The numerical concentration of insects in the atmosphere is such that frequently at short range there is less than one individual insect at a time in the pulse volume, whereas at long range several individuals are present in $V$ at the same time (we note in passing that it is not surprising that such targets cannot be easily referenced visually). The analysis of the variation of the signal-to-noise ratio for a homogeneous population with almost horizontal elevation in this case is in $r^{-4}$ and $r^{-2}$ for short and longer ranges, respectively (Figure 4.9). In the $r^{-4}$ region, we can estimate the individual $\sigma$ (when the insect is on the beam axis). When the population is homogeneous and has a preferential orientation, the isoecho lines in horizontal polarization are elliptical around the radar, the minor axis of the ellipse indicating the direction

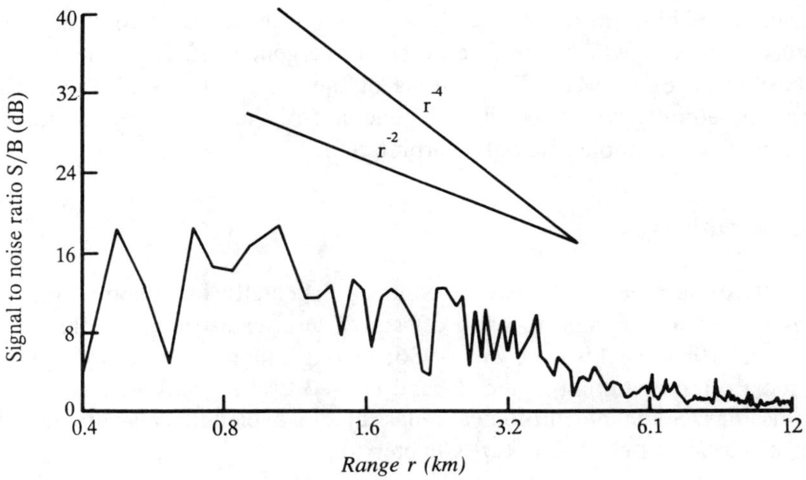


**Figure 4.9** Signal-to-noise ratio *versus* distance $r$ for insects observed in SW France with a $K_a$ band radar in 1987. Also shown are the slope of $r^{-2}$ and $r^{-4}$ curves.

of preferential orientation. The measurement of $\sigma$, the polarization ratio ($Z_{DR}$), and the characteristic frequencies of modulation of $\sigma$ form signature elements for radar identification of species of insects.

### 4.5.2 Insects and Birds

Birds, which sometimes are sufficiently numerous to interfere with radar observation of air traffic at low levels around airports, have their own velocity of movement (10 to 20 m $s^{-1}$), which is too large in relation to that of atmospheric flow in the lower layers to be used as a tracer of air movements. On the other hand, bird echoes are a useful signal in ornithology; radar offers a means of studying the distribution of birds in the atmosphere, their seasonal migration to their resting and nesting sites, their diurnal migration between the rest areas and supply areas, and their local migrations that can reach an altitude of 6 km for certain species [Eastwood, 1967].

The distribution of birds in the atmosphere is the result of various stimuli, necessities, or opportunities. Annular echoes are observed that, starting from a point, spread out to diameters exceeding 30 km and are due to birds leaving the rest area with the first rays of the sun. In the proximity of coasts, line echoes are created by birds flying in ascending zones generated by the coastal circulation. Line echoes are also created by birds gathering in local fronts induced by currents of cold air diverging close to the ground behind cumulonimbus clouds. The main cause of local concentrations of birds in the boundary layer is the presence of food; that is, concentrations of insects, which originate in the convergence and ascending zones, with the highest value of $C_n^2$. The distributions of these three types of radar scatterers therefore are naturally correlated. The dependence of the insects on the polarization allows us to resolve ambiguities of interpretation.

### 4.5.3 Observations

Since 1970, some radars have been developed especially for entomological observation as part of a campaign to ward off agricultural catastrophes [Schaefer, 1976; Drake *et al.*, 1981; Dickison *et al.*, 1986; among others]. These are light mobile radars, most often operating in the *X* band ($\lambda$ = 3.2 cm), well adapted to entomological missions. Some meteorological radars at all wavelengths also are occasionally used for observations of insects (cf. Chapter 5).

## 4.6 ARTIFICIAL TRACERS

### 4.6.1 General Properties

When no natural target can be detected, air movements can be observed by deliberate creation of clouds of reflecting objects in the atmosphere. This technique is widely used in military operations as a passive "countermeasure," either to create false targets or to form a target screen concealing the objective. The targets in question, called *chaff*, have very low terminal fall velocity and follow the air motions. There is an extensive literature on the subject, which also provides various techniques for particular applications: particulate tracers, gas tracers, absorbent tracers, tracers for the measurement of winds in the upper atmosphere, and so on. Different shapes of reflectors may be used. Their dimensions are matched to the radar wavelengths. This matching covers one or more frequency bands.

The procedure has been adapted to meteorological observation for studying the medium- and small-scale circulation of air in a clear atmosphere under circumstances where the observation of wind by current methods is not sufficiently representative; this is the case in particular for the study of the movements associated with convective structures and for the measurement of the horizontal divergence over a small range or circulation across a velocity discontinuity.

The most commonly used tracers for the observation of the atmosphere are resonating dipoles manufactured with the help of metallized nylon wires or glass fibers with terminal velocities of fall less than 1 m $s^{-1}$ for nylon and 0.3 m $s^{-1}$ for glass fiber. They have a very small diameter, on the order of 10 $\mu$m (a cylindrical strand 1 cm in diameter contains up to $10^5$ dipoles). These therefore are simple, inexpensive tracers.

The physical length of the resonating dipole ($l$) is not exactly the half-length of the radar wave in a vacuum or a multiple of $\lambda/2$ (refer to the theory of wire antennas). The main corrective term for $l$ is the ratio of the length of the conductor to its diameter; for length-diameter ratios of $10^3$ and $10^4$ (values corresponding to metallized fibers), we have $l \approx 0.480\ \lambda$ and $l \approx 0.485\ \lambda$. The band width (at $-3$ dB) of the resonating dipole is equal to approximately 10% of the central frequency.

Under optimal conditions, as for a dipole adjusted to the length $l$ oriented perpendicular to the direction of propagation and parallel to the polarization plane ($E$) of the incident wave, the effective backscattering cross section has a maximum and is equal to [Nathanson, 1969]

$$\sigma_M = 0.86\ \lambda^2 \tag{4.14}$$

For $N$ dipoles of random orientation, on averaging eq. (4.14) over $4\pi$ steradian, the effective cross section becomes

$$\sigma = 0.18\ N\lambda^2 \tag{4.15}$$

This formula and the next formula are valid when the average spacing between dipoles is greater than $2\lambda$. For smaller spacing, the dipoles are not electromagnetically independent. However, in the atmosphere, the dipoles do not have a random orientation and tend to align horizontally during their fall. Therefore, the choice of the radar's plane of polarization is very important, horizontal polarization being the most favorable for efficient detection.

For $N$ dipoles orientated randomly in the horizontal plane and observed with a radar with horizontal polarization at a low angle of elevation, we have

$$\sigma_H = 0.22\ N\lambda^2 \qquad (4.15)$$

When the dipoles are all identical and therefore have the same effective backscattering cross section, the equivalent radar reflectivity factor is proportional to their numerical concentration in the observed volume. Under suitable observation conditions, a single dipole placed in the optimum position, can be detected at a range on the order of 10 km.

### 4.6.2 Applications to Atmospheric Observation

In the atmosphere, a cloud of dipoles tends to scatter and disperse as a result of turbulent or shear movements. Thus, the numerical concentration of an isolated cloud decreases over time whereas its volume increases. The result is an alteration of the definition of the point lowering the accuracy of its localization. In this case, a minor error is committed if the geometric center of the cloud or the barycenter of the signal distribution is used for the measurement. This scattering effect can also be used for studying turbulent diffusion in air.

With noncoherent radars the movement and the evolution of the cloud of tracers is observed over time, and from its successive positions, the speed and direction of the air movements are derived. The same procedure is used with a doppler radar as with hydrometeors in which the speed of fall is uniform (or negligible). Furthermore, because the concentration of tracers decreases over time, the spacing between dipoles may be several meters or tens of meters. The measurement of the velocity spectrum therefore is restricted to the wave number below the reciprocal of the minimum spacing of the dipoles (Lhermitte, 1969-b). In other words, the tracer does not reproduce the small scales (filtering effect).

Depending on the type of measurement to be carried out, tracers are dispersed in the atmosphere continuously or discontinuously from one or more points. Various dissemination vectors may be used. For example, a balloon or a rocket, under which is attached a capsule of dipoles equipped with a small explosive charge that is fired with a suitable delay, makes it possible to obtain a point scattering at a chosen altitude. With an aircraft the tracers can be distributed in altitude at a number of predetermined sites; for example, as a function of the wind relative to a particular cloud. At low levels, the use of a scaled-down model aircraft is a convenient solution to obtain a scattering of dipoles at short range at precise points. In the convective boundary layer, manual scattering at ground level is also possible for minimum horizontal wind; only a fraction of the dipole is carried upward. From a practical point of view, it is not easy to obtain a homogeneous scattering of the tracers. The fibers tend to form packets, which fall at a much higher speed than that of the isolated elements.

The technique of artificial tracers has been used to measure winds at altitude (up to 80 km), gradients in strong shear zones, and deformation profiles with the help of a dispersion perpendicular to the lines of flow. It is efficient for the observation of the air circulation around isolated cumulonimbus [Fankhauser, 1971; Marwitz, 1973]. The release of a packet of tracers can be used in "cloud modification experiments" to precisely locate with radar the seeding points or any other simultaneous event [Summers, Mather, and Treddenich, 1972]. The use of a doppler radar for tracking artificial tracers makes it possible to observe turbulent movements and especially the small-scale circulation of air at low levels [Lhermitte, 1969a and 1969b; Chernikov *et al.*, 1969; Hildebrand, 1977; Kropfli and Kohn, 1978].

The artificial tracer technique therefore is very convenient for studying the lower convective layers; released close to the ground, the dipoles carried by the updrafts associated with a natural or artificial thermal cell, disperse throughout the convective volume, marking its top and side boundaries [Sauvageot *et al.*, 1982]. In a cell with closed circulation, the demarcation can be perceived for several tens of minutes, as long as the convection remains active, and disappears rapidly when the cell breaks up. The dipoles therefore can be used not only as tracers of movement but also to study diffusion processes and turbulent mixing.

Tracers are used occasionally in studying clouds and precipitation; however, their use can be justified by the search for an increase in the level of echoes, which may be necessary for dynamic or trajectographic observations. In this connection, an interesting property of dipoles is that they are not dragged by precipitation: the average increase in the velocity of fall due to rain is very small compared to their terminal velocity of fall and may be neglected in applications relating to the observation of air movements by radar. This property is due to the significant difference in the drag forces experienced by the fraction of the dipole situated inside and outside the drop; the separation occurs after a fall of a few centimeters [List, Low, and McTaggart-Cowan, 1974]. This conclusion is valid only if the fibers are appreciably longer than the drop size, which is generally the case, and in the absence of icing.

Wire dipoles also have another application, a modification of the electric field underneath and in thunderstorms. In fact, in the presence of a significant electric field, the dipole is orientated in the direction of the field and at its ends produces an ionization by the corona effect, which modifies the conductivity of the air. The intensity of the current generated is proportional to the length of the dipole and the square of the field. It seems that the value of the field thus can be lowered sufficiently to oppose the creation of the high values necessary for discharges (except if the cloud of dipoles has much larger dimensions in the sense of the field than in the perpendicular sense; in this case, a field concentration is obtained that favors the formation of lightning) [Holitza and Kasemir, 1974; Kasemir *et al.*, 1976]. In addition, a radar receiver or a sensitive radiometer can detect at short range the microwave radiation produced by microdischarges due to the corona effect; this is not lacking in interest because the radiation intensity is proportional to the total current due to the microdischarges [Rust and Krehbiel, 1977].

# Chapter 5
# Introduction to the Study of Some Meteorological Structures by Radar

## 5.1 INTRODUCTION

### 5.1.1 Diversity of Meteorological Structures

Meteorological structures and the processes they generate occupy a remarkably broad range of scales: between a cloud droplet of a few microns and a synoptic disturbance, the space of the phenomena covers more than ten orders of magnitude. In addition, the movement of the atmospheric fluid, the mechanisms influencing the transformation of water (cf. Figure 2.1) and radiative transfers are of such complexity that the quantities describing the state of the atmosphere can be distributed in an infinite number of ways in space and time. Globally, the atmosphere is a turbulent flow whose kinetic energy is distributed according to a continuous spectrum within which the energy transfers are produced between different scales. However, without always knowing exactly why, there exist identifiable structures within this continuum. The main ones are shown in Figure 5.1 with their characteristic space and time scales.

### 5.1.2 Movements of the Atmosphere

To try to find one's way in the meteorological atmosphere, that is, to understand the relationships among the phenomena involved in different scales as well as their relative importance and to define the role of radar as a means of studying and observing the atmosphere, a simplified schematic description of the mechanical energy cycle associated with the general atmospheric circulation (Figure 5.2) can be used.[1] This approach finds some justification in the close relationships that exist between the atmospheric transfer mechanisms and energy conversion and the significant structures from a meteorological point of view observed through radar.

[1]The term *general atmospheric circulation* is used to denote, in a broad sense, the complete statistical description of the atmosphere movements around the earth.

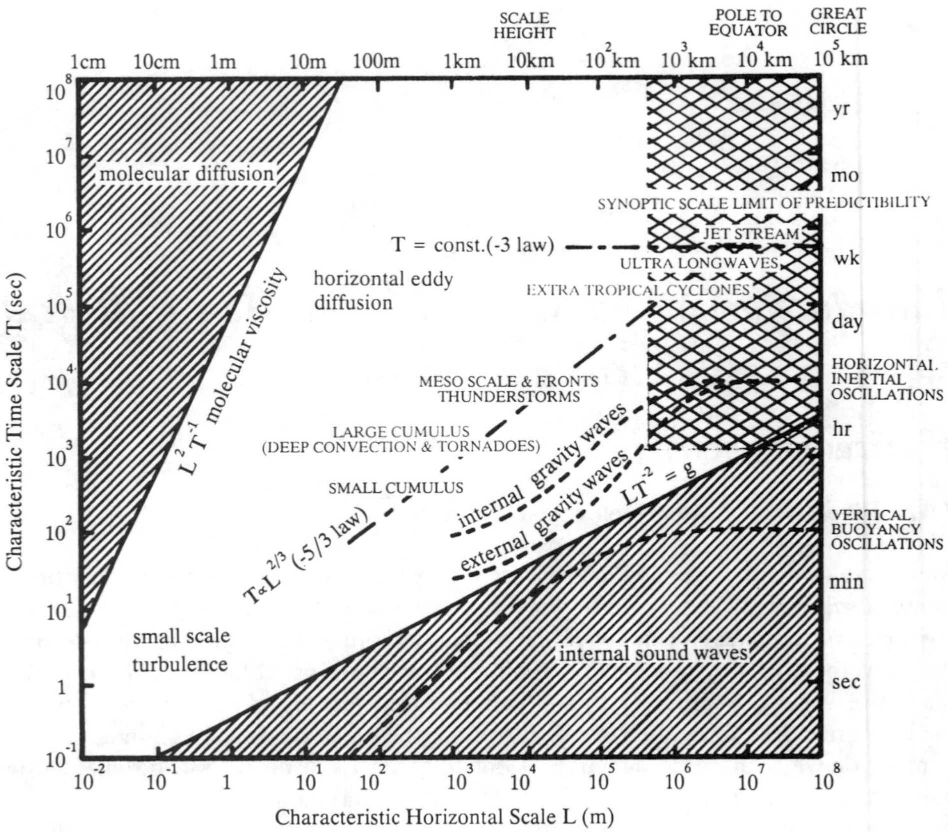


**Figure 5.1** Space-time domain of the characteristic atmospheric phenomena. The nonhatched area represents most of the kinetic energy involved, with a predominance of the extratropical cyclones, ultralong waves, and jet streams. The cross hatching marks the area occupied by the phenomena typically solved by general or macro-scale circulation models (according to [Smagorinsky, 1974]).

The origin of circulation is the solar energy flux that, converted into heat, maintains the heat sources of the atmospheric thermal machine. Vertical thermal gradients are established between the base of the atmosphere, where the main energy is brought, and the upper troposphere, which behaves as a heat sink. At the same time, the differences in heat flux as a function of latitude (differences in solar flux at ground level between the equatorial and polar regions) and also as a function of the geographical distribution of the thermal and radiative properties of the ground (tropical forests, polar ice, seas, deserts) and the atmosphere (aerosols, gases, clouds) create horizontal thermal gradients on a planetary scale in the stratified stable regions that occupy most of the volume of the atmosphere.

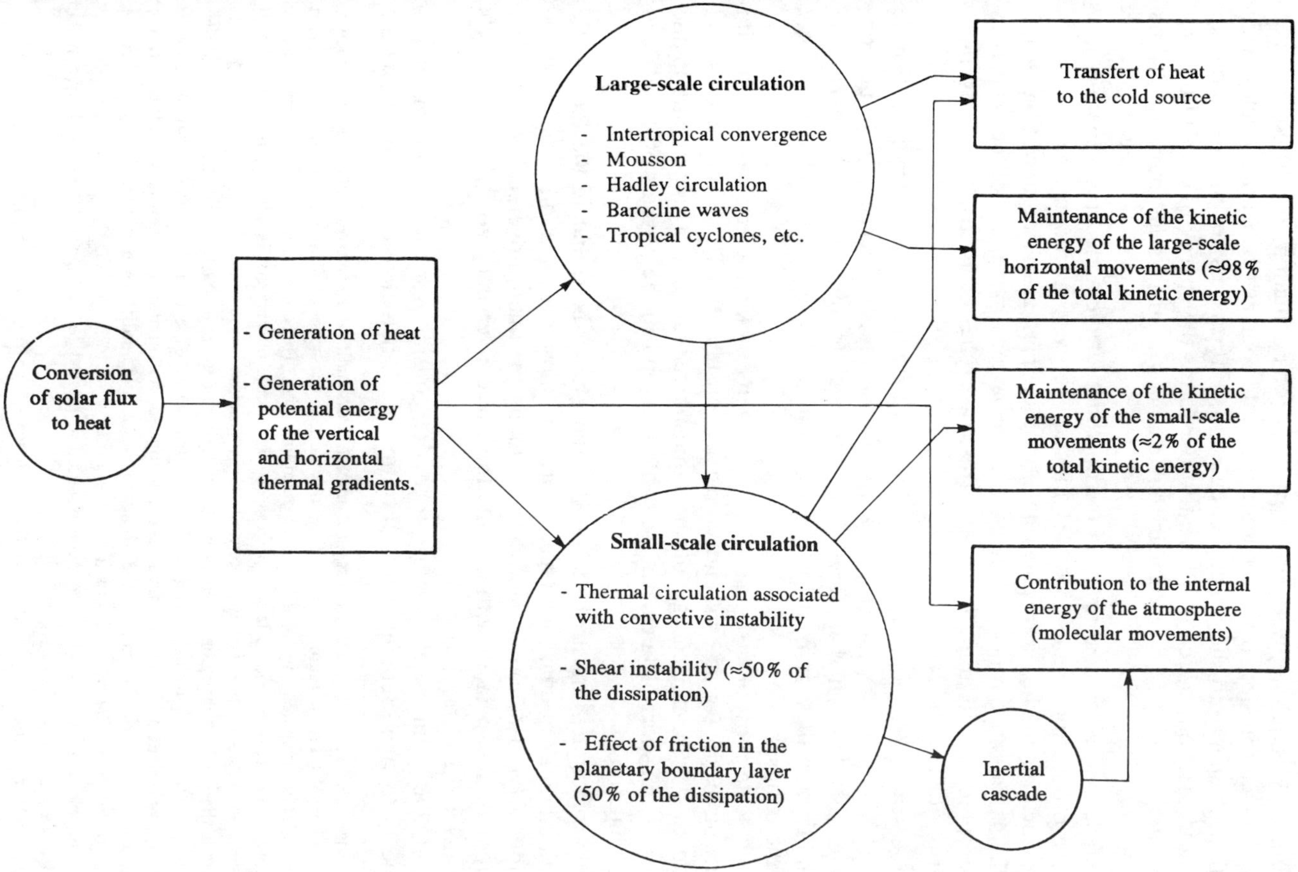


**Figure 5.2** Simplified diagram of the transfers of mechanical energy associated with the general circulation of the atmosphere. The circles surround processes, whereas the rectangles enclose their results.

The potential energy stored in these thermal gradients is continually released in the form of kinetic energy of the movement (or circulation) on different scales operating the heat transfers from hot sources to cold sources (the Carnot cycle). This contribution of energy to the general circulation compensates for the losses due to dissipation by friction, which permanently transforms the energy of the fluid movements to internal energy of molecular movements via small-scale movements and their degradation.[2]

Overall, the mechanical result of the operation of the atmospheric thermal machine is the production of the necessary kinetic energy to maintain general circulation. The rotation of the earth plays an important part in the organization of the general circulation movements (conservation of the angular movement of the air masses and geostrophic equilibrium). In broad terms, two types of atmospheric movement can be distinguished:

1. Large-scale movements associated with the horizontal thermal gradients. These are characterized by the rise of warm air and the subsidence of cold air connected by a horizontal air flow going from high to low pressure; the horizontal wind field of these movements represents more than 98% of the kinetic energy of the atmosphere.
2. Small-scale movements (less than a few tens of kilometers) associated with vertical thermal gradients and convective instability. These movements add little to the kinetic energy of the atmosphere but play a major part in the vertical transport of latent and sensible heat and in the physics of clouds and precipitation.

At low latitudes, most of the kinetic energy can be found in the large-scale thermal circulations of quasi-stationary character whose organization is related to the distribution of the heat sources and sinks. Among the main movements of this type is the circulation associated with the intertropical convergence zone,[3] and Hadley's circulation,[4] and the monsoons.[5]

At middle latitudes, the major process responsible for the conversion of potential energy to kinetic energy is the relaxation of the barocline instability,[6] by virtue

[2]The degradation or cascade of energy to smaller and smaller scale movements is a consequence of the second law of thermodynamics, according to which, in the absence of an external force, a system tends toward a state of increasing disorder (maximum entropy). The contribution to the internal energy of the atmosphere due to this degradation process of small-scale movements is very small compared to other physical processes involved in the internal energy balance.

[3]This is the convergence zone created by the meeting of the southeast and northeast trade winds.

[4]A direct, thermally driven circulation, symmetric with respect to the meteorological equator, containing on average at a low level a trade current (toward the equator) between about 30° of latitude and the equator, an ascending current in the tropical zone, a poleward flow in altitude, and a subsidence at about 30° of latitude. The whole forms a "Hadley cell."

[5]The monsoon is a seasonal large-scale circulation related to the thermal contrast between the land and the sea.

[6]Hydrodynamic instability created by the existence of a strong horizontal temperature gradient and a wind shear in a statically stable atmosphere.

of which the planetary scale horizontal thermal gradients break into large-scale "waves" associated with extratropical cyclone disturbances (Section 5.5). Among the large-scale mechanisms, a particular case is that of certain thermal circulations developing above warm seas, such as the tropical cyclones[7] (cf. Section 5.6). The potential energy of the vertical gradients is liberated by the convective instability that develops locally when the conditions are favorable, independent of the large-scale structures or within these structures. It is directly converted to kinetic energy of small-scale movements occupying a spectrum ranging from microturbulence to the scale of a storm (Sections 5.2, 5.3, and 5.4). Finally, some small-scale movements take their energy from the degradation of large-scale movements by different processes. These include the vertical disturbances of the horizontal flow due to orography or shearing instability (Section 5.7). These mechanisms are particularly active in the planetary boundary layer and the stably stratified flows in the presence of wind shear.

*A priori,* such a multiplicity of structures and movements dictates more of a statistical approach than a deterministic approach. However, the main practical problems or applications relating to the meteorological atmosphere depend on the knowledge and forecasting of medium- and small-scale structures.

### 5.1.3 The Area of Radar Application

The problems in which radar is involved or can be involved are essentially related to three concerns:

1. The acquisition of new knowledge on phenomena or atmospheric structures;
2. The systematic measurement over the entire atmospheric space of various useful physical quantities for operational meteorological forecasting models (current or future): three-dimensional velocity, reflectivity, refractive index field, which are best tackled by arrays of doppler radar;
3. Detailed short-term forecasting of local meteorological evolution (visibility, wind, rain, hail, tornado, lightning) by extrapolation after a few tens of minutes to a few hours of direct observation as well as immediate control of local situations for the benefit of a large number of different sectors of activity: aeronautics, agriculture, and telecommunications, to name a few.

Current knowledge far from covers the phenomena accessible to radars and their applications. Therefore considerable work is to be done in this area. However, because of their capability and the variety of problems to be solved, conventional or doppler radar, used singly or in network, have been involved in a large number

[7]This mechanism is the convective instability of the second kind. Contrary to the barocline waves that are supplied from already existing horizontal temperature gradients, tropical cyclones create their own warm zone by absorbing the latent and sensible heat coming from the ocean surface and transforming the latent heat to sensible heat in the convective regions where precipitation is developed.

of evaluations and research for basic and applied operations covering practically all aspects and for all types of structure, from large-scale circulations to the fine structure of clouds and precipitation. A correspondingly extremely large number of publications involve radar. Therefore, it is hardly possible, and certainly of little use, to collect all the results concerning such a vast domain in one book.

In this chapter, we limit ourselves to an introductory presentation of significant meteorological structures, which have been covered by the most important studies. Each type can be found in the different parts of Figure 5.2. For each type of phenomenon we indicate only what the radar sees without making any explicit reference to the large number of specific problems that can be handled from the information obtained. The main references are given.

## 5.2 CONVECTION IN THE PLANETARY BOUNDARY LAYER

The possibility of detecting turbulent variations in the refractive index of air by radar as well as by using artificial tracers (dipoles) has found an important application in the study of the dynamic processes of the boundary layer in clear air.

### 5.2.1 The Convective Boundary Layer

Diurnal heating at ground level by solar radiation when the atmosphere is clear results in the development of a convective layer at a low level above the ground that thickens during the course of the day. In fact, this layer consists of three distinct regions, identified by the thermodynamic properties of the air and their effect on convective movement.

Immediately above the ground is a layer with a thickness of a few tens of meters, in which the decrease in temperature is greater than the dry adiabat; therefore, it is in a state of convective instability. The temperature and humidity "pulses," which mark the nascent convective elements, are manifested as positive fluctuations; the ascending air is warmer and more humid than that which surrounds it. The deviations have a strong rapidly varying amplitude; the convective activity has a turbulent and diffuse character.

Above this is a thicker region in which the vertical decrease in temperature is moderately stable. The temperature pulses from the superadiabatic layer are still positive at the bottom of this layer; the convective elements receive a vertical acceleration proportional to the temperature difference between the cell and the surrounding air. During its upward movement, the convection manifests a more and more marked cellular organization, the elements reducing in number but increasing in size and persistence; the cells are bounded by a "cutoff" characteristic of the average value of the thermodynamic quantities and the refractive index at the transition of the boundary with the external air (called a variation of the *top hat* type). However, because of the cooling due to the expansion of the ascending air and the entrainment of environmental air in the cell, the amplitude of the temperature difference with its surrounds decreases with height until it becomes indistinguishable, close to the level of hydrostatic equilibrium. Above this, the temperature pulse becomes negative and its amplitude increases, whereas the ascending current trails off until the amount of vertical movement of the cell is cancelled (phenomenon of overshoot of the hydrostatic equilibrium). The air of cells, which is warm and humid below the equilibrium level, is cold and humid above it.

Finally, at the top of this region is a layer in which the surrounding air has a stable vertical temperature gradient that marks the limit of the convective field; the amplitude of the temperature and humidity fluctuations is large here. The boundary between the air of the cells and the relatively calm overlying air is a turbulent mixing zone at the level of which the convection erodes the upper layer (entrainment of air from the upper layer increases the thickness of the convective layer).

### 5.2.2 Observation of the Convective Field

When the winds are sufficiently weak, observation by highly sensitive radar, thanks to the fluctuations of water vapor, reveals quite clearly, in this convective boundary layer at up to the more than 50 km range, thermal cells of horizontal dimensions lying between 1 and 3 km, several hundreds of meters in height, with very variable persistence, sometimes reaching 30 min [Hardy and Ottersten, 1969; Konrad, 1970, Harrold and Browning, 1971]. The echoes are produced by variations in the refractive index created by the turbulence in the transition zone between the air transported by the ascending convective movement and the surrounding air with different properties into which the cell penetrates. The echoes therefore are more intense at the boundaries of the cell and particularly at the vertex; the almost continuous wavy line formed by the meeting of echoes produced by different adjacent cells marks the upper limit of the convective layer. When the cells are separated, the echo of their vertex takes on a characteristic inverted U shape. The movement of air in these cells is a convective circulation of the Rayleigh-Benard type with a central ascending flow. As with the cumulus fields, such structures can be observed throughout the year over land or sea; they are distributed either in a random fashion in a weak wind field or in parallel alignments in the presence of vertical wind shear. They move with the average winds at the level of the convective layer. The ascendant vertical velocities increase during the progression of the cells toward the vertex of the convective layer, where their value can reach 1.5 m $s^{-1}$, until the ascending air crosses its equilibrium level with respect to the surrounding air. Overall, the convective layer in clear air consisting of the entire population of cells has different average properties from those of cells taken individually and in particular its average rate of thickening above the ground does not exceed 4 to 5 m $min^{-1}$.

The upper limit of the convective layer in clear air, as can be observed by radar, sometimes has variations in thickness of a few hundreds of meters, which identify zones of higher convective activity due to local effects (overheating area, orography). If the conditions are favorable for subsequent development of a thicker cloud convection, these zones of higher convective activity may become source areas of cumulus congestus and cumulonimbus. It therefore seems worth locating them for the short-term prediction of certain types of local storms [Harrold and Browning, 1971].

Occasionally, the spatial distribution of the convective cells makes it possible to identify different types of meso-scale structures such as transversal or longitudinal alignments and subsident regions limited by convective zones, or local phenomena such as coastal circulation. For example, convection lines are observed in clear air associated with sea or land breezes [Meyer, 1971]. This circulation is due to the contrast in temperature between the air situated over land and that over the sea due to the differences in thermal inertia between the ground and water. A horizontal pressure gradient results perpendicular to the coast as well as a closed circulation containing ascending and subsiding lines.

Observation of the planetary boundary layer in clear air by radar is also possible by using artificial tracers (Section 4.6), when the fluctuations in the refractive index of air are not detected by the radar. This technique and the results obtained from it have been discussed in Section 4.6.2. We saw that dipoles can be used to define a natural or artificial convective volume of air as long as the cell concerned remains active [Rowland, 1976; Sauvageot *et al.,* 1982]. Dropped at a large number of points, situated, for example, in a horizontal plane within a convective boundary layer, the tracers first spread out following a distribution strongly influenced by the instantaneous arrangement of the cells. However, the dipoles tend to occupy all the space and after times much longer than the duration of the average life of the convective elements, approximately homogeneous distributions are obtained. These return echoes have the advantages of spatial continuity and high intensity; they lend themselves well to the observation of dynamic fields by doppler radar and the identification of particular structures [for examples, see Kropfli and Kohn, 1978; Doviak and Berger, 1980].

Figure 5.3 shows an example relating to a tropical convective boundary layer observed by a system of two doppler radars during the COPT experiment in the north of the Ivory Coast by Eymard and Weill [1988]. The tracer used is chaff dispersed by a light airplane flying at a height of 600 m. The study made it possible to analyze the structure of the boundary layer (which appears similar to that observed at moderate latitudes) and its relationship with the flow on a larger scale.

Convection in clear air can also be detected at very short ranges by acoustic pulse sounders or sodars (Little, 1969; Weill *et al.,* 1980). Simultaneous observation by sodar and radar provides a complementary description of the phenomena in the sense that, in the observation of convection in the lower layers of the atmosphere,

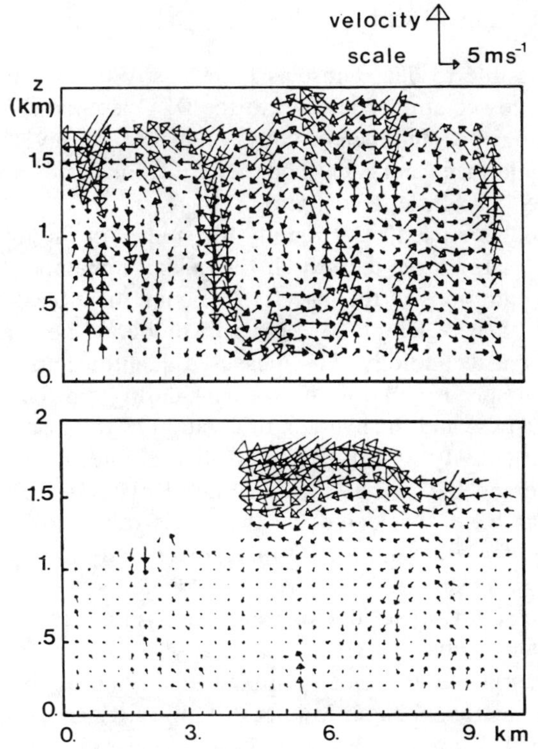


**Figure 5.3** Vertical cross sections of the wind field in a tropical convective boundary layer observed by a system of two doppler radars on 24 May 1981. The wind tracer is chaff. $Z_i$ is the temperature inversion height. The vertical scale of distance and velocity are multiplied by a factor of 3 (from [Eymard and Weill, 1988]).

radar is sensitive to the fluctuations in humidity, whereas monostatic sodar responds to fluctuations in temperature [Eymard and Weill, 1982]. The observations on small-scale cells in the convective boundary layer can be summarized as follows:

- The top of the convective field, where strong contrasts between temperature and humidity are found, is seen by the radar and by the sodar;
- The lower part of the convective field is observed differently: the doppler acoustic sounder sees the plumes where the temperature fluctuations are manifested. These plumes display the ascending flow, which is warmer than the environmental air that makes up the central part of the convective cell; the presence or absence of radar echoes in this region depends on the strength of the humidity fluctuations found there;
- At the lateral limits of the cells, the mixed air of the top is transported to the lower levels; the echoes are weaker there.

### 5.2.3 The Aerobiological Field

Radar makes it possible to detect insects quite easily (cf. Section 4.5). These are omnipresent in the lower atmosphere up to the 0° C isotherm. This detection is of interest to aerobiology (or entometeorology) and meteorology, two fields that have a large number of interfaces and cannot but benefit from a multidisciplinary approach (which till now has been little developed).

Entomologists have used radar mainly for studies of the migration of dynamic insects of relatively large size, the search for radar "signatures," and evaluation of the numerical concentrations of the insects. The equipment used often involves non-coherent radar with a wavelength of 3 cm and moderate power. The results show the efficiency of radar technology. The measured quantities are the distribution over time and space, the speed of the group by considering the displacement of the reflectivity characteristics, and the average orientation of the insects deduced from the variation in the reflectivity as a function of the azimuth (Schaeffer, 1976; Drake *et al.,* 1981; Drake and Farrow, 1983; Riley and Reynolds, 1983; among others).

Meteorologists have considered the insects as tracers (cf. Section 4.5). Contrary to the dipoles, these are not totally inert tracers but are naturally present in the lower layers of the atmosphere. They are of undeniable operational interest because of their quantitative significance (thickness of the layer concerned) and the durations over which they can be observed.

Their distribution often is closely linked to the thermodynamic conditions of the lower layers, in particular to atmospheric stability [Lhermitte, 1966; Campistron and Sauvageot, 1974; Campistron, 1975; Takeda and Murabayashi, 1981; Mueller and Larkin, 1985; among others]. The main types of distribution are

- Homogeneous layer. The echoes decrease in number and power as a function of height. The vertical distribution varies slowly over time. This organization is related to a homogeneous convection of weak intensity in the terrestrial boundary layer.
- Stratiform layers. The vertical distribution of the number and the power of the echoes is maximal at different levels, forming horizontal thin sheets varying between 100 and several hundreds of meters of thickness. These stratifications sometimes last for several hours; they are situated at the limit of stable levels (inversion or isothermy) and always above the convection layer.
- Convective structures. The echoes sometimes are assembled in "packets" or cells distributed in an irregular manner. Frequently, their base is at ground level and their top corresponds to that of the convective layer; these structures are related to the convective zones and display the organization of dry phase convection. Echoes are strong in the ascending regions (air rich in insects coming from the ground) and weak in descending regions (for the opposite reason).
- Clear zones. Large areas producing few echoes of insects are sometimes observed in the scattering layer. These zones correspond to the location of small

cumuli (humulis or mediocris) scattering very weakly (diameter of the droplets $< 20\ \mu$m) [Campistron and Sauvageot, 1974].

The diurnal evolution of the thermodynamic conditions in the lower layers of the atmosphere creates relatively complex and evolving echo distributions that involve several characteristics of the structures just defined at the same time. Figure 5.4 demonstrates this. The observation was made at fixed vertical elevation (HTI) with a 0.86 cm wavelength radar in a flat region in summer in clear air. It represents a typical case of the thickening of a convective layer above the ground (Figure 5.5).

We conclude this section by stating that radar offers a considerable potential for entomological research and for "entometeorology."

### 5.2.4 Pollution and Plumes

In general, atmospheric pollution consists of particles too small to be detected from their microwave backscattering. However, radar can make an important contribution to the problems of treating pollution by virtue of the information it provides on the state of the environment. This can be coupled with short-term forecasting, treated in Chapter 2, concerning the evolution of the boundary layer, the dynamics of which govern the dispersion of the pollutants and the measurement of the precipitations involved in the atmospheric "washing" process. Thus, doppler radar lets us analyze the flow dynamics at a polluted site such as the island effect of a large town [Kropfli and Kohn, 1978].

When the pollution is in the form of a cloud of particles of sufficient size, direct detection by radar is possible. In this case, the concepts developed for natural clouds can be used. Gaseous, particulate or thermal pollution often is generated at a point and develops in the form of plumes in which diffusion is the important "parameter." Various studies have been carried out on this subject, based on the dispersion of chaff tracked by radar in a nonperturbed natural medium [Hildebrand, 1977; Moninger and Kropfli, 1982].

Local atmospheric perturbations due to massive emissions of heat into the atmosphere have also been studied. The dynamics of dry thermal plumes and the surrounding flow have been observed by radar using chaff thrown out at the base and around the ascending column [Sauvageot *et al.*, 1982]. In particular, the chaff lets us precisely determine the contours and the height at which the perturbation culminates. With radar we can prove the existence of snow induced by the emission of heat and humidity from plumes of humid cooling towers from nuclear power stations in perturbed atmosphere over ranges greater than 20 km leeward of the towers (Figure 5.6) [Sauvageot, 1987]. It has also been shown that, in the absence of natural hydrometeors, radar makes it possible to measure saline drizzle present in the trails [Sauvageot, 1989].

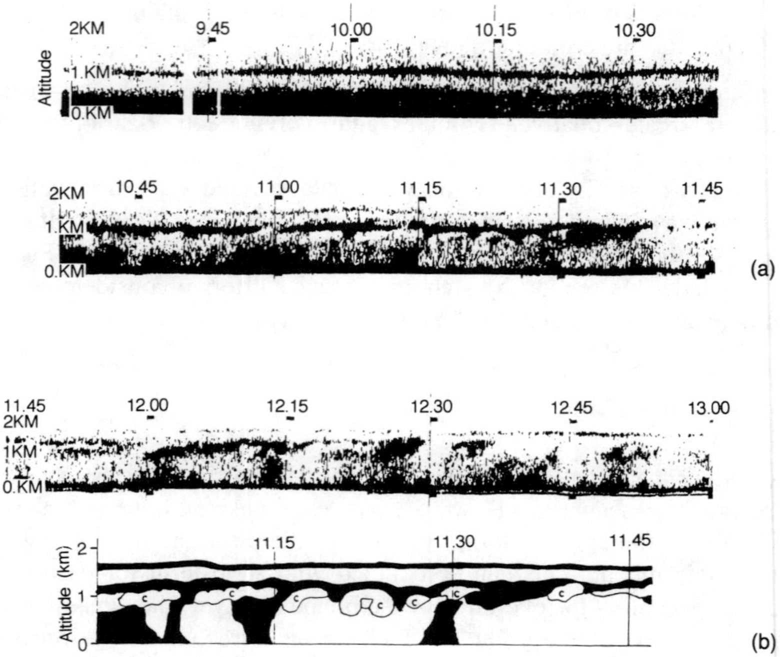


**Figure 5.4** Height-time record of insect echoes on 23 June 1971 obtained with a 0.8 cm radar. (a) The organization of echoes is characterized at 0900 UT (universal time) by a continuous layer between the ground and approximately 600 m and by a stratification towards 1100 m, at the level of a stable layer situated between 1000 and 1350 m below a temperature inversion (between 1350 and 1500 m). A few point echoes can be seen between these two regions and up to 1800 m. Then, the lower layer thickens out, and the stratification is reinforced. The structure of the lower layer then becomes more irregular, breaks up, and forms upward protuberances. After 1045 UT, the echo stratification at 1100 m disintegrates. It shows outgrowths pointing downwards. After 1215 UT, only "packets of echoes" are left at the stratification level and in the interval that separates it from the lower level at the beginning of the observation. In the same period, clear regions are found between 800 and 1200 m containing few or no echoes. (b) A diagramatic representation given to facilitate referencing. The letter *C* marks cumulus clouds. Photographic observations and in situ by aircraft show that these clear spots coincide very precisely with the cumulus; the distribution of echoes carries the imprint of condensed zones. When the cumulus reaches the inversion, which itself is a layer containing few echoes and spread out horizontally, its upper limit disappears and large clear spaces are created. After 1045 UT, the echoes situated above the main stratification reassemble and form a new stratification that lasts without undergoing any significant disturbances until 1300 UT. This stratification is found just at the top of the inversion and it defines its upper limits (from [Campistron and Sauvageot, 1974]).

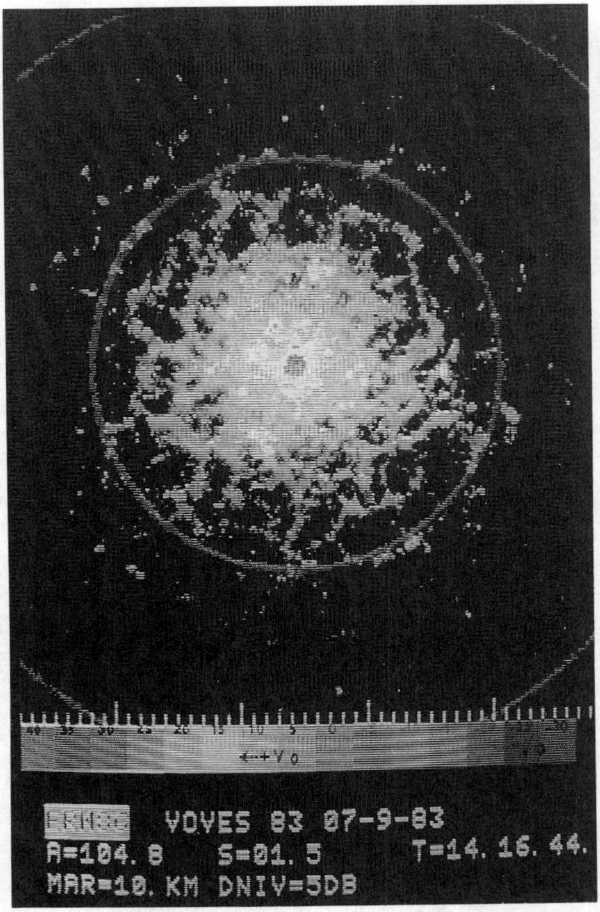


**Figure 5.5** Panoramic (PPI) display of $Z_e$ (in dB $Z_e$) for a convective boundary layer in clear air observed with a 0.86 cm wavelength radar on 7 September 1983 at 1416 UT above a flat ground about 100 km south of Paris. Elevation is 1.5°. Range markers are at 10 km intervals. $Z_e$ of echoes is between 10 and −20 dBZ.

## 5.3 DEEP CONVECTION AND THUNDERSTORMS

The formation of convective clouds of considerable vertical extent result from situations favoring first of all the storage of latent energy in the atmosphere and then the sudden liberation (or relaxation) of this energy in a few principal areas. These clouds of the cumulus and cumulonimbus type are isolated or grouped together according to various configurations in the cold fronts of extratropical disturbances, in squall lines, tropical clusters, and the band structures of cyclones. They are accompanied by intense and localized precipitation.

**Figure 5.6** Radar reflectivity factor for a cooling tower plume observed on 9 February 1983 at 0824 UT near Dampierre-en-Burly (France): (a) longitudinal section; (b) transversal section at $X = 11.5$ km. Cooling towers are at $X = 0$. The wind is from the northeast. The plume generates a precipitation of snow reaching the ground between 10 and 20 km, markers ring in the azimuth 230° (from [Sauvageot, 1987]).

Considerable research has gone into studying deep convective systems because of their importance for human activity and their large-scale circulation. Several reviews of the very large numbers of articles on this subject have been published [in particular, Kessler, 1983, in 3 volumes; and Ray, 1986]. We therefore confine ourselves to a summary.

### 5.3.1 The Convective Cells

The description of the structures of deep convective clouds, in particular those of severe storms, are based mainly on the concept of the cell[8] as a dynamic entity, characterized by a specific convective current that has associated with it a precipitation structure also called a *cell*. The latter is the main element in the characterization of the structure by radar.[9] A distinction is made between two types of cells: ordinary cells and supercells. The ordinary cells are the basic elements of most deep convective clouds. They are characterized by a process consisting of three stages [Byers and Braham, 1949]:

1. A first stage of cumuliform growth marked by the establishment and intensification of the updraft accompanied by the rapid increase in height of the cloud top and echo intensity;
2. A maturity stage with the coexistence within the same cloud of a specific updraft and a specific downdraft[10] in the presence of strong and localized precipitation;
3. Finally, a decaying stage during which only a downdraft exists with precipitation tending to generalize to the entire cell while its intensity falls off.

Ordinary cells drift horizontally approximately at the velocity of average winds at their level. Their lifetime is a few tens of minutes. A main factor governing the duration of the phenomenon is the size of the cell; in fact, the entrainment of the

---

[8]Generally speaking, the concept of cell, always tied to that of convection, covers a broad range of phenomena (in size and lifetime). This varies from the small precipitation generator characteristic of stratiform clouds (cf. Section 5.5) with a size of a few hundred meters for precipitation intensities of a few mm $h^{-1}$ produced by a thin ascending convection, to the "deep convection" cell contained in severe storms. Convection cells on different scales are also observed in clear air (cf. Section 5.2).

[9]Most of our knowledge relating to the structure of the systems has been obtained by using radar as the main means of observation and measurement by referring to the organization of the precipitation. The precipitation in fact is a very significant meteorological quantity; it represents the culmination of the combined action of microphysical and dynamic processes acting in the clouds and is quantitatively related to the transfers of latent heat.

[10]Several causes combine to create downdrafts; the main ones are the entrainment of air by precipitation, the evaporation of precipitation in regions where the cold and dry air of the middle and upper levels erodes the sides of the cloud because of a shear of the horizontal wind in terms of velocity and direction (this evaporation cools the air locally and therefore increases its density), and the presence of fluctuations in pressure of dynamic origin close to the sides of the cloud because the cell behaves as an obstacle to atmospheric flow [see, for example, Ramond, 1978].

environmental air by the updraft that normally acts as a brake against the convection (the mixture reduces the temperature contrast between the external air and the air inside the cloud and reduces the humidity of the latter) becomes less significant as the horizontal cross section of the cloud increases and the external area-volume ratio is smaller. More generally speaking, the ordinary cell dies because the convective circulation causes vertical transfers that tend to cancel the initial instability of the air or, in other words, because the ordinary cell evolves in a closed system.

Sometimes, in the presence of favorable conditions, storm cells evolve to a more perfected organization, that of the supercell [Browning and Ludlam, 1962; Browning, 1964]. These conditions lead to an explosive development of the convection and the creation in a giant cell of an intense three-dimensional circulation, which in crude terms consists of an updraft and a downdraft (Figure 5.7). This organization is maintained in an apparently quasi-stationary form for durations that can last more than an hour (even occasionally several hours), during which the storm covers considerable distances at a velocity close to that of moderate winds, producing intense precipitation of water and hail and strong gusts of wind. The supercells often

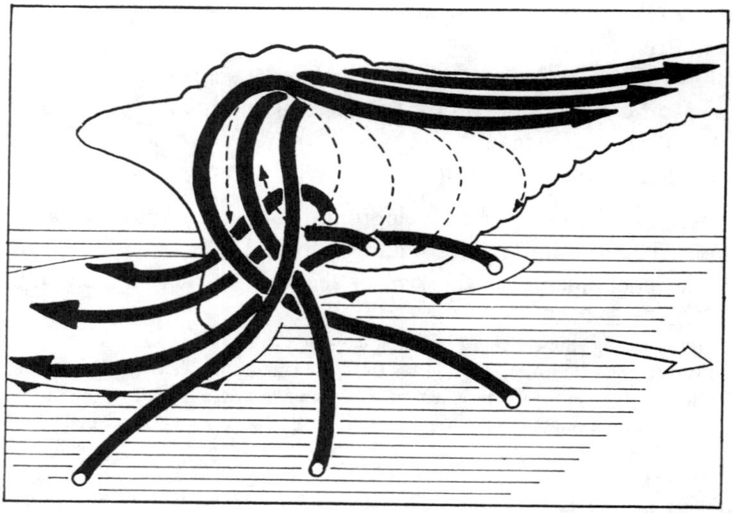

**Figure 5.7** Three-dimensional circulation of the air in a supercell with respect to a reference moving with the cloud. The ascending air, converging close to the surface, enters the front right-hand-side sector (with respect to the direction of the average horizontal movement of the cloud) and leaves the cell at an altitude flowing into the anvil, at the front, following the direction of the high-level winds. The origin of the descending current is an airflow entering on the right-hand side of the cell, passing around the ascending air, and then diverging close to the ground where it creates a secondary cold front. The dotted curves represent the trajectory of precipitation particles (according to [Browning, 1964]).

have special characteristics making it possible for them to be identified by radar if the observation resolution is sufficient (Section 5.3.2).

The quasi-stationary nature of the circulation of the supercell is the subject of controversy and must be understood as an approximation for the overall properties of the cell such as they appear when observed with equipment whose resolution is limited. The supercell has a finer structure, which is not stationary, as is shown by observation of the fluctuations over space and time of the precipitation that develops there [Changnon, 1977]. However, the persistence of the average properties (size, velocity, radar structure) over a long period of time is an essential difference compared to ordinary cells, which are not stationary. This difference is because the supercell is an open system in which a continuous convective circulation is maintained (Figure 5.8). The instability responsible for these movements is regenerated by a change in the air crossing the cell vertically. The conditions favorable for this circulation result in particular from the organization of horizontal movements of the atmosphere, which are on a larger scale than that of the cell and favor its supply, and its progression to zones where the air is potentially unstable.

The horizontal movement of the supercells often presents significant deviations from the average direction of the winds in the troposphere. In the northern hemisphere, most cells drift to the right of the mean winds and more rarely to the left.

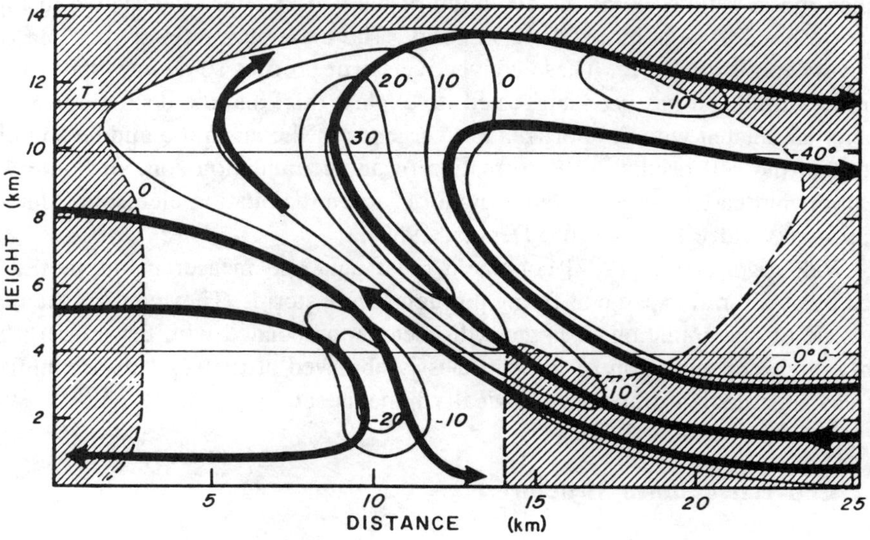


**Figure 5.8** Streamlines relative to the storm and isopleths of vertical velocity (m $s^{-1}$) within a vertical section along the direction of movement of a supercell hailstorm. The storm motion is from left to right. The unshaded area corresponds to radar reflectivity factor in excess of 30 dBZ (from [Browning and Ludlam, 1962]).

The opposite occurs in the southern hemisphere. Occasionally, a supercell can break up into two supercells, one drifting to the right and the other to the left [Fujita and Grandoso, 1968]. The causes of such deviations have not been completely elucidated; for one part at least, they are probably due to an interaction in the proximity of the ground between the region of cold descending and divergent air associated with the precipitation zone of the cell and the convergent current supplying the updraft; in fact, the downdrafts have an effect on the delimitation of the zone of warm and humid air pumped by the updraft and create a violent thermal contrast that increases the instability appreciably in the zone where the two currents meet.

The leading edge of the horizontal flow of air resulting from the spreading of the downdraft above the ground is called the *gust front*. It is observed in all storms when the downdraft has a significant intensity but it is particularly violent and sometimes destructive in supercell storms and in storms organized in line. The temperature gradient at the transition between the environmental warm and humid air and the almost saturated cold air from the storm can be above 10° C over some tens of meters and can affect the propagation of the radar beam (quasi-specular reflection, superrefraction).

Figure 5.9 shows the typical vertical profiles of the radar reflectivity in the cellular part of cyclone systems and in the precipitation of convective storm cells. The profiles relating to severe storms obviously vary considerably from one cell to another and as a function of the evolution of the cell. The profiles shown in Figure 5.9 correspond to cells in a state of maturity. The maximum reflectivities can reach and overstep 60 dB*Z* here. These convective storm profiles sometimes have a maximum in altitude. It has been suggested, to explain this, that the decrease in vertical velocity associated with the horizontal divergence of the air in the middle and upper regions of the cell results in the formation of an accumulation zone of liquid water that in certain cases can favor the growth of large precipitation elements [Marshall, 1953; Sulakvelidze *et al.*, 1965; Dennis, 1980].

Waldvogel *et al.* [1987] compared radar data and measurements in situ collected by an aircraft capable of flying through severe storms. They confirm the reality of a zone of accumulation of large hydrometeors associated with the updraft; however, a high concentration of ice particles is observed at the level of accumulation whereas supercooled drops are practically nonexistent.

### 5.3.2 Convective Storm Structure

*Medium Scale*

The properties and structure of severe convective storms (duration, intensity, nature, and spatio-temporal distribution of the precipitation) essentially depend on the type

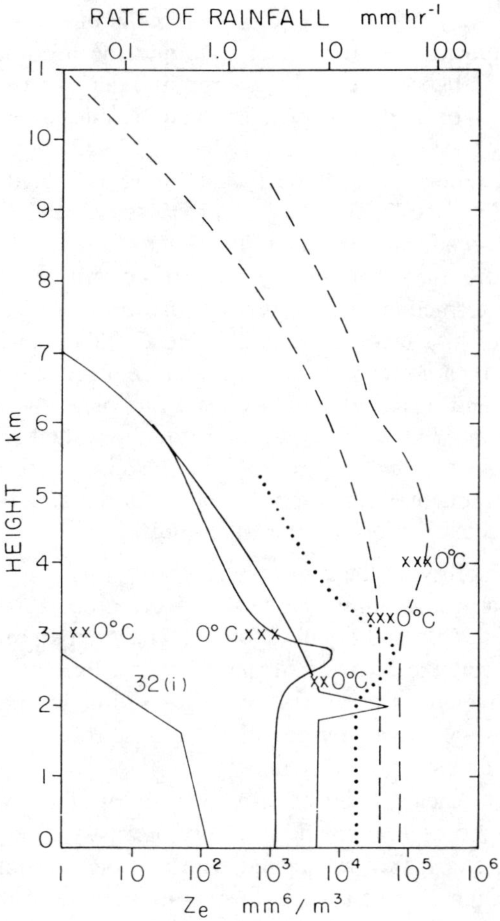


**Figure 5.9** Variation of equivalent radar reflectivity factor ($Z$) with height. Full lines = widespread cyclonic precipitation. Dotted line = cells embedded in widespread rain. Dashed line = heavy showers or thunderstorms. The rate of rainfall scale assumes $Z = 200\ R^{1.6}$ and is approximately applicable beneath the 0° C level on each profile (from [Harrold and Austin, 1974]).

of cells they contain: they may have one or more cells of a single type or a mixture of the two; this composition evolves over time and the lifetime of the storm. The structure of storms has been covered by a large number of case studies by conventional radars and single or dual doppler radars and includes attempts at classification [Browning and Ludlam, 1962; Atlas, 1963; Marwitz, 1972; Newton and Fankhauser, 1975; Lhermitte and Gilet, 1975; Kropfli and Miller, 1976; Browning and Foote, 1976; Foote and Knight, 1977; Brandes, 1977; Kessler, 1983; among others], (see also Section 5.4 on tornadoes). However, for severe storms, as for cells, the delimitation among the different types is very clear in certain cases and very vague in others; there exist intermediate structures. Furthermore, the classification criteria adopted are imperfect and incomplete and liable to subclassification. The organization and evolution of a convective storm are closely related to the corresponding particular conditions and differ appreciably from one place to another. For the sake of convenience, storms with ordinary cells are distinguished from storms with supercells. They are either isolated or grouped in clusters, possibly in line.

The main environmental conditions involved in the structure of severe storms and favoring their intensification are the following:

1. Convective instability of the atmosphere.
2. Blocking of warm and humid air between the ground and a stable layer not very high up with cold and dry air above it. The strength of the storm is directly related to the available quantity of humidity at a low level in which energy is released by convection (latent heat of condensation and possibly latent heat of freezing). The stable layer, by delaying the onset of convection, promotes the accumulation of this "instability energy."
3. A strong vertical shear of the horizontal velocity of the wind. Ordinary cells with a low degree of organization usually are associated with a weak shear, whereas severe storms are associated with a strong vertical shear of the horizontal wind. In the case of severe storms, an increase in the shear of the wind direction is observed at low levels, creating a countercurrent opposite to that of the movement of the entire cell, which ensures the supply of humid air to the updraft (Figure 5.8).
4. The nature of the dynamic mechanisms involved in the triggering of the convection; that is, the liberation of energy over a limited extent.

The storms formed from a single ordinary cell have a short lifetime. This is frequently the case of orographic storms developing in a mass of stagnant air in the absence of an intense and sheared wind field.

The most common storm is formed by a system of several ordinary cells. Analyzed by radar under adequate resolution conditions, it contains two to four cells at any given time. Its evolution results from the combination of individual cells it contains; each one passes through the same phases with a certain time shift with respect to each other. The cells therefore have a shorter lifetime than that of the entire system (Figure 5.10). Consequently, particularly during the growth and maturity phase, the cloud activity has variations with a short period (similar to pulses) that are reflected by the spatio-temporal fluctuations of the precipitation intensity. The new cell is formed next to the existing cloud with which it tends to merge during its development, becoming its center while the previously formed cells gradually decay. The period for new cells to form is a few minutes up to about 15 min; they can be identified by radar for several tens of minutes (on the order of 30 min) and up to several tens can be counted during the life of a storm. During the growth phase of the cell, the maximum echo is situated in the upper part of the cloud associated with a powerful ascending current (20–30 m $s^{-1}$). It intensifies and extends downward; then, the maximum echo zone descends while its value decreases and the cell spreads out and becomes uniform. However, the progression of the convective echoes toward the bottom of the cell should not be interpreted simply in terms of the trajectory of the precipitation; in particular, the velocity of the progression of the echo can be higher than the velocity of fall of the precipitation because, the particles continuing to grow during their fall, the reflectivity of the cell evolves simultaneously at all levels.

Often, new elements are formed on the right-hand side of the storm with respect to the direction of movement whereas the older elements decay on the left-hand side, which helps shift the trajectory of the entire system to the right of the moderate winds [Browning and Ludlam, 1962]. In general, the velocity of the entire cloud is slower than that of the cells it contains, which corresponds approximately to the average wind at the level of their ascending currents.

The most common supercell cloud (that which deviates to the right of the moderate winds), seen from the radar in horizontal section, presents a circular or elliptical shape with a dimension of approximately 10–30 km and an extension in altitude in the form of an anvil stretching over a distance from 100 to a few hundred kilometers on the lee side of the cell. An important radar characteristic of storm convection supercells is the presence of a region of weak echoes; this marks the location of an ascending current in which the intensity is such (the vertical component can reach 40 m $s^{-1}$) that the scattering particles are driven away from it. This weak echo region is surmounted by a vault or overhang of appreciably stronger echoes and surrounded, either totally or partly, by precipitation. The anvil, the weak echo region, and the vault partly result from the coupling between the updraft passing vertically through the cell and the strong horizontal current found at altitude [Dessens, 1960; Ludlam, 1980; Marwitz, 1972; Browning and Foote, 1976].

At low levels, the strong echoes created by the precipitation are situated at the rear of the cell with possibly an extension in the form of a hook. The vault marking the updraft is situated in the front part of the cell up to a height corresponding to one-half to two-thirds the total height of the cloud and a horizontal dimension of a

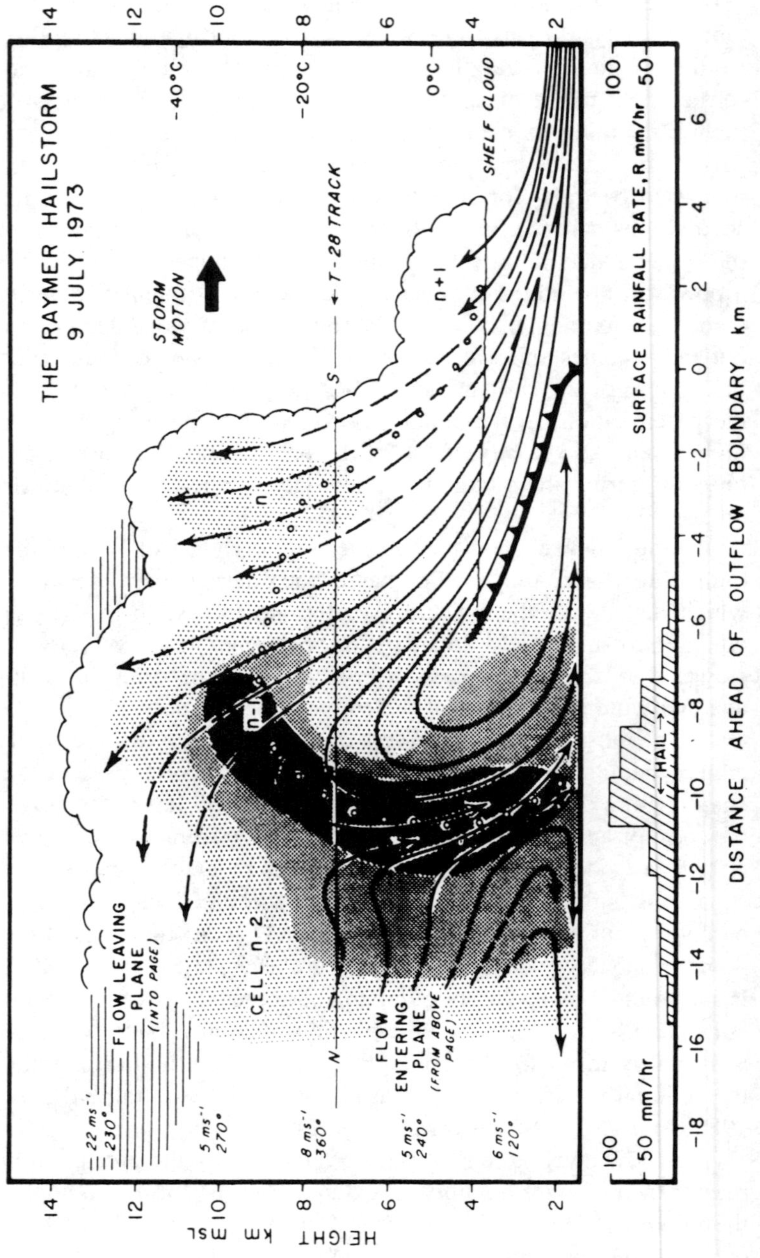
THE RAYMER HAILSTORM
9 JULY 1973
STORM MOTION
T-28 TRACK
SHELF CLOUD
-40°C
-20°C
0°C
CELL n-2
n-1
n
n+1
FLOW LEAVING PLANE (INTO PAGE)
FLOW ENTERING PLANE (FROM ABOVE PAGE)
22 ms⁻¹ 230°
5 ms⁻¹ 270°
8 ms⁻¹ 360°
5 ms⁻¹ 240°
6 ms⁻¹ 120°
N
S
HAIL
HEIGHT km msl
SURFACE RAINFALL RATE, R mm/hr
50 mm/hr
100
DISTANCE AHEAD OF OUTFLOW BOUNDARY km
-18
-16
-14
-12
-10
-8
-6
-4
-2
0
2
4
6
14
12
10
8
6
4
2

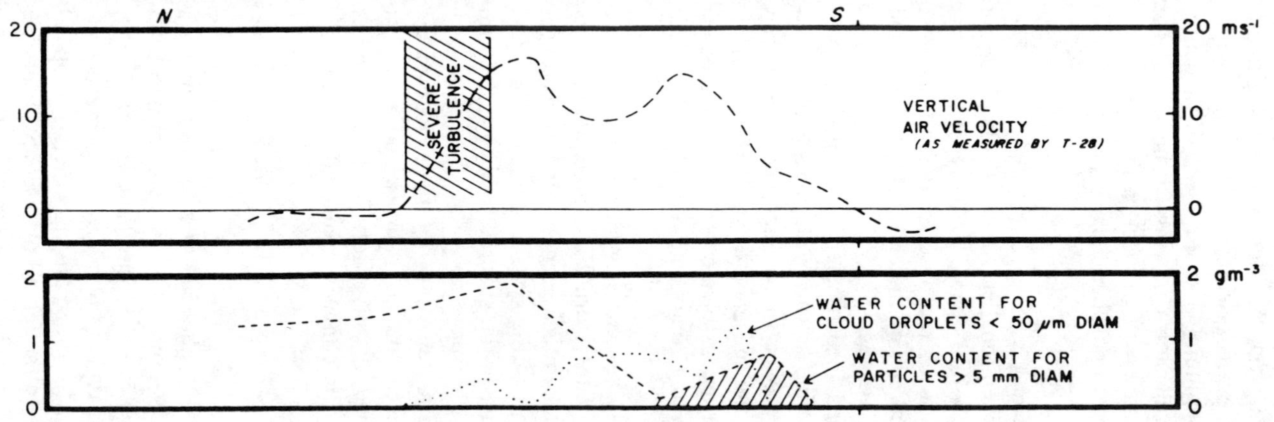


**Figure 5.10** Schematic model (top) of an ordinary multicell hailstorm in Colorado (Raymer, 9 July 1973) showing a vertical section along the storm's direction of travel through a sequence of evolving cells. Solid lines are streamlines of flow relative to the moving system; they are shown broken on the left side of the figure to represent flow into and out of the plane and on the right side of the figure to represent flow remaining within a plane a few kilometers closer to the reader. The open circles represent the trajectory of a hailstone during its growth from a small droplet at cloud base. Actually the airflow in each cell has been drawn relative to the individual cell and, since the developing cells $n + 1$ and $n$ traveled more slowly (5 m $s^{-1}$) than either the mature cells (7 m $s^{-1}$) or the storm as a whole (10 m $s^{-1}$), the streamlines in the young cells would have had a stronger component from the south relative to the storm as a whole. This explains why in the model the trajectory of the growing hailstone crosses over the streamlines during its early growth as shown in the figure. Lightly stippled shading represents the extent of cloud and the three darker grades of stippled shading represent radar reflectivities of 35, 45, and 50 dBZ. The temperature scale on the right side represents the temperature of a parcel lifted from the surface. Winds (m $s^{-1}$, deg) on the left side are environmental winds relative to the storm based on soundings behind the storm. Surface rainfall rate averaged over 2 min intervals during the passage of the storm is plotted below the section. The horizontal line NS through the section at 7.2 km shows the track of the T-28 penetration aircraft, smoothed data from which are plotted at the foot of the figure. Although the T-28 data were not quite synchronous with the data in the vertical section, a comparison of the T-28 updraft velocity measurements with the flow pattern in the vertical section shows that the agreement is reasonably good (from [Browning *et al.*, 1976]).

few kilometers. The zone of maximum echo is situated in the middle. These characteristics are maintained with little change throughout the lifetime of the supercell. Isolated storms with a single supercell therefore are associated with quite clearly defined radar characteristics. However, at the beginning and at the end of their life, they have a multicellular structure. The supercell storm moving to the left of the average winds has a certain symmetry of structure with respect to that moving to the right in terms of the direction of the average tropospheric shear vector.

*Fine Scale*

Within the medium-scale cells of the storm, even smaller structural elements are found that doubtless are present in forms and to various degrees in all types of storms but are not revealed unless the spatial resolution of the observations allows this. These elements have dimensions of a few hundreds of meters up to 2–3 km and appear as local maxima of the reflectivity or velocities.

Some high-resolution observations obtained in particular by doppler radar at vertical elevation have made it possible to describe storms with considerable fluctuations in the reflectivity and vertical velocity over short ranges and sometimes considerable heights; these fluctuations reflect the presence of pockets of precipitation particles with larger sizes locally. They are associated with heterogeneous structures of the currents within an overall ascending zone [Battan and Theiss, 1970; Battan, 1975].

The small-scale structural elements are also associated with minor convective towers, which are formed mainly on the sides of the storm. The precipitation they generate creates local reflectivity maxima. They retain their identity for periods longer than 10 min while they cross the most active part of the system and can be used as tracers to estimate the local movement of the air. In the supercells, they have been called *hot points* [Browning and Foote, 1976].

Generally speaking, small-scale cellular structures are observed close to the summit of the storms. They either reflect the decay of the main ascending current of the cell that, close to the upper limit of the cloud, dissolves into small-scale eddies or instabilities specific to the interface between the convective air and the clear air of the upper levels.

### 5.3.3 Squall Lines

Storm cells sometimes are organized in line a few hundreds of kilometers along a frontal or pseudo-frontal zone. As with isolated storms, the lines are formed from an arrangement of one or two types of cells [Newton, 1963; Smull and Houze, 1985]. At tropical latitudes the squall lines are a main type of organization of deep convection [Houze, 1977; Roux, 1988; Chalon *et al.*, 1988, among others]. Tropical squall lines have a relatively simple organization, although all the aspects of this are not clearly understood. They consist of an active and intense convection line approximately 10–50 km wide followed by a vast anvil of stratiform character that can extend over several hundreds of kilometers. The convection line of approximately north-south orientation can be several hundreds of kilometers long (Figure 5.11); it moves from east to west more rapidly than the surrounding winds at a speed of 30–60 km $h^{-1}$. Its lifetime can be several days and its trajectory greater than 2000 km. The powerful individual cumulonimbi that make up the intense convection line usually last less than 1 hr. They are born at the front of the system, develop very rapidly, then having produced a shower whose intensity sometimes exceeds 150 mm $h^{-1}$, are stratified in the anvil at the rear of the line. The generation of the line by the front is associated with the progression of the cold density current (or pseudo-cold front) created by the divergence above the ground of the cold air carried downward by the precipitation. In semiarid regions, where the ground is dry, the gust front creates very dense dust storms. The renewal process of the cells is rapid and contributes to the propagation of the line although the cells advance relatively little. Figure 5.12 shows an example of the dynamic organization of a squall line retrieved from measurements with dual-doppler radars carried out during the COPT 81 experiment in the north of the Ivory Coast [Roux, 1988; see also Chalon *et al.*, 1988, for the estimate of the energy and matter flows in the same squall line].

### 5.3.4 Microbursts

Microbursts refer to an intense small-scale downdraft that, when it diverges above the ground, produces an outburst of violent and destructive wind. The term *downburst* was proposed by Fujita, who made the first analyses of the phenomenon [Fujita and Byers, 1977] and later reviewed the subject [Fujita, 1985]. He divides downbursts into two categories, depending on the horizontal scale of the zone perturbed by the violent winds close to the ground:

- Macrobursts, in which the outburst wind zone stretches horizontally over more than 4 km with a duration of 5–30 min;
- Microbursts, which stretch over at least 4 km on the ground but can be more violent than the macrobursts. For the sake of convenience, most authors only use the term microburst and do not make a distinction between the two scales.

According to Hjelmfeld [1988] who has studied the structure and life cycle of the microbursts observed by doppler radar in Colorado, they can be isolated or organized in lines with an average lifetime of 13 min and 1 hr, respectively. The low-level reflectivity of the associated precipitation is similar for the two types of structure and stretches from 20 to 60 dBZ. The average maximum differential radial velocity observed is 24 m $s^{-1}$ over a distance of 3.1 km. Figure 5.13 shows an example of the perturbation of the wind field at low level created by a microburst observed by doppler radar. See also the observations and modeling of Tuttle *et al.* [1989].

Microbursts constitute a major danger for air traffic in airport terminals because of the wind shears they induce at low levels. A problem therefore arises of detecting

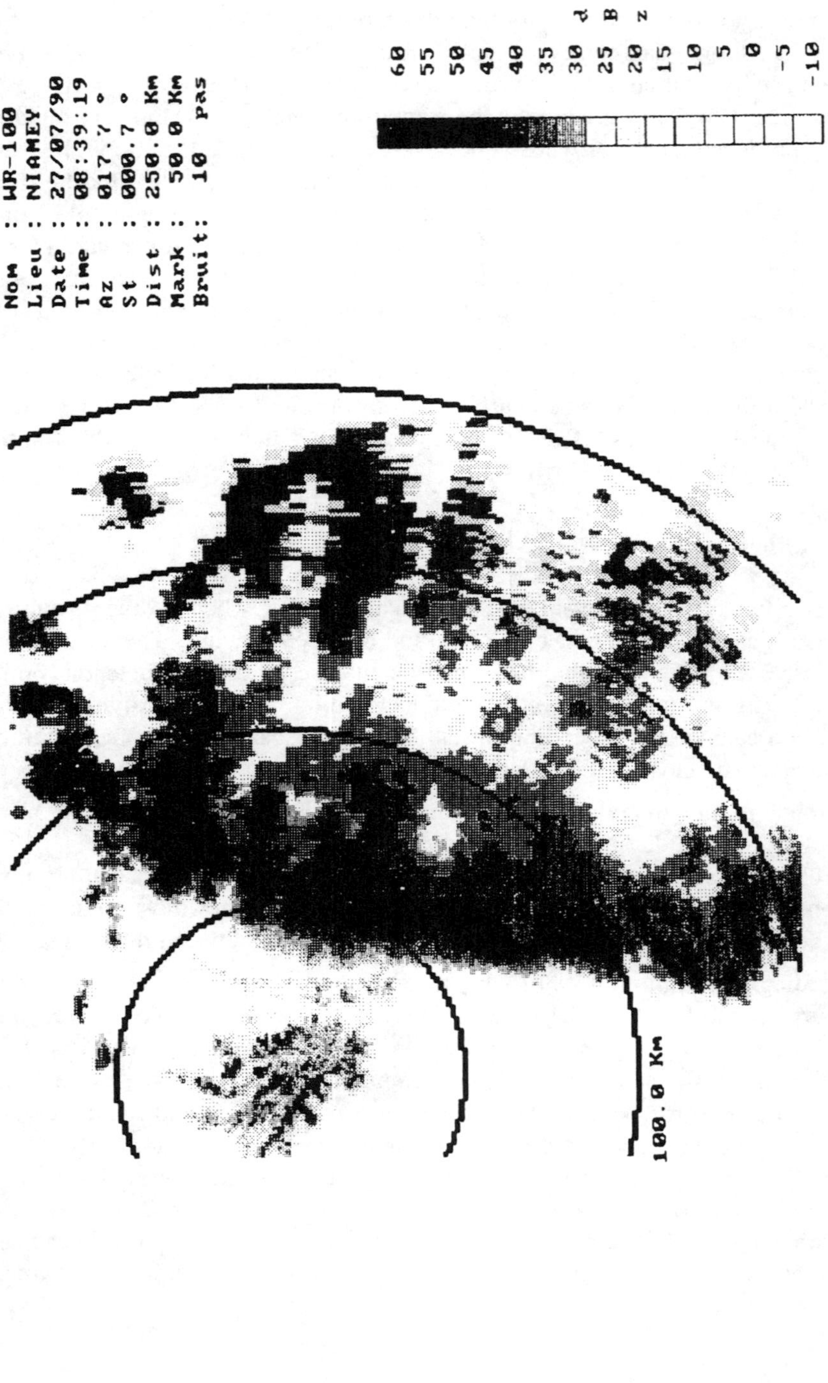


**Figure 5.11** Radar reflectivity factor (in dBZ) in a sahelian squall line observed with a 5 cm wavelength radar from Niamey (2°10 E, 13°28 N) in Niger.

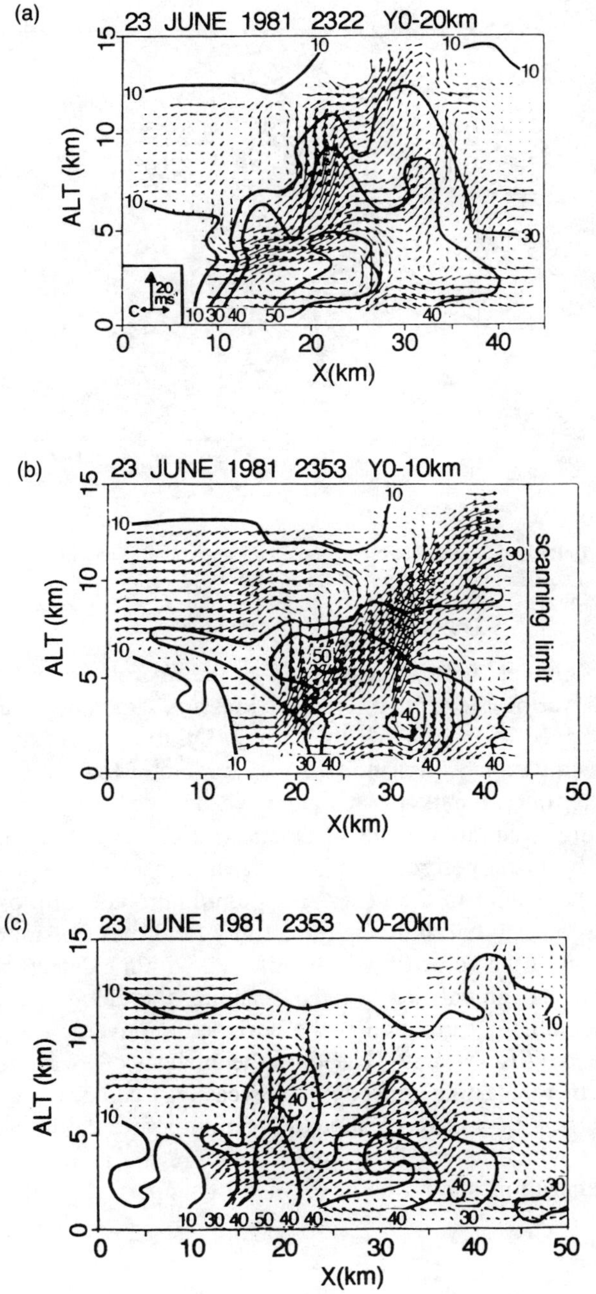


**Figure 5.12** Vertical cross sections of cell-relative wind and reflectivity (in dB*Z*) of tropical squall line observed in Korhogo (9°24 N, 5°36 W) in the northern Ivory Coast, using the dual doppler radars RONSARD. The cross sections are oriented toward the northeast; that is, perpendicular to the mean squall line orientation (from [Roux, 1988]).

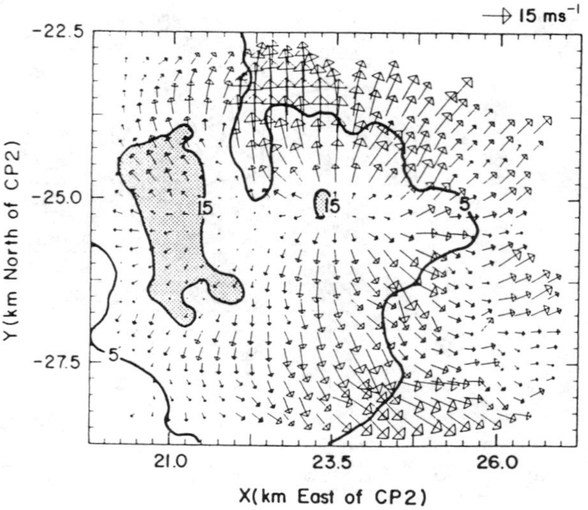


**Figure 5.13** Horizontal winds and reflectivity factor in horizontal plane at 50 m for a microburst observed near Denver on 14 July 1982 at 1445 MDT using dual doppler radars (from [Hjelmfelt, 1988]).

them in time to give an alert. At the origin of the microburst, we find precipitation associated with various forcing processes (pressure gradient, evaporation), which accelerates the speed of the air that they carry. On the ground, the microburst may be humid if precipitation is present or dry if it is not. However, in humid regions, practically all the microbursts are accompanied by intense precipitation, the generating cloud being a cumulonimbus. Detection aimed at providing an alert, to be effective, must be made before the microburst has reached the ground. Therefore a precursor must be sought in the generating cumulonimbus. In most cases, we have available for this a perturbation in the reflectivity field (descendant reflectivity cores) and a perturbation of the velocity field manifested within a detectable medium acting as a tracer of the movement. For a radar scanning the cloud at a low angle of elevation, the observable perturbation is the horizontal convergence around the downdraft. Roberts and Wilson [1989], based on the analysis of 31 microbursts, proposed a very short-term forecasting procedure (nowcasting) using a single doppler radar. The radar precursors they observe appear 2–6 min before the impact with the initial surface outflow, which leaves little time to analyze the data and indicates that the identification algorithms should be developed to support the alert decision.

### 5.3.5 Hail

The conditions determining the birth and growth of hail in storms are poorly known. Doubtless, several development processes can be involved successively or simultaneously within the same storm. Also development processes that are partly different correspond to different types of storm structures. A large number of phenomenological models have been proposed in an atttempt to explain all or some of the phenomenon [see the review of knowledge and abundant bibliography on hail, storms, and hail suppression by Foote and Knight, 1977, and Kessler, 1983]. These models usually are based on assumptions concerning the relationships between particular conditions characterizing the storm or its environment (for example, the intensity of the vertical wind shear) and the occurrence of hail. In many cases, faced with the difficulty of direct detection of hail in the storm, a statistical study is simply made of the correlation between the assumed causes and the final result (i.e., the properties of the precipitation at the ground), neglecting the chain of processes involved in the storm. However, the natural variability of the phenomena is such that an approach of this type leads to uncertain conclusions of little practical value.

Both the direct physical study of hail and the hail-producing storm and experimental studies relating to the suppression of hail by various "artificial action" methods on clouds (Section 5.8) make it imperative to solve the problem of remote detection for the observation and measurement of the formation and evolution of the hailstorm in space and time (size, concentration, environmental conditions). Radar provides an important resource in this connection and its possibilities are analyzed in detail in Section 2.5.4.

### 5.3.6 Electrical Activity of Storms

The passive and active detection conditions of lightning by radar have been discussed in Section 2.5.5, and the main results have been indicated. Radar in its conventional form (i.e., using a single and mobile directional beam) only fortuitously detects the discharges within a cell or a group of storm cells. Thus, many studies of the electrical activity of storms combine radar as a means of observing the dimensional structure of storms with other techniques of observing discharges in the storm. The latter mainly consist of techniques of radioelectric localization by detection of sferics [for example, Richard and Auffray, 1985, from tridimensional interferometry] and techniques providing access to the distribution of the charges in this storm and the transfers due to the lightning. Such observations can be made with the help of simultaneous measurements at multiple points of the electrostatic field on the ground and its variations.[11] The charge centers can be localized in space by this method with a precision of a few hundred meters [Jacobson and Krider, 1976; Krehbiel *et al.*, 1976]. The use of doppler radar arrays adds an extra dimension to this type of study.

The comparison of such observations with radar data suggests making a joint examination of the variation in intensity of the precipitation echoes and the electrical

[11]The measured field values are compared with the calculated values which would be produced by a hypothetical distribution of electric charges whose magnitude and position are varied until the best fit is obtained. The greater the number of measurement points compared to the number of unknown parameters of the chosen charge distribution, the better is the method.

activity in storms. For example, a relationship has been sought between $L$, the number of sferics per minute, $A$ the area (in $km^2$) occupied by the echoes corresponding to a radar reflectivity factor higher than a given level, and $r$ the range (in km) between the radar and the storm [Marshall and Radhkant, 1978]. For $A > 30$ dBZ, the obtained formula is

$$L = 2.7\, A^{1.64} r^{-1.62}$$

Equations of similar form obviously could be obtained by considering other values of the level determining $A$ or an echo volume rather than an area. The preceding relationship represents only an attempt made by neglecting important parameters for electrical activity, such as the altitude of the summit of the echoes and its temperature or the type of storm structure (dimension and number of cells, fine structure characteristics). Also a relatively intense electrical activity that can produce discharges, sometimes seems to develop in certain convective clouds with weak radar reflectivity where only snow precipitation is observed (for example, in old cumulonimbus in the dissipation stage) or even on the outside of the cloud. The value of this type of relationship, insofar as it can be refined and validated (possibly by taking into account radar information other than the reflectivity such as the polarization of the echoes or the dynamic fields), derives from its ability to make radar an indirect measurement instrument of the electrical activity of clouds.

## 5.4 TORNADOES AND VORTEXES

One of the most violent and most destructive byproducts of the storm is undoubtedly the tornado (sometimes called a *waterspout,* when it is formed above the surface of water). It appears as a vortex with its axis usually vertical and small diameter, developing over a considerable height and moving at very high velocities. Its observation is difficult because of the sporadic nature of the phenomenon and its relatively brief duration. Radar provides remarkable opportunities for scanning the space with a view to detecting tornadoes and to studying the conditions of their formation and evolution with, however, some difficulties concerning the spatial resolution of the observations.

### 5.4.1 General Characteristics

Observations carried out in the past with the help of conventional radars have shown that tornadoes frequently are formed close to the rear right-hand side of the "mother cloud" (with respect to the direction of movement of the cloud) possibly related to the hook echoes (mentioned in Section 5.3.2). However, this is not a general rule, and tornadoes have been observed at other points in the cloud. The same cloud can successively produce several tornadoes; occasionally, a tornado can last several hours, moving with a cloud over long distances.

Observation by doppler radar shows that tornadic storms are characterized by high values of the velocity spectrum variance, sometimes exceeding 4 $m^2\ s^{-2}$ with tangential shear values greater than $4 \times 10^{-2}\ s^{-1}$. Doppler radar has made it possible to establish that the tornado is preceded by the formation of a mesocyclone vortex, which develops aloft at the middle level of the cloud with diameter of a few kilometers (typical value is 5 km). The mesocyclone concentrates and intensifies; its diameter decreases while it expands toward the lower part of the cloud. A little later, it appears under the base of the cloud in the form of a compact vortex that progresses downward, possibly down to the ground, with its diameter sometimes being more than ten times shorter than that of the mesocyclone vortex [Arakawa, 1967; Fujita, Bradbury, and van Thullenar, 1970, among others]. When the tornado disappears under the cloud, the vortex dissipates at any height.

The main mechanism capable of producing an intense rotation with a vertical axis is the horizontal convergence of air already in slow rotation. In fact, if the angular moment is conserved, the velocity of a particle around an axis is inversely proportional to its distance to the axis. The most marked effect is produced under the storm with a strong horizontal convergence associated with an intense and localized ascending current of air in the cloud; instead of entering directly into the ascending current, the air converges in a spiral with an increasing angular velocity. The ascending movement in itself is not sufficient to create the phenomenon and most convective clouds pump the air below them without any significant rotation effect. When the rotation is sufficiently intense, the centrifugal forces create a depression at the center of the vortex in which, as a result, a condensation is produced, often over all or part of its height [see Davies-Jones, 1983, for a detailed presentation of tornado dynamics].

### 5.4.2 Identification of Vortexes by Radar

*Signature of the Vortex in the Reflectivity Field*

Attempts have been made in vain to find special features in the configuration of the connective storm echo contours that could constitute a signature or a precursor of the tornado and thus the basis for its detection or its very short-term prediction: a ring echo, a hole that could correspond to the eye of the whirlwind. However, storms producing tornadoes frequently have a maximum reflectivity factor greater than the average, but some examples run counter to this. Overall, the prediction and identification of tornadoes with the help of conventional radar on the basis of the reflectivity field are hardly encouraging; at the most, there is a slightly higher probability of observing an already established tornado where there are certain peculiarities in the shape of the echo contours (such as hook echoes) than where these peculiarities are absent.

*Signature of the Vortex in the Velocity Field*

The identification of tornadoes from discontinuities in the velocity field by using doppler radar has a better chance. The first doppler radar constructed in the United States for meteorological use, a continuous-wave 3 cm radar, was intended also for measuring velocities in tornadoes [Brantley and Barczys, 1957; Smith and Holmes, 1961]. Pulse systems appeared on the scene a few years later.

The use of pulsed doppler radar for observing tornadoes poses the problem of the spatial resolution of the observations in a particularly acute manner as well as that of the velocity and range ambiguities expressed by formula (1.13) because of the high values of the maximum velocities produced by the vortex and the very strong velocity gradient associated with it. In fact, an intense vortex movement centered in the radar beam will produce a broad velocity spectrum centered on zero and containing approximately equal positive and negative components. With an inadequate repetition frequency, the aliased spectrum will entirely occupy the range of measurable velocities and cannot be interpreted. Therefore we must use radars with high repetition frequency and limit the observation range to moderate ranges.

Under optimum observation conditions (i.e., in the absence of range and velocity ambiguities and with a radar with perfect resolution), a vortex of the tornado type would produce a signature in the velocity field with characteristics like those illustrated in Figure 5.14 [according to Donaldson, 1970]. Here, a vortex is considered in which the distribution of the tangential velocities following a radius perpendicular to the axis of rotation is that of Rankine's vortex [1901], which seems appropriate for tornadoes and the parent mesocyclones of the tornado: a central part in solid rotation whose tangential velocity increases proportionally to the radius $r$ up to a value $r = R$; that is, $u = C_1 r$, with $C_1 = u_{max}/R$. Beyond this, the flow regime is a potential vortex with a tangential velocity that decreases with the reciprocal of the radius: $u = C_2/r$, with $C_2 = u_{max}R$. A doppler radar scanning the vortex, with vertical axis horizontally, differentiates between two velocity peaks with opposite signs; the diameter of the mesocyclone deduced from the observation is that of the solid rotation part, or $2R$. Although the theoretical limits of the potential vortex ($r > R$) are at infinity for increasing values of $r$, they become practically indistinguishable from the surrounding circulation. The distinguishable part of the potential vortex extends the limits of the total mesocyclone to twice the central "solid" part.

In practice, the detection of vortexes in general and tornadoes in particular, with the help of a doppler radar, depends on the relative size of the vortex and the sampling volume of the radar as well as their respective positions; for a particular vortex, these conditions change with the observation range [Donaldson, 1970; Brown,

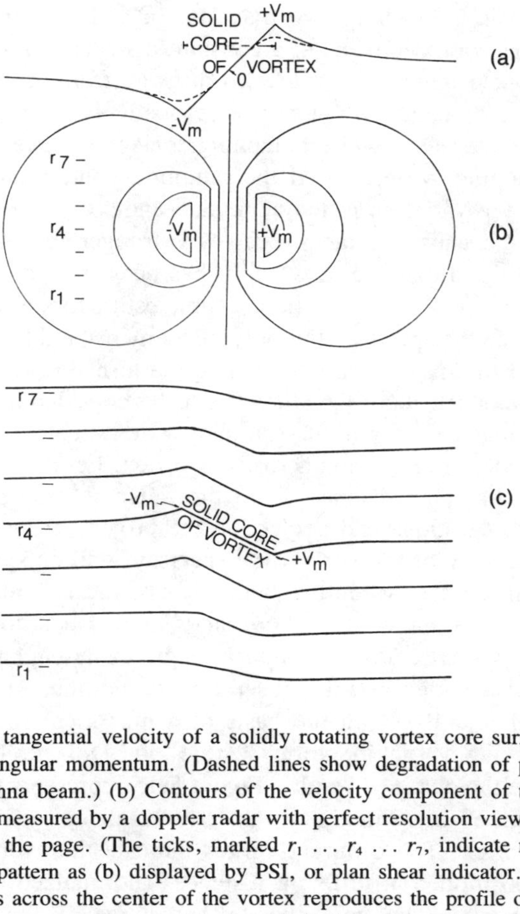


**Figure 5.14** Profile of tangential velocity of a solidly rotating vortex core surrounded by a region of constant angular momentum. (Dashed lines show degradation of peaks as observed by a wide antenna beam.) (b) Contours of the velocity component of the vortex in (a) which would be measured by a doppler radar with perfect resolution viewing the vortex from the bottom of the page. (The ticks, marked $r_1 \ldots r_4 \ldots r_7$, indicate ranges from the radar.) (c) Same pattern as (b) displayed by PSI, or plan shear indicator. (Note that the PSI arc which cuts across the center of the vortex reproduces the profile of tangential velocity of the vortex, except for a reversal of polarity) (from [Donaldson, 1970].)

Lemon, and Burgess, 1978]. In general, the tornado as such has a horizontal dimension less than that of the radar sampling volume. Therefore, we cannot directly determine the tangential velocity profile by scanning the vortex at constant elevation; in fact, a doppler velocity spectrum is measured with respect to the radar, weighted by the reflectivity distribution.

The radial and vertical distributions of the reflectivity in tornadoes, in particular the peculiarities that could be due to centrifugal effects on the droplets and to the presence of debris carried from the ground are not known. Observations have not allowed any significant variations to be revealed. If the radar's sampling volume is centered on the vortex, the positive and negative extremes of the spectrum correspond to the maximum tangential velocities in the tornado apart from instrumental errors; the spectrum is broad with two maxima. In addition, it is symmetric with respect to zero if the vortex remains at a constant distance from the radar. If it has a translation component that moves it toward or away from the radar, the spectrum is shifted to positive or negative velocities, respectively. If the sampling volume moves away from the tornado (while remaining centered in direction), the spectrum narrows while remaining symmetric. If the sampling volume moves away laterally from the tornado (by a variation in the azimuthal angle of the beam), the spectrum becomes asymmetric because the radar receives a stronger contribution from the circulation corresponding to the side to which it has been moved.

As mentioned in Section 3.1.2, the real-time estimators of doppler velocity do not explicitly calculate the spectrum but only its moments and therefore do not allow the determination of the maximum velocities in the tornado. To detect the vortexes by using such instruments, their signature has to be considered on the basis of the variation of the average velocity in the sampling gates situated at fixed range while the beam moves at constant elevation across the vortex. Let us consider the sampling volume situated at the same distance from the radar as the center of the vortex; following these considerations and neglecting the projection on the beam direction of the translation velocity of the vortex, the spectrum will be symmetric and so the average velocity will be zero when the sampling volume is centered on the vortex. The spectrum will be asymmetric on both sides of this position, and the average velocity will have two extrema with opposite signs. The dependence between the doppler signature of a vortex and the observation conditions of the radar has been analyzed [Brown *et al.*, 1978] on the basis of a numerical simulation study of a doppler radar observing a Rankine vortex [Zrnic' and Doviak, 1975]. Figures 5.15 and 5.16 summarize the obtained results. Figure 5.15 represents the average doppler velocity as a function of the azimuthal distance (normalized by the beamwidth) from the center of the vortex. The different curves indicate the average velocity in the sampling volume positioned at different distances (normalized by the pulse depth) of the center of the vortex. The most obvious characteristic is that the extreme values of the average doppler velocity are approximately situated at a half-width of the beam on both sides of the vortex when the sampling volume is centered at the same distance as the vortex. When the sampling volume is shifted in distance by 1, 2, and 3 pulse depths, the azimuthal separation of the extreme values increases whereas the amplitudes decrease. The marked fluctuation of the average velocity as a function of the azimuth rapidly disappears at a distance of a few pulse widths of the tornado.

Figure 5.16 also shows the average doppler velocity as a function of the azimuthal distance (normalized by the beamwidth) counted from the vortex center for a sampling volume situated at constant distance for different values of the ratio of the beamwidth to the radius of the solid part of the vortex. The separation between the extreme values does not significantly depend on the value of this ratio. On the contrary, the amplitude of the "signature" that governs its detectability is what is affected. Now, the signature can be resolved only if the amplitude of the extremes

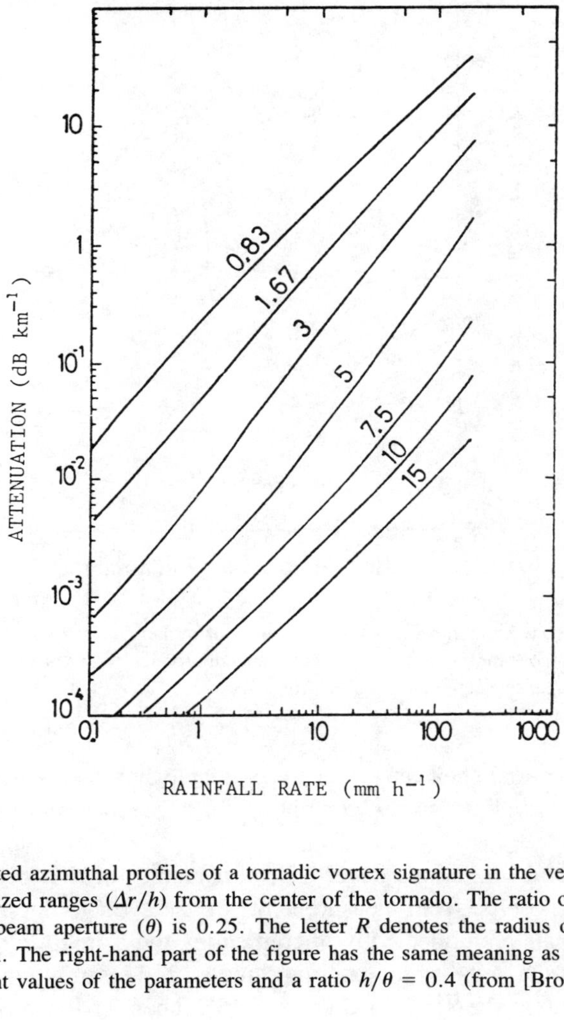


**Figure 5.15** Simulated azimuthal profiles of a tornadic vortex signature in the velocity field at various normalized ranges ($\Delta r/h$) from the center of the tornado. The ratio of the pulse length ($h$) to the beam aperture ($\theta$) is 0.25. The letter $R$ denotes the radius of the region in solid rotation. The right-hand part of the figure has the same meaning as the left-hand side for different values of the parameters and a ratio $h/\theta = 0.4$ (from [Brown *et al.*, 1978]).

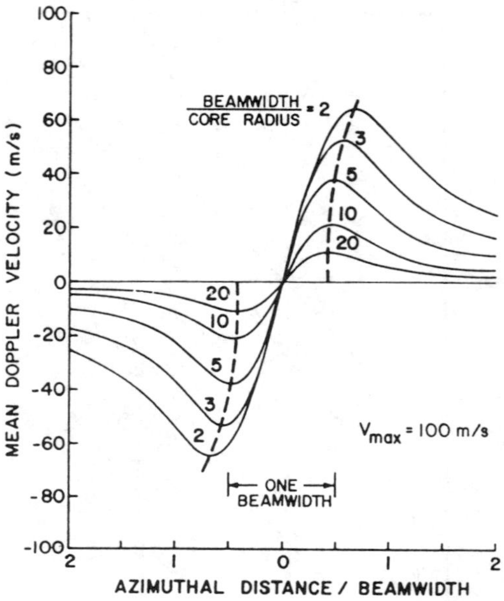


**Figure 5.16** Simulated azimuthal profiles through center of tornadic vortex signature for various beamwidth to core radius ratios. Maximum tangential velocity in vortex is 100 m $s^{-1}$. Ratio of pulse depth to beamwidth is 0.15 (from [Brown *et al.*, 1978]).

reflecting the tangential shear is appreciably greater than the surrounding shear produced by the mesocyclone vortex generating the tornado; a typical threshold value is $10^{-2}$ $s^{-1}$.

Other practical limitations have to be considered for the identification of vortexes of the tornado type from measurements of the average doppler velocity. First of all, when the data are collected by incrementing the azimuth by finite values, the extreme values of the velocity may not be sampled if the sampling interval is greater than a beamwidth (in the sense of Figure 5.15). Then, if the rotation of the radar antenna is rapid compared to the sampling time, the apparent width of the beam is increased. Thus, for a maximum tangential velocity and a given vortex dimension, the amplitude of the extreme values decreases when the antenna rotation velocity increases.

It follows from this analysis that the characteristic signature of the vortex is only qualitative and does not lend itself to a direct quantitative interpretation. The amplitude of the peaks of the characteristic curve is directly proportional to the size of the vortex and the maximum tangential velocity; it is also influenced by the relative position in distance of the sampling volume, producing an ambiguity in interpretation. Furthermore, the fitting of a profile of measured doppler velocities to a theoretical curve does not let us determine the size or the maximum velocity of the vortex that produced a given curve unless the size of the vortex can be determined independently, in which case an estimate of the tangential velocity is possible. The kinematic and dynamic three-dimensional structure of tornadic storms has been investigated in a few studies using doppler radar systems [Heymsfield, 1978; Eagleman and Lin, 1977; Ray *et al.*, 1975 and 1980], which widen the results obtained with a single radar, in particular concerning the signature.

### 5.4.3 Application to Warning Systems

One of the most attractive applications offered by doppler radar is its use as a warning against tornadoes in regions where this phenomenon is frequent. The detection of the characteristic signature of the mesocyclone vortex in altitude from its first appearance in the cloud indicates the commencement of air rotation.

Just the observation of a wind shear, indicated by a variation in velocity similar to those mentioned earlier, is not sufficient to identify a vortex. Several criteria, which together could form a basis for identification, have been suggested [Donaldson, 1970; Lemon, Burgess, and Brown, 1978]:

- A significant tangential shear in an almost horizontal plane over an azimuthal range of approximately one beamwidth (at least 20 m $s^{-1}$ for a vortex of the tornado type) with maximum doppler velocities of opposite signs on the azimuthal variation curve for the average velocity (after elimination of the component due to translation);
- A vertical extension of the region with shear perturbation greater than a horizontal diameter and for vortexes of the tornado type, a shear disturbance not extending over a horizontal range greater than 1 km (which indicates a small-scale vortex and not a shear line);
- A persistence in the shear disturbance at the same height for at least half the time necessary for one revolution of the vortex.

The use of similar criteria has made it possible to identify 10 vortexes of which 8 then led to the observation of tornadoes [Brown *et al.*, 1978]. However, a large number of questions remain unsolved.

Certain tornadoes give no detectable signature, even at short range. Some of the detected signatures show no rotation on the ground; that is, some vortexes do not "descend." The problem is to know the signature complement that distinguishes tornado mesocyclones from those that are not, what distinguishes long-lasting mesocyclones capable of producing multiple tornadoes from less persistent mesocyclones that produce only one, and how to predict changes or branching in the trajectory.

Well-documented warning times between the first detection of the mesocyclone vortex at altitude and the first contact of the tornado with the ground range from 5 to 60 min with a typical value of approximately 20 min; a delay on this order makes it possible to put out a general alert so that the most urgent precautions can be taken in the most threatened zone. Information on the movement of the generator storm lets us forecast approximately its progression by extrapolating from its past trajectory. The maximum ranges for detecting a tornado and a mesocyclonic vortex of average size with a doppler radar of 0.8° beam aperture are less than approximately 100 and 200 km, respectively.

## 5.5 EXTRATROPICAL CYCLONE DISTURBANCES AND STRATIFORM CLOUDS

### 5.5.1 Structure of Extratropical Cyclone Disturbances

The major synoptic meteorological structure at middle latitudes is the frontal depression system or extratropical cyclone. In the general circulation of the atmosphere, it plays a key part in converting the potential energy of horizontal thermal gradients to kinetic energy (Section 5.1). The succession of certain types of clouds and precipitation in a defined order during the movement of extratropical cyclones has been identified by the first observers. A phenomenological description of the average ensemble characteristics of the system was developed by the Norwegian meteorological school at the beginning of the twentieth century [see, for example, Godske *et al.*, 1957]. The Norwegian model is organized around the concept of a front[12] as a discontinuity of air mass but does not discuss the medium-scale structure of the systems. Ideas have undergone profound changes with the development of the concept of the barocline wave that associates cyclogenesis with a preexisting large-scale instability (i.e., baroclinic instability[13]), which is the primary process; cyclogenesis not being its cause but its consequence [see Bluestein, 1986]; and Keyser, 1986, for a review on these notions]. The study of the medium-scale and small-scale structure really started only with the use of radar. In fact, the presence of snow and rain in extratropical cyclone systems over a very large range and often over a considerable height, which can reach 10 km in altitude, lends itself to measurement by conventional and doppler radars.

[12]In meteorology, the front is the boundary or the transition zone between two air masses of different density (and temperature). Other quantities exhibit characteristic variations across frontal zones, wind, humidity, the shape of the clouds, and the precipitation. A distinction is made between several types of fronts, the main ones being on a planetary scale: polar fronts, arctic fronts, intertropical fronts, and within extratropical cyclone systems, warm, cold, and occluded fronts.

[13]Baroclinic instability is a hydrodynamic instability arising from the existence of a meridional temperature gradient (and hence a thermal wind) in an atmosphere in quasi-geostrophic equilibrium and possessing static stability.

*Air Movement*

The structure of extratropical disturbances is associated with the rise of warm air imprisoned by the frontal surfaces, which leads to the formation of extensive thick altostratus and nimbostratus layers in the vicinity of the warm front and cumuliform clouds above the cold front. The average vertical component of the air velocity estimated from the convergence at low levels is only a few centimeters a second.

The combined study of these systems by radar, satellite, radiosounding, and pluviometric measurements on the ground shows that their structure is determined by air movements on a subsynoptic scale [Palmen and Newton, 1969]. In crude terms, the organization of this system results from three main air currents whose arrangement (Figure 5.17) and curvature develop during the various phases of the "life" of the disturbance. The system is fed by a warm and humid current that converges at low levels to the warm sector and rises slowly above the warm front. At higher levels, a rapid, cold, and dry flow passes from the rear to the front of the system in the direction of its movement. In the front, a third current of middle tropospheric air descends toward the lower layers under the warm current [Browning and Harrold, 1969].

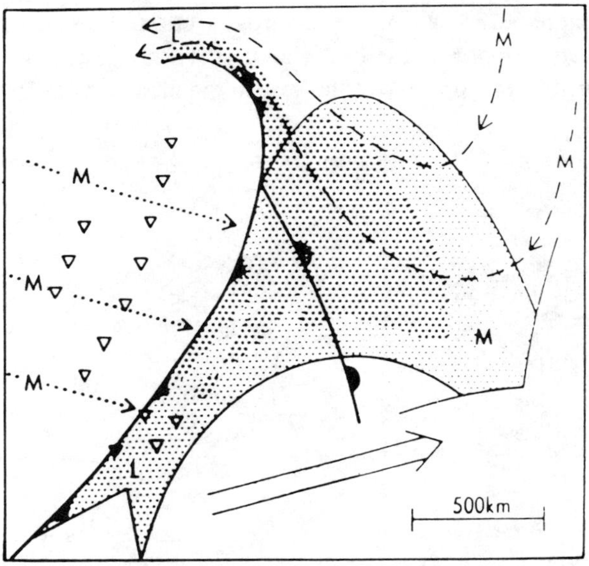


**Figure 5.17** Conceptual model of the air circulation and the distribution of precipitation in the partially occluded extratropical cyclone. The letters indicate that the air in movement is situated at middle (*M*) or low (*L*) levels. The hatched area denotes the area occupied by widespread precipitation whereas triangles denote areas of shower-type precipitation (from [Browning and Harrold, 1969]).

Radar observation lets us achieve a resolution compatible with the study of the medium-scale and small-scale dynamic structures. A large number of investigations, essentially carried out with the help of doppler radar operating in isolation by conical scanning (VAD mode; Section 3.3.1) have been involved in this type of study. They reveal in a fragmentary manner, a complex circulation, particularly in the vicinity of the frontal surfaces resulting from the overlapping dynamic and thermodynamic processes related to the conversion and partial degradation of the energy within the system: deep convection or convection confined to a sheet presenting an organized structure on various scales, wave movements with different wavelengths including dynamic instability waves (Section 5.7) in certain stable zones subjected to a wind shear, *et cetera*. [Browning, Harrold, and Starr, 1970; Browning and Harrold, 1970; Boucher and Ottersten, 1971; Borresen, 1971; Bryant and Browning, 1975; Gilet *et al.*, 1980; Heymsfield, 1979; Testud *et al.*, 1980, Hobbs *et al.*, 1980; among others]. This circulation still has been only investigated and understood very partially because of the lack of observation data. The installation of doppler radar arrays covering large areas, combined with other means of measurement in situ and by remote sensing, seems to be the only way of advancing the phenomenological analysis of these systems and to understand their global mechanisms.

Figures 5.18 and 5.19 illustrate the capacity of the doppler radar to observe frontal zones. Figure 5.18 is a height-time cross section of the turbulent energy measured with a 3 cm doppler radar by conical scanning of the VAD type (Section 3.3.2), in a warm-frontal zone above the sea, in the presence of strong winds [Bryant

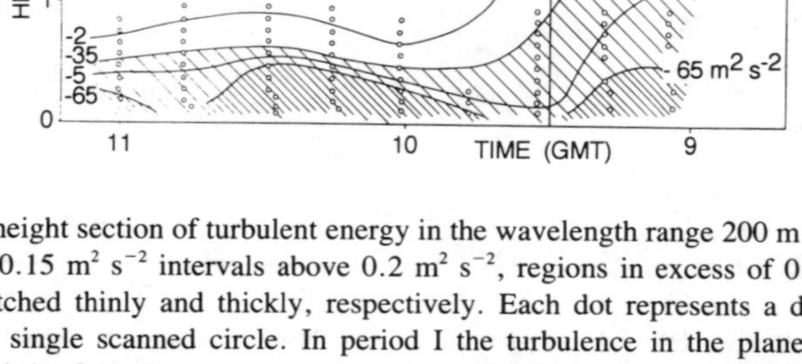


**Figure 5.18** Time-height section of turbulent energy in the wavelength range 200 m to 2.5 km. Isopleths are at 0.15 $m^2\ s^{-2}$ intervals above 0.2 $m^2\ s^{-2}$, regions in excess of 0.35 and 0.65 $m^2\ s^{-2}$ are hatched thinly and thickly, respectively. Each dot represents a data point associated with a single scanned circle. In period I the turbulence in the planetary boundary layer reached the free shear layer, whereas in period II the two zones of turbulence were separated by a placid layer. The warm front was traveling at 58 km $h^{-1}$ and so this figure may be regarded as a spatial cross section spanning a total distance of about 150 km (from [Bryant and Browning, 1975]).

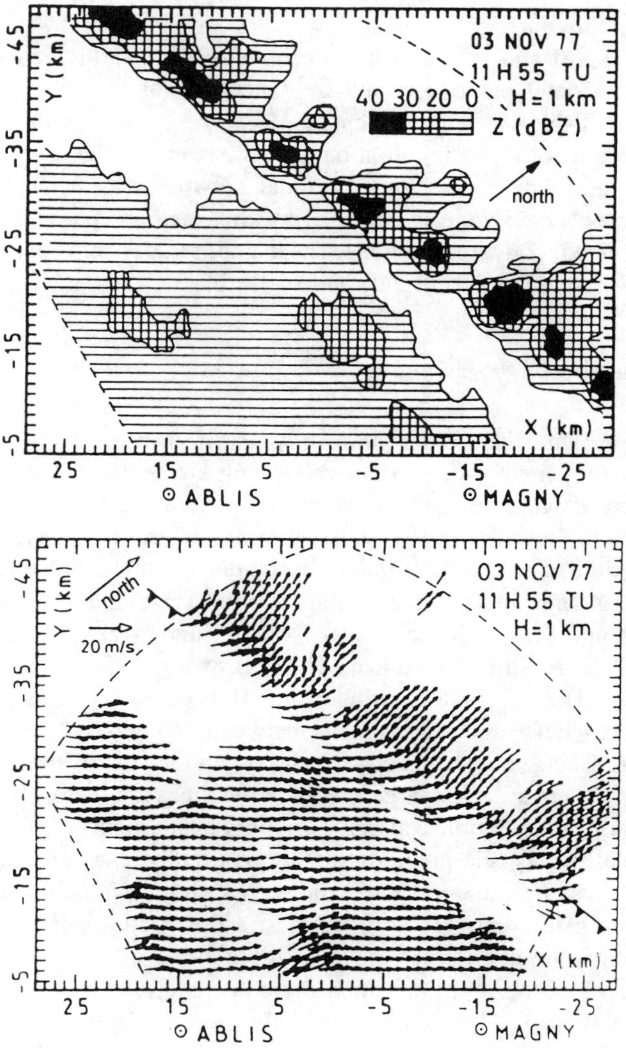


**Figure 5.19** Dual doppler radar analysis of a precipitation band associated with a cold front on 03 November 1977 south of Paris. The upper part represents the radar reflectivity factor (in dBZ) for a horizontal level of height 1 km. The lower part represents the field of horizontal air velocity relative to a reference moving with the frontal cells in the same plane as the upper part (from [Campistron and Sauvageot, 1981]).

and Browning, 1975]. The results show the differences in the structure of the turbulent field between the planetary boundary layer (up to 600 m) and the so-called free shear zone above it.

Figure 5.19 (a) and (b) show the average velocity fields and the reflectivity fields in the horizontal plane obtained by a network of two 5 cm doppler radars (Ronsard radars; Table A4.5 in Appendix 4) carrying out simultaneous and coordinated observations of a precipitation band at the leading edge of a cold front. The implementation and data processing of a radar network is heavy but a detailed description of the observed zone is obtained by this method, particularly where precipitation is detected: convergence zone, shear line, relationships between the movement field and the precipitation field, and so forth.

### *Horizontal Structure of Precipitation*

The precipitation is not distributed uniformly within these systems. Radar observation shows that they are formed from overlapping elements of different scales, one element of a given scale always enclosing one or more elements of smaller scale. These substructures are associated with variations in the intensity of precipitation over space and time. The results of radar observations in this field can be summarized by distinguishing three main scales, which are the large and small mesoscale precipitation areas and the small-scale cells [Nozumi and Arakawa, 1968; Kreitzberg and Brown, 1970; Austin and Houze, 1972; Browning *et al.*, 1973; Mielke and Houghton, 1977; Hobbs, 1978; Businger and Hobbs, 1987]. The large mesoscale precipitation areas have dimensions lying between $10^3$ and $10^4$ km$^2$, and they last from 1 to 4 hr. The average velocity of the ascending air in these regions is approximately 20–30 cm s$^{-1}$. The large precipitation areas are themselves formed from several groups of precipitation cells clustered over an extent of 50 to $10^3$ km$^2$ and forming the small mesoscale precipitation areas. Finally, the cells making up these groups occupy areas less than 50 km$^2$, and the ascending velocities associated with them range from a few tens of centimeters to a few meters per second. Cells and small and large precipitation areas move with the average horizontal velocity of the winds at the level of the layer in which the convection develops (or active layer). This velocity is typically 1 km min$^{-1}$, usually more rapid than the synoptic characteristics. This organization therefore shows the absence of uniformity in the vertical movements in the system.

The characteristics of medium-scale structures that are large and small precipitation areas usually tend to be organized in bands in extratropical cyclones. Such bands have been observed in many regions. For example, bands, 35–70 km wide with spacing from 50 to 100 km have been identified in occluded systems on the west coast of the United States [Elliott and Hovind, 1964]. A band organization has also been observed in the northeast and the northwest of the United States [Kreitzberg and Brown, 1970; Hobbs and Locatelli, 1978; Businger and Hobbs, 1987], in Japan [Tatehira, 1968; Nozumi and Arakawa, 1968], in Europe [Browning and Harrold, 1969; Gilet *et al.*, 1980; Campistron and Sauvageot, 1981; Testud *et al.*, 1980, reviewed by Hobbs, 1978]. Typically, these bands have a width of 50 km with a rather strong dispersion of the observed values. Comparison of the results obtained by different authors makes it possible to distinguish several types of band.

The warm frontal bands are parallel to the surface warm front, mainly situated above and ahead of it. These bands consist of groups of thin convective cells localized at the top of the cloud in what is called the *precipitation-generating layer*. The cells are held up by updrafts of approximately 1 m $s^{-1}$ and produce long characteristic snow trails (Section 5.5.2). The precipitation forms there from crystals in a partly supercooled medium (as long as the updraft lasts). The stratiform cloud between the precipitation bands consists of droplets and there is no precipitation. The convection in the generating cells is fed by the potential instability released at an altitude by the large-scale ascending motion. The cells move at the velocity of the winds at the bottom of the generating level; they therefore progress more rapidly than the synoptic characteristics of the system. The bands situated in the most advanced part of the system are less active.

The warm sector bands occupy the warm sector, with a surface that runs approximately parallel to the cold front. The intensity of these bands is very variable and can reach that of a squall line of the midlatitudes. This therefore is a convection developing over a larger depth associated with a convergence at lower levels. The precipitation there is stronger and more localized than in the warm front bands.

The surface of cold frontal bands is also parallel to the cold front. Two subcategories can be distinguished: a narrow deep-convection band associated with the surface of the front and wider bands forming above the frontal surface behind the preceding ones, consisting of generating cells due to localized convection over a small thickness at the top of the cloud. As with the other types of bands associated with generating cells, the broad bands move with the wind at the level of generation and therefore advance across the system. The narrow, deep-convection band moves as the trace of the front in surface with which it is associated.

These different types of bands, within which it seems subdivisions can be distinguished, have unequal intensities; not all may be present or identifiable in a single cyclone. Because of the complexity of the configuration of the precipitating systems, we cannot always identify subsynoptic structures from a single radar scan; sometimes we must use observations extending over a certain time interval and, even then, identification often remains subjective.

### 5.5.2 Stratiform Precipitation

*Generating Cells and Precipitation Trails*

Short-range and high-resolution radar observation clearly shows that precipitation in clouds of a stratiform nature is always associated with convection. The generating cells confined to a thin layer, in the generation layer at the top of the cloud, produce precipitation trails that bend and stretch out under the influence of the wind shear (Figure 5.20). The partial overlapping of the trails can lead to a certain uniformity at low levels. However, such precipitation of low intensity, continuous and uniform appearance, results from nonuniform processes in space [Marshall, 1953; Gunn and Marshall, 1955; Douglas, Gunn, and Marshall, 1957; Wexler and Atlas, 1959; Browning and Harrold, 1969; Sauvageot, 1974; Carbone and Bohne, 1975; Henrion and Sauvageot, 1977; Henrion, Sauvageot, and Ramond, 1978; Hobbs, 1978; Ludlam, 1980; Sauvageot *et al.*, 1982; Sauvageot, Auria, and Campistron, 1984]. This type of structure is found in extratropical cyclone disturbances, in warm front bands, the broad bands of the cold front, and also sometimes in other types of bands during the dissipation stage. A similar structure is also observed in the stratiform clouds on the edge or at the rear of an extratropical cyclone or in certain cloud systems due to local effects with no direct relationship to the cyclone wave of the general circulation.

The main causes leading to the formation and maintenance of unstable layers at altitude are (1) the liberation of a potential instability by a larger-scale ascending motion, and also (2) the release of the latent heat of freezing or deposit of water vapor from the surrounding medium during the growth of the crystals in the cell or in the trail (3) radiative transfers, the consequence of which is a heating of the base of the cloud layer and a cooling at the top, and finally (4) the temperature contrast resulting from an advection of cold and dry air above the saturated layer. Several causes can be involved at the same time. Therefore, the concept of the generating cell encompasses phenomena that, from a physical and dynamic point of view, have quite significant differences. Sometimes, the demarcation between the generating layer and individual cells is imprecise.

Figure 5.21(a) is a typical example of the microscale structure of stratiform clouds showing the generating cells and the trails of crystals in the presence of vertical shear of the horizontal wind [Sauvageot *et al* 1984]. It was obtained from the data of a 35 GHz doppler radar. Volumes where the precipitations are initiated are situated at the upper end of the trails where the region of ascendant air is located (Figure 5.21(b)). Figure 5.21(d) shows the doppler velocity spectrum in the generating zone and in the trail below the convectively active level. The difference between average doppler velocities is 1 m $s^{-1}$ and, above all, the standard deviation $\sigma_v$ is four times larger in the cell than in the trail. The shape of these cells is strongly influenced by the wind shear [Sauvageot *et al.*, 1982].

Certain generating cells of the warm frontal bands apparently can remain active for more than 1 hr and create continuous trails longer than 100 km. On the other hand, small cells are observed in certain clouds in thin layers of the altocumulus or cirrocumulus type with a short lifetime. The generating cells and particularly the long-lasting cells frequently have a structure on a finer scale, consisting of overlapping subelements, resulting in the creation of a single precipitation trail [Sauvageot *et al.*, 1982].

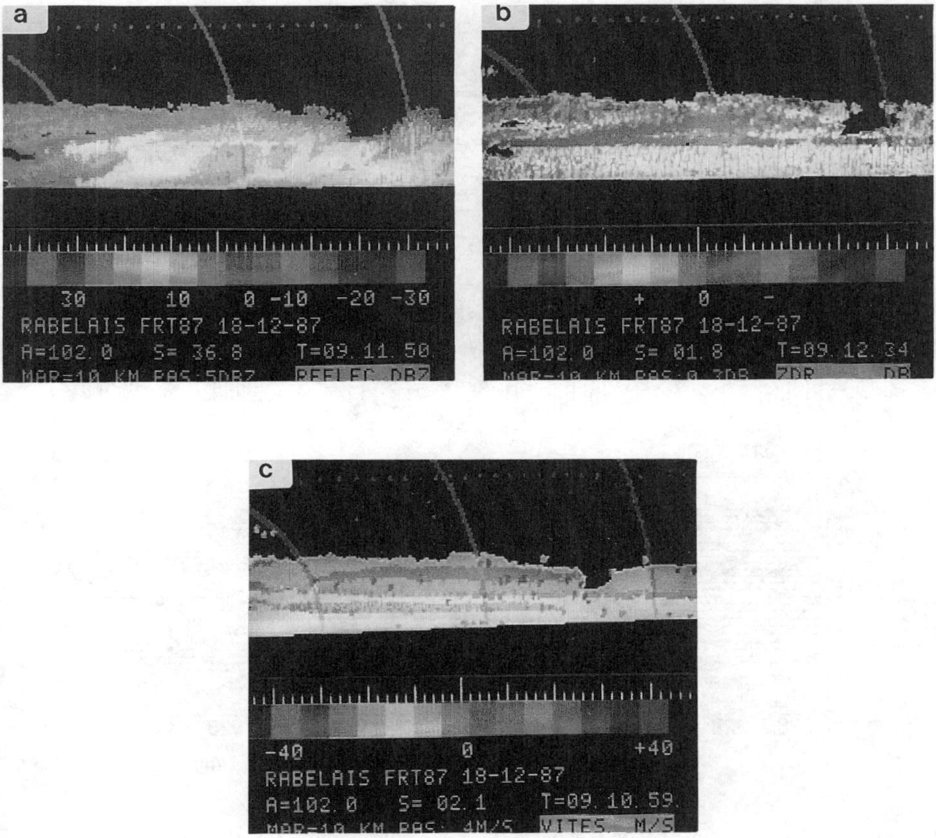


**Figure 5.20** RHI display of (a) reflectivity factor $Z_e$, (b) differential reflectivity $Z_{DR}$, and (c) radial velocity for a stratiform cloud observed on 18 December 1987 from Brest in the west of France with the 35 GHz RABELAIS radar. Azimuth is 102° (i.e., toward the east). Range markers are at 10 km intervals. $Z_e$ is in dBZ in 5 dB steps, $Z_{DR}$ is in dB in 0.3 dB steps, velocities are in m s$^{-1}$ between $-40$ and $+40$ m s$^{-1}$ in 4 m s$^{-1}$ steps. The 0° C level is at 3.4 km of altitude. The "bright-band" is just below this level.

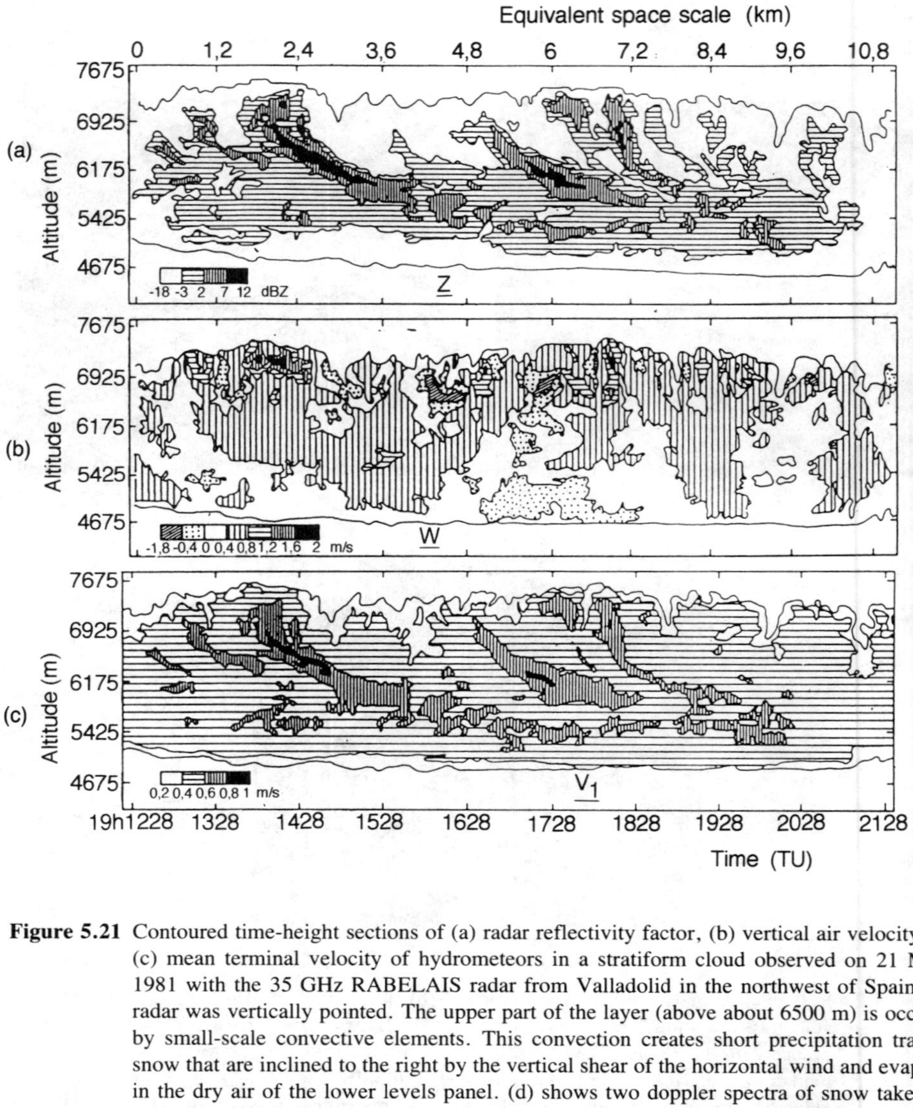


**Figure 5.21** Contoured time-height sections of (a) radar reflectivity factor, (b) vertical air velocity, and (c) mean terminal velocity of hydrometeors in a stratiform cloud observed on 21 March 1981 with the 35 GHz RABELAIS radar from Valladolid in the northwest of Spain. The radar was vertically pointed. The upper part of the layer (above about 6500 m) is occupied by small-scale convective elements. This convection creates short precipitation trails of snow that are inclined to the right by the vertical shear of the horizontal wind and evaporate in the dry air of the lower levels panel. (d) shows two doppler spectra of snow taken ($d_1$) within a precipitation generating cell and ($d_2$) below the generating level, in the precipitation trail. In ($d_1$) the spectrum width shows the effect of turbulence in the generating cell ($\sigma$ is 4.6 times the value in $d_2$). In ($d_2$) the spectrum has the narrowness typical of snow (from [Sauvageot *et al.*, 1984]).

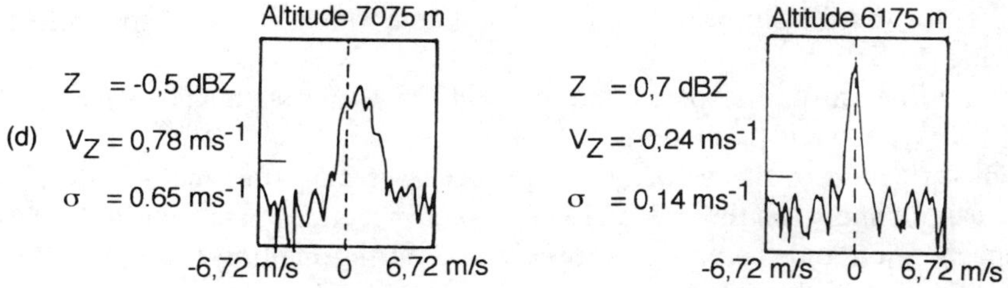


**Figure 5.21** (continued)

The bottom limit of the stratiform or cumuliform cloud echoes is usually defined not by the base of the cloud (condensation level) but by the precipitation, which creates considerably larger echoes than the cloud medium itself. When it is not joined to ground echoes, it therefore presents a complex, somewhat anarchic contour. In certain stratiform clouds, vertical outgrowths even sometimes are observed at the base of the echoes in the form of "stalactites." This phenomenon is a consequence of the evaporation of crystal precipitation trails that, by locally creating an unstable temperature gradient, destabilize the air and induce a thin convection at the base of the echoes [Harris, 1977].

### *The Bright Band*

We saw in Chapter 2 that the melting of ice particles results in appreciable modifications of their attenuation and backscattering cross section. In natural precipitation, these modifications create a region of discontinuity of the radar properties below the 0° C isotherm, called the *bright band* or *melting band* (Figure 5.20). This band is particularly clear when the particles have a low velocity of fall and when melting takes place over a limited height; it particularly affects snow precipitation in extratropical cyclone systems, of which it is one of the main radar characteristics, but it is also observed in old cumulonimbus clouds during the decaying stage and in showers of compact ice particles of small diameter. It appears in Figure 5.9, which contains a few typical profiles of the vertical variation of the equivalent reflectivity factor $Z_e$ calculated from the measured reflectivities.

The bright band consists of a marked maximum of the radar reflectivity in a layer 100–400 m thick below the 0° C isotherm. The maximum separates an overlying, low-reflectivity region (corresponding to the solid precipitation) from a belowlying region with greater reflectivity (corresponding to rain). Several effects combine to determine the observed variation:

1. First of all, the variation of the term $|K|^2$ appearing in the expression for the backscattering cross section, which differs approximately by a factor 5 (i.e.,

7 dB) for the transition from ice to water and which increases the reflectivity during the fall;

2. A modification of the precipitation spectrum that also tends to increase the reflectivity toward the bottom;
3. A modification of the velocity of fall that is greater after melting, at least in the case of snow (on the order of 2–6 m $s^{-1}$), which tends to reduce the particle concentration and therefore the reflectivity during melting (Figure 5.22); this modification in the velocity of fall is accompanied by a change in the slope of the precipitation trails when they are inclined;
4. Finally, a form effect during melting that can act on the polarization and the intensity of the backscattered signal.

Overall, the increase in reflectivity between the levels immediately above the 0° C isotherm and the maximum in the melting layer is 10–15 dB and the decrease between the maximum and the lower levels is 5–10 dB.

The extent of the modifications affecting the precipitation spectrum across the melting layer has given rise to conflicting analyses. Certain authors base their interpretation on an aggregation of crystals above the maximum and a disintegration of the drops below it. The quantitative nature of this effect is considered uncertain or insignificant by others [du Toit, 1967; Ohtake, 1969].

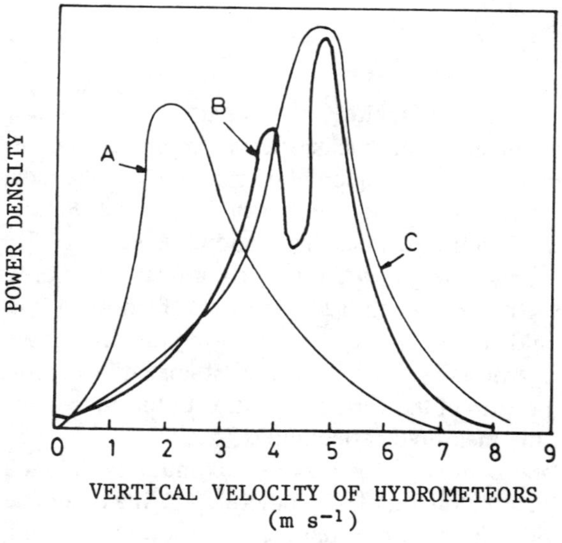


**Figure 5.22** Velocity spectrum at three levels in a bright band. The curves *A*, *B*, and *C* are 100, 250, and 400 m below the 0° C isotherm, respectively. The spectra are averaged over intervals of 5 mm (from [Dyer, 1968]).

A turbulence and small-scale disturbances of the dynamic field also have been observed at the level of the melting zone. They are attributed to temperature fluctuations due to cooling, caused by transfers of the latent heat of melting that follow the spatio-temporal variations of the intensity of the precipitation and move horizontally with the trails. Disturbances of the mesoscale wind field are observed there as well [Atlas *et al.*, 1969]. The melting layer of the precipitation also has the peculiarity of attentuating the radar radiations more strongly (see Sections 2.3.2, 2.3.3, and 2.4.3 on the attenuation by melting snow).

*Kinematics of Precipitation Trails*

The formation of precipitation from small volumes of vapor that are generating cells has various implications for the interpretation of radar observations and for the measurement of hydrometeor size distribution. Usually, the cells behave as point sources with constant altitude moving with the wind at their level. They form precipitation bands or trails in which the slope, which depends on the vertical shear of the horizontal component of the wind, is highly variable; in certain cases, it can be such that the precipitation resembles a cloud stratification or exhibits multiple changes of inclination. The length of the band depends on the length of time over which the precipitation has been produced (a very long trail indicates that the generating mechanism inside the cell is stable both in terms of its functioning and its humidity supply; see Figure 5.20). A trail with no generator and in which the upper end is poorly defined indicates that production has ceased and that the generating cell has disappeared or has not been detected by the radar.

Over a time interval, during which the fall velocities and the profile of the horizontal velocities as a function of the height do not change, the generator and the trail move as a whole; and all the points of the trail have the velocity and the direction of the generator. In particular, the velocity of the ground trace of the precipitation trail is the same in terms of magnitude and direction as that of the generating region. The distribution of the horizontal winds below the cell have an effect only on the shape of the trail (slope and orientation).

In a moving reference frame tied to the generator, the trajectory of a particle coincides with the precipitation trail [Gunn and Marshall, 1955]. Consider a moving reference frame $Gxyz$ with the generator ($G$) as its origin of coordinates where $Gx$ is horizontal and has the direction of the movement of the generator but opposite sign; $Gz$ is vertical and directed downwards, $Gy$ is perpendicular to $xGz$. Let $v_l(z)$ be the velocity of fall of a particle ($P$) with coordinates $x$, $y$, $z$; and $\mathbf{u}_a(z)$ be the horizontal velocity of the air with respect to the mobile reference frame. The term

$\alpha(z)$ is the angle formed by the plane $xGz$ and the direction of $\mathbf{u}_a(z)$. In this reference frame, the equations of the trajectory are

$$x = \int_0^z \frac{u_a(z)}{v_l(z)} \cos\alpha \, dz$$
$$y = \int_0^z \frac{u_a(z)}{v_l(z)} \sin\alpha \, dz \qquad (5.1)$$

Therefore, if $\mathbf{u}_a(z)$, $\alpha(z)$, and $v_l(z)$ are known, the trajectory of the precipitating elements can be determined.

Let us assume that, at least approximately, $\alpha(z) = 0$ and that $v_l$ is constant between the generator and a level $z$; and let us denote the horizontal velocities of the air with respect to the ground at the generator level and at level $z$ by $V_g$ and $V(z)$, respectively. Because radar observation is made from a fixed point on the ground, the horizontal space relating to the precipitation trail is related to time by $dx = V_g \, dt$. According to (5.1), we have

$$\frac{dz}{dt} = v_l(z) \frac{V_g}{V_g - V(z)} \qquad (5.2)$$

Knowing $V_g$ and $V(z)$ (from wind sounding), by measuring the slope of the trail $dz/dt$ at level $z$, we can calculate the approximate value of $v_l$.

In the simple case of a constant vertical shear of wind and with $\alpha = 0$ (or by considering projections only in the vertical plane parallel to the direction of the generator) and $v_l$ constant at all levels, the trajectory of the particles is parabolic in the space with respect to the ground or in the moving reference frame. It is the same for the precipitation trail seen by the radar (Figure 5.23).

A variation in the velocity of fall of the precipitation is reflected by a change in the slope of the trail with respect to that which it would have had with constant velocity of fall. This is what usually happens on crossing the melting level. The variation in the velocity of fall in this case is localized over a small thickness and the slope has a discontinuity.

Equation (5.2) shows the effect of the horizontal velocity of the air $V(z)$ on the value of the slope of the precipitation trail $dz/dt$. Therefore, the slope of the trail provides information on the relative value of $V(z)$ with respect to the horizontal velocity $V_g$ of the generator and on its variations as a function of height. The sign of the vertical wind shear $dV(z)/dz$ is related to that of the second derivative of the trajectory $d^2z/dt^2$.

The precipitation produced by the generator consists of particles in which the dimensions and velocities of fall are distributed according to a spectrum. In the presence of wind shear, a mass spectrometer or sorting effect is developed in the inclined trail; large elements arrive at the ground faster than small elements leaving the generator at the same time, and their trajectory has a greater slope [Marshall, 1953].

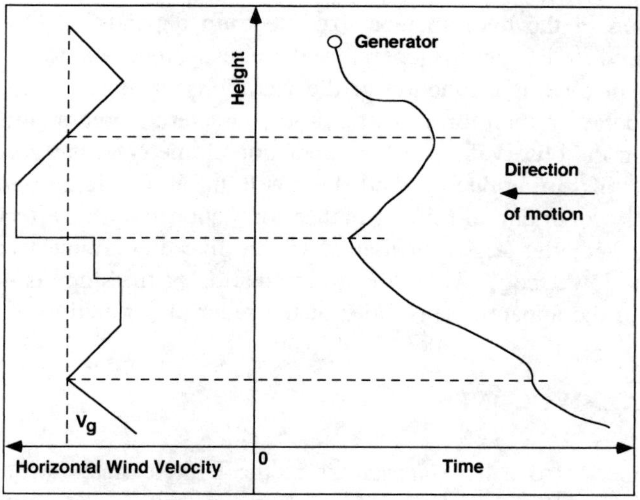


**Figure 5.23** Relation between the vertical wind velocity profile (left part) and the shape of the precipitation trail (right part) for a constant terminal fall velocity.

The distance between the vertical passing through the generator and the point of impact at the ground of a particle in fact is proportional to the velocity of fall or more exactly to

$$\int_0^z v_t(z)\,\mathrm{d}z$$

The result is a spreading of the precipitation that, on the ground, can amount to several kilometers or tens of kilometers. Therefore, in the presence of shear and with $\alpha(z) = 0$ at all levels, a fixed point on the ground situated in the plane of the trajectory successively receives all the elements of the hydrometeor size distribution in order of decreasing or (more rarely) increasing dimension following the direction of the slope. However, at the ground level, the precipitation lasts longer than the time for the generator to pass, and its intensity is lower than at the generator level.

If $\alpha$ is appreciably different from 0, the sorting effect also is produced, following a different direction from that of the movement of the entire precipitation and therefore a fixed point on the ground only receives some of the size spectrum of this. If, in addition, and this is the general case, several neighboring generators are active at the same time, the precipitation on the ground is determined by its respective contribution; the spectrum received at a given point in time at a point in space is discontinuous and consists of lines from different point sources, the width of the lines being related to the dimensions of the source and the duration of capture.

The composition of the hydrometeor size spectrum depends on the nature of the precipitation produced by each generator, their respective positions, and the wind field. It varies in time as a function of the variations of these different parameters with a certain delay. It therefore is difficult to characterize precipitation produced at an altitude from the observation just by their granulometry at the ground.

The effect of sorting due to wind shear is found in precipitation from all types of clouds, stratiform or cumuliform. Another consequence of this effect is that, when the trail is detected, the largest elements, which give the strongest echoes, always are found at the lower edge. Thus, the determination of the slope is not affected by the variations in the apparent sensitivity of the radar as a function of the range.

## 5.6 TROPICAL CYCLONES

Tropical cyclones, also called *hurricanes* in the United States (when the wind is greater than 33 m $s^{-1}$) and in the Caribbean and *typhoons* in the west of the Pacific Ocean, are characterized by a thermal circulation containing a spiral convergence of warm air toward the center of the system at low levels, and a region of ascending air close to the center surrounded by a descending region. The air diverges at the top. The system has a diameter ranging between 100 km and more than 1500 km. The mechanism of generation and maintenance of the warm zone close to the center is independent of any existing horizontal thermal gradients. The positive temperature anomaly close to the center of the system results from the heating and humidification of the convergent air of the low levels by flows of water vapor, sensible heat coming from the surface of the sea above which it flows, and the release of latent heat associated with the very thick vigorous convection in the convective ring surrounding the eye of the cyclone and in the spiral bands. The central warm zone behaves as a reservoir of potential energy that is continuously transformed into kinetic energy by the thermal circulation. This kinetic energy is responsible for the generation within the system of high-speed tangential winds and their maintenance despite the dissipation by friction. The velocity of the winds decreases proportionally to the square root of the radial distance from the center of the eye [see review by Anthes, 1982].

Observation by coastal radars, instrumented aircraft, and airborne radar has made it possible to describe the mesoscale structure of precipitation in tropical cyclones. The closer the convective cells or towers forming the base of the system are to the center, the higher they are. They are organized on several different scales; they form mesoscale bands arranged according to a spiral configuration on the scale of the disturbance as a whole. The distribution usually has a high degree of symmetry, the curvature of the spirals making it possible to approximately locate the center of the system; but it also can be temporarily asymmetrical. The organization of the tropical cyclone is found to be extremely efficient in the production of precipitation and kinetic energy [Parrish *et al.*, 1982; Barnes *et al.*, 1983; Marks, 1985].

Figure 5.24 shows a panoramic observation sequence of the precipitation echoes of a tropical cyclone.

The eye of the cyclone is a depression zone of partially clear, slowly descending air, in which the diameter, which measures only a few tens of kilometers, can vary during the movement of the system. It is surrounded by a vertical wall of intense precipitation (which can reach 100 mm $h^{-1}$), called the *eyewall,* in which the echoes (with a radar reflectivity factor reaching 45–50 dBZ) frequently have a bright band below the 0° C isotherm, indicating the presence of icy precipitation [Tatehira and Itakura, 1966; Hawkins and Imbembo, 1976]. The pressure can be less than 900 mb in the eye, with winds reaching 100 m $s^{-1}$ at the periphery. Sometimes, a structure containing a double eye is observed (the eye is surrounded by a cloud ring and then by a concentric zone of clear air [Hoose and Colon, 1970, diameter of the double eye 13 and 46 km; Marks, 1985]).

The study of the wind field by radar in tropical cyclones has been carried out by considering the movement of identifiable structures in the radar reflectivity field over time. Since about 1980, observations with airborne doppler radar, making it possible to bring the radar closer to its objective, with a view to improve the resolution of the measurements, have also been made [Marks and Houze, 1984; Jorgensen and Marks, 1984]. The data have made it possible to establish a three-dimensional model of the circulation. The strongest winds are observed close to the base of the

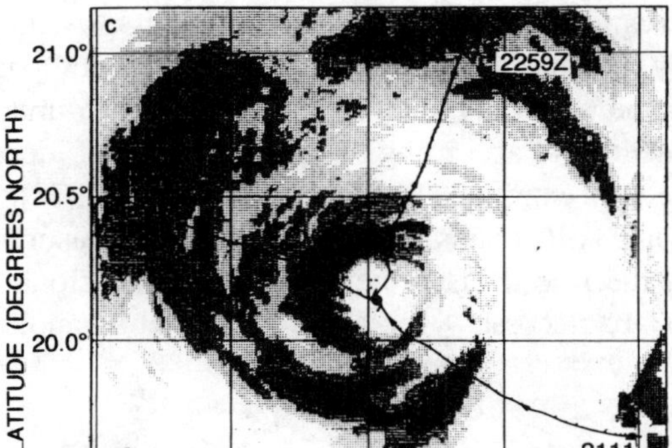


**Figure 5.24** Time composite of the horizontal distribution of precipitation in Hurricane Allen between 2111 and 2259 GMT on 6 August 1980 from airborne radar data. The reflectivity contours are at 25, 30, 35, 40, and 45 dBZ indicated by increasing shades of gray. The solid lines indicate the flight track of the aircraft during the period of the time composite and the hurricane symbol at the storm center. The cross hatching indicates areas not covered by the composite (from [Marks, 1985]).

clouds on the periphery of the eye. Their velocity decreases rapidly with height in the middle troposphere; in the upper troposphere, the wind field is asymmetric to the center of the cyclone with a divergent air flow (from the updraft in the spiral bands) rotating in a direction opposed to that of the air situated between the bands so that the tangential velocity around the cyclone is on average close to 0 at these levels.

Frequently the convective elements forming the cyclones that reach the intensity of hurricanes develop tornadoes. According to Gentry [1983] tornadoes develop around the center, within a radius of 50 km. Beyond this distance they are observed mainly to the east and northeast of the center.

Measurements of precipitation by radar in tropical cyclones have been compared with those of a pluviometric ground-based array and granulometric measurements have been made in situ with a view to investigate a specific *Z-R* relationship with no clear result [Willis and Spahn, 1977]. Finally the radar, in the particular form of a coherent system using a HF frequency of 15 MHz has been used to observe the condition and the movements of the sea surface below a tropical cyclone, up to distances of 3000 km [Maresca and Barnum, 1979].

Cyclones are one of the most destructive meteorological structures. The practical problem they pose is the operational tracking of their state and forecasting their evolution. The quality of radar observations depends on the distance of the cyclone. Now, cyclones form over oceans where most of their trajectory is situated so that certain global observations, first carried out by radar, are currently obtained with greater efficiency by meteorological satellites: thermography, surveillance of the cyclogenesis, tracking the trajectory and the evolution of the dimensions of the system, the apparent movement field according to the movement of identifiable structures on the images etc. [see, for example, Shenk and Rodgers, 1978]. Radar, in particular airborne doppler radar, still retains all its value by the measurement of the intensity of precipitation and the fine measurement of three-dimensional velocity and reflectivity fields, which are necessary for advancing phenomenological knowledge on tropical cyclones.

## 5.7 TURBULENT STRATIFICATIONS AND SHEAR INSTABILITY

High-sensitivity decimetric radar quite frequently detects extensive and thin horizontal echo sheets; several sheets can be observed at the same time at different levels. The entire height of the troposphere is affected by this phenomenon. The backscattering is produced by a disturbance of the average gradient of the potential refractive index created by a turbulent mixture and exhibits a weak dependence on the wavelength characteristic of the turbulence ($\lambda^{-1/3}$) [Hardy *et al.*, 1966].

In a general manner, there is little doubt that stratified echo layers correspond to zones of stronger static stability in which the variations of some meteorological quantities are strongly marked, in particular the humidity and wind shear and thus also the average potential refractive index. At low levels, the stratified layers are frequently associated with temperature inversions limiting an underlying convective activity, which disturbs the stable layer (Section 5.2). In the presence of a strong humidity gradient, turbulence with scarcely perceptible intensity can suffice to create echoes caused by variations of the humidity. Although this is not necessary for the creation of scattering layers, a strong shear often is observed at the level of the stratifications associated with a marked turbulence, in particular at altitude. The thickness of the scattering layers often is only a few tens of meters and rarely exceeds a few hundreds of meters.

Frequently, when observations are made with sufficient spatial resolution, the echo layers in clear air reveal a fine structure of a wavy nature. Very detailed observations of the structure and development of the waves has been made with a vertically pointing frequency-modulated continuous-wave 10 cm radar [Gossard and Richter, 1970; Gossard, Richter, and Atlas, 1970; and Gossard, Jensen, and Richter, 1971] (this equipment has a radial resolution of 1–2 m and high sensitivity; the minimum cross section that can be detected at 1 km is $3.7 \times 10^{-6}$ $cm^2$ for a point target). Figure 5.25 gives an example of this phenomenon, which is interpreted in terms of gravity waves associated with the internal instability of the stratified flows or Kelvin-Helmholtz dynamic instability [Hicks and Angell, 1968; Hicks, 1969; Reed and Hardy, 1972; Browning *et al.*, 1973; Metcalf, 1975].

Such waves are sometimes observed at the top of thin stratiform cloud layers, in particular in the upper troposphere at the level of the cirrus clouds when the condensation marks the crest of the waves although the troughs remain clear. They develop in the layers of hydrostatically stable air in the presence of a wind profile presenting a vertical velocity shear with an inflection point. They are also found in the ocean, under similar flow conditions. These waves, the existence of which has been known for a long time, appear in isolated groups of a few units, directed transversely to the vertical wind shear, and move with the average wind at their level. The unstable nature of the phenomenon is reflected by a rapid increase in the amplitude of the initial wave form that, after a time of less than 30 min, can break down into a turbulent flow of smaller and smaller scale.

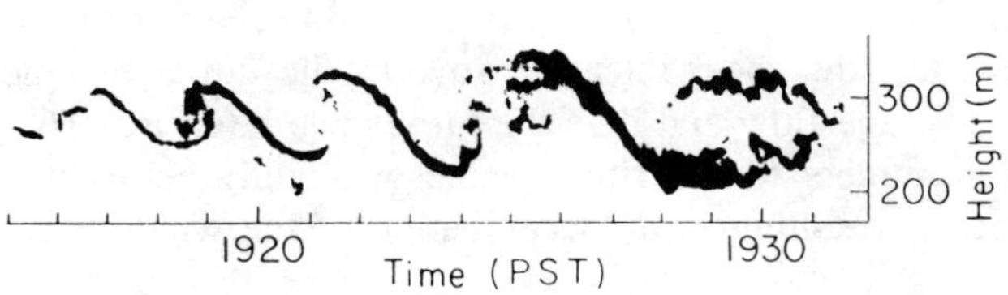


**Figure 5.25** Clear air echo from a breaking Kelvin-Helmholtz wave on 6 August 1969 observed at vertical elevation by a FM-CW *S*-band radar (from [Metcalf, 1975b]).

The phenomenon develops simultaneously on several scales, the smallest being several orders of magnitude less than the largest, within which they are produced, the most rapidly increasing principal wavelength depending on the thickness of the unstable layer [Scorer, 1978]. The necessary condition for the initiation of the instability is that the value taken locally by the gradient Richardson number (Ri) should be less than or equal to the critical value:

$$\mathrm{Ri}_c \approx 0.25 \tag{5.3}$$

with

$$\mathrm{Ri} = g\beta \bigg/ \left(\frac{\mathrm{d}u}{\mathrm{d}z}\right)^2 \tag{5.4a}$$

where $\mathrm{d}u/\mathrm{d}z$ is the vertical gradient of the horizontal velocity of the fluid,

$$\beta = \frac{1}{\theta}\frac{\mathrm{d}\theta}{\mathrm{d}z}$$

is the static stability of the layer, and $\theta$ the generalized potential temperature,[14] dry or saturated depending on the case [Browning, 1971]. We also can write

$$\mathrm{Ri} = N^2 \bigg/ \left(\frac{\mathrm{d}u}{\mathrm{d}z}\right)^2 \tag{5.4b}$$

where $N$ is the Brunt-Vaisala natural oscillation frequency of the fluid.

Ri is an indicator of the ratio of the stabilizing forces of the static stability to the forces inducing the turbulence contained in the wind shear. High positive values of Ri correspond to a stable flow in which the disturbances induced by the shear are damped by the static stability of dominant influence. On the contrary, for small positive values of Ri, these disturbances develop with an intensity that depends on that of the shear until the unstability conditions have been destroyed by the turbulent mixture, which tends to homogenize the properties of the medium in which it appears. Low values of the shear and the static stability may be sufficient for waves to develop. However, the shear intensity governs the turbulent kinetic energy extracted from the average flow and thus the turbulence intensity. The local lowering of the Richardson number to its critical value or slightly below it is due to a disturbance of the flow resulting in an accentuation of the wind shear at the level of the stable layer.

---

[14]Characterizing a particle of air brought adiabatically to an arbitrary reference level of pressure $p_0$, usually chosen equal to the average value in the layer considered.

The evolution of the phenomenon is summarized by Figure 5.26 without showing its small-scale structure: the main wave develops from a thin stable layer; it is amplified and then breaks down. A movement is created, called a *cat eye pattern, plait,* or *roll,* in which the air has a closed circulation. This is completely identical to that of cloud rolls created by the combined action of the static instability and wind velocity vertical shear [Sauvageot, 1976]; in particular, the ratio of the "wavelength" of the phenomenon (or of the spatial periodicity in the direction of the average wind) to the height of the roll is close to 2.7. In the roll, the stable potential temperature gradients are destroyed by this circulation, which returns and inverts the transported air volume and mixes it. Neutral or weakly unstable gradients are created locally and temporarily (the overall effects on the temperature profiles, with a more complex analysis, are different). The breakdown therefore makes possible a transition to a convective circulation in the roll. This circulation can last for some time. Then, the air spreads out horizontally in a sheet the thickness of which depends on the amplitude of the waves and transport across stable limits during the falling-off phase: a strong static stability at the beginning will produce a rapid falling-off and thus little thickness in the intermediary sheet and *vice versa*. In all cases, the final mixed layer is four or five times thicker than the original layer. The average wind shear is destroyed within the turbulent layer. At the limits, two thin layers with very stable temperature gradients marked by a strong wind shear are created in which new waves can develop. In Figure 5.26, the thick lines indicate regions of strong heterogeneity of the refractive index that produce the radar echoes; the fine lines represent current surfaces at the limits of the layer concerned. Within the roll, the turbulence destroys the index gradients whereas at the limits, in the air film that undergoes maximum displacement, the contrasts are accentuated. The radar therefore sees only the limit of the rolls. In the final phase, a thin scattering zone lasts for some time at the limits of the mixture layer. Finally, note that, during the breakdown phase, the vortexes are clearly two-dimensional; they probably produce a three-dimensional turbulence only at small scales. In the observation of this phenomenon as in that of convection, the small-scale turbulent component detected by the radar, which is assumed stationary and isotropic, therefore serves to trace the larger-scale anisotropic and nonstationary movements.

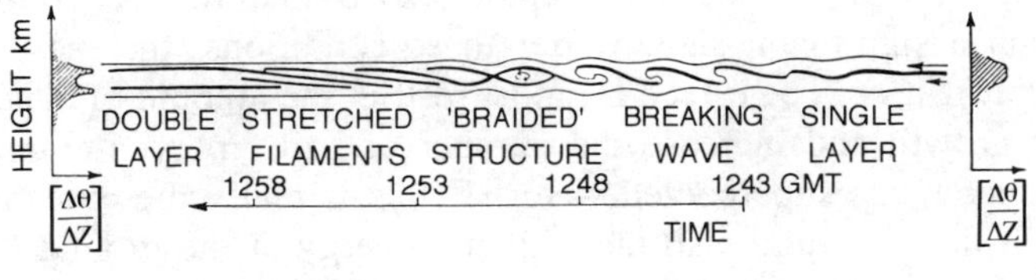


**Figure 5.26** Schematic representation of the life cycle of an individual Kelvin-Helmholtz billow based on experimental data. Thick lines correspond to the detectable clear air radar echo. Schematic vertical profiles of potential temperature ($\Delta\theta/\Delta z$) are indicated before and after the occurrence of the phenomenon (from [Browning and Watkins, 1970]).

## 5.8 EXPERIMENTAL MODIFICATION OF CLOUDS AND PRECIPITATION

We have known for many years that it is possible in certain cases to intervene with the physical processes responsible for the evolution of clouds and the formation of precipitation with a view to artificially modifying their course. The large number of attempts in this direction have been aimed at mainly the dispersion of low stratiform clouds or ground fog to increase visibility, increasing precipitation to combat droughts or increase water reserves at a particular site (the reservoir of a hydroelectric dam, for example), the suppression of hail, or reduction in the intensity of tropical cyclones [see, for example, the reviews of Dennis, 1980; or Todd and Howell, 1985].

The general principle for most of these interventions consists of dispersing in the cloud a substance that can act in a catalytic manner on the physical state of the hydrometeors and thus modify their granulometry. In warm clouds (i.e., at positive Centigrade temperatures), hydroscopic particles are used, such as sodium chloride, capable of rapidly reaching a precipitable size and so initiating the process of growth by coalescence. In cold clouds (i.e., at negative Centigrade temperatures), the substances used either make it possible to stimulate the freezing of drops by local cooling of the air of the cloud (this is a thermodynamic action used, for example, to treat supercooled fogs by dispersion of carbon dioxide snow or liquid propane) or themselves act as the freezing nuclei.

The best-known *seeding* agent and the one most frequently used to create artificial freezing nuclei to increase precipitation and reduce hail is silver iodide (AgI) [Simpson, Brier, and Simpson, 1967; Weinstein and MacCready, 1969; among others]. The method of seeding in freezing nuclei is based on the finding that, in convective clouds, the condensation resulting from the ascending air remains partly and temporarily in the form of supercooled liquid droplets of nonprecipitating sizes. Introduced into this medium, the freezing nuclei are assumed to first rapidly grow by deposition (because of the difference of the equilibrium water vapor pressure with respect to water and ice), then start the growth process by coalescence, and so initiate a precipitation (cf. Section 2.2.1). Hail is associated with an extreme case, where, in large cumulonimbus clouds, the large quantities of available supercooled water (due to the thickness of the cloud) are captured by a small number of frozen elements that thus reach a significant size. Under these conditions, the massive seeding of freezing nuclei, if it is to be effective, must restrict the amount of supercooled water available for growth and increase the number of hail embryos entering into competition to capture this supercooled water. If this is really the case, the size of the hailstones will decrease and also their kinetic energy at the ground level, which is the cause of damage to property and crops (the kinetic energy of the hailstone is proportional to approximately $D^4$, where $D$ is the diameter).

Some encouraging results have been obtained for local increase of precipitation, in particular in orographic systems. However, many aspects of these methods are poorly known. Considerable research remains necessary to understand the conditions of their implementation in the natural medium and evaluate their expediency. However, such an effort is largely justified because of the enormous potential economic rewards and the scientific interest in the studies to be carried out.

The evaluation of the efficiency of the operations for increasing precipitation and reducing hail involves looking for a correlation between the cause (the seeding) and the effects it produces. The difficulty here is due to natural variability of the phenomena, which is very large for rain, even larger for hail, and in all cases larger than the effects to be measured. A large number of wide ranging research programs have been targeted at this problem of evaluation, with little result. Often we must adopt a randomization procedure involving treating half a sample of clouds selected with apparently similar properties, the other half acting as reference. Broadly speaking, the evaluation is carried out in two different ways. The simplest consists of a control made at the end of the sequence of processes we seek to modify; for example, ground measurement of the precipitations (intensity, granulometry, distribution over time and space). The other approach is based on direct observation in the cloud of the physical consequences of the intervention.

In all these operations, radar is involved, before the intervention, in the characterization of the situation and the choice of the sample, then in carrying out the operations of treatment (demarcation of the zones to be treated), guiding the planes making the measurements in situ, and finally in the acquisition of the base data for statistical analysis and modeling. Radar has become an almost indispensable tool for this type of experiment.

Conventional or doppler radar has also been used to control the tests on treating tropical cyclones. The modification of tropical cyclones with the help of seeding freezing nuclei is based on the assumption that the introduction of an agent capable of increasing the frozen fraction of the clouds situated around the eye of the system, and so increase the amount of liberated latent heat, should make it possible to increase the dimensions of the central warm zone, spread the low-pressure region, and therefore reduce the intensity of the winds at the base of the eyewall. Various attempts were made between 1961 and 1969 but the reductions in observed wind have not been significant [Black, Senn, and Courtright, 1972].

Many parameters have been extracted from radar observations to evaluate the efficiency of the treatment; for example, geometric parameters, such as the rate of growth and the maximum height reached by the echoes, to provide evidence of a larger amount of latent heat of freezing being liberated. However, most often, the radar variable used is the reflectivity factor $Z$, because of its high sensitivity to the variations in the size of the hydrometeors. Its interpretation in terms of physical quantity is based on the concepts discussed in Chapter 2. It therefore suffers from the disadvantages discussed there, but perhaps here for an additional reason: the seeding is aimed at locally modifying the granulometry or the thermodynamic phase of the hydrometeors and these modifications influence the coefficients of the $Z$-$X$

relationships under unknown conditions. This is why, in experimental programs for modifying clouds and precipitation, multiparameter radar measurements are used more and more, doppler for access to the dynamics, double wavelength, and more recently polarization diversity for access to the microphysics of the hydrometeors. Microwave radiometric measurements are also used to estimate the quantities of liquid water. The large number of references to these problems can be found for hail and its suppression in the monograph edited by Foote and Knight [1977] and for the increase of precipitations in the monograph edited by Braham [1986].

# Appendix 1
# Bibliographical Note

The original literature on the theory and application of radar to the observation and study of the atmosphere has been subject to a large number of articles, which can be found in all the specialist reviews in meteorology and the physics of the atmosphere. However, the main part has been and continues to be published in the following:

- The reviews of the American Meteorological Society (AMS), in particular: *Journal of Applied Meteorology* (entitled *Journal of Meteorology* until 1961), *Journal of Atmospheric and Oceanic Technology, Journal of Atmospheric Sciences,* and *Monthly Weather Review*.
- Certain reviews of the American Geophysical Union (AGU): *Journal of Geophysical Research* (*Atmospheres*) and *Radio Sciences*.
- European reviews, mainly: *Annales Geophysicae* of the European Geophysical Society (EGS) and the *Quarterly Journal of the Royal Meteorological Society*.

An important source for original work is also the Proceedings of the Conferences on Radar Meteorology of the AMS (the 24th is for the year 1989 and is 753 pages long).

The main edited works are on radar techniques, Skolnik [1962], Nathanson [1969], the compilation edited by Skolnik [1970], and the collection of 36 titles on radar published by Artech House. For electromagnetic propagation in the atmosphere, one can consult Bean and Dutton [1968] and Hall [1979]. A general presentation on the application of microwaves to remote sensing is given by Ulaby, Moore, and Fung [1981].

The best-known texts on the meteorological applications of radar are those of Atlas [1964] and Battan [1973]. More recently, a book by Gossard and Strauch

[1983] on observation of clear air and clouds and a book by Doviak and Zrnic' [1984] giving more attention to doppler measurements have been published.

In this book, to keep the list of references to a reasonable length, it was necessary to omit references to articles of essentially historic interest. We also have refrained from mentioning articles in reports or reviews of restricted circulation often difficult to procure.

# *Appendix 2*
# *Units and Symbols*

The units used in this book conform to the international system except when more or less common usage dictates a departure.

System International (SI) Units

| *Quantity* | *Units* | *Symbol* | *Definition* | *Dimensions* |
|---|---|---|---|---|
| Length | Meter | m | | L |
| Mass | Kilogram | kg | | M |
| Time | Second | s | | T |
| Electric intensity | Ampere | A | | I |
| Temperature | Kelvin | K | | $\theta$ |
| Density | | | kg $m^{-3}$ | M $L^{-3}$ |
| Velocity | | | m $s^{-1}$ | L $T^{-1}$ |
| Acceleration | | | m $s^{-2}$ | L $T^{-2}$ |
| Force | Newton | N | kg m $s^{-2}$ | M L $T^{-2}$ |
| Pressure | Pascal | Pa | N $m^{-2}$ | M $L^{-1}$ $T^{-2}$ |
| Energy, work | Joule | J | kg $m^2$ $s^{-2}$ | M $L^2$ $T^{-2}$ |
| Power | Watt | W | J $s^{-1}$ | M $L^2$ $T^{-3}$ |
| Voltage | Volt | V | W $A^{-1}$ | |
| Electric charge | Coulomb | C | A s | |
| Impedance | Ohm | Q | V $A^{-1}$ | |
| Electric capacity | Farad | F | A s $V^{-1}$ | |

## List of Symbols

| Symbol | Meaning | Usual Units |
|---|---|---|
| $a$ | Radius of the drops or equivalent spheres | $\mu$m or mm |
| $a$ | Amplitude of the signal scattered by a point target (see $e(t)$) | |
| $a_g$ | Rate of attenuation by gases | dB $km^{-1}$ |
| $a_n$ | Rate of attenuation by clouds | dB $km^{-1}$ $g^{-1}$ $m^{-3}$ |
| $a_\eta$ | Rate of attenuation by rain | dB $km^{-1}$ $mm^{-1}$ $h^{-1}$ |
| $a_t$ | Total rate of attenuation by particles | dB $km^{-1}$ |
| $A_e(\theta, \phi)$ | Effective (or apparent) aperture area of the antenna | $m^2$ |
| $A_0$ | Amplitude of the bipolar video | |
| $A_\eta$ | Physical (or geometrical) aperture area of the antenna | $m^2$ |
| $A(t)$ | Amplitude of the bipolar video as a function of time | |
| $B_n$ | Noise band | MHz |
| $B$ | Pass band at 3 dB | MHz |
| $C_n^2$ | Structure constant of the fluctuations in the refractive index | $cm^{-2/3}$ |
| $C_V^2$ | Structure constant of the fluctuations in velocity | $m^2$ $s^{-2}$ $m^{-2/3}$ |
| $C_T^2$ | Structure constant of the fluctuations in air temperature | $°C^2$ $m^{-2/3}$ |
| $c$ | Velocity of light | m $s^{-1}$ |
| $C_d$ | Drag coefficient | |
| $d$ | Antenna diameter | m |
| $D$ | Diameter of drops | $\mu$m or mm |
| $D_0$ | Median volume diameter | $\mu$m or mm |
| $D(\theta, \phi)$ | One-way antenna directivity ($D_0$ on the beam axis) | |
| $D_n(r)$ | Structure function of the refractive index | |
| $e$ | Partial pressure of water vapor | h Pa |
| $E$ | Electric field intensity | |
| $E(t)$ | Instantaneous voltage of the signal scattered by a distributed target | |
| $f_n^2(\theta, \phi)$ | Normalized one way radiation pattern | |
| $f_0$ | Transmitted frequency | MHz or GHz |
| $f_c$ | Intermediate frequency | MHz |
| $f_d$ | Doppler frequency ($< 0$ for $V_r > 0$) | Hz |
| $f_r$ | Repetition frequency | Hz |
| $F_n$ | Noise factor | dB |
| $g$ | Acceleration due to gravity | m $s^{-2}$ |
| $G(\theta, \phi)$ | One-way antenna gain ($G_0$ on the beam axis) | |
| $G_0$ | Gain in power on the antenna axis ($= G$) | |
| $G(\theta, \phi)$ | Gain function of the antenna for a path | |
| $h$ | Spatial extent of the pulse ($= c\tau$) | m |
| $H$ | Intensity of the magnetic field | A $m^{-1}$ |
| Im( ) | Imaginary part of | |
| $k$ | Absorption coefficient of a material | |
| $k$ | Number of independent samples or order of a sample | |
| $k$ | Wave number | |
| $\|K\|^2$ | Dielectric factor | |
| $k_c$ | One-way coefficient of attenuation by clouds | dB $km^{-1}$ |

## List of Symbols

| Symbol | Meaning | Usual Units |
|---|---|---|
| $k_p$ | One-way coefficient of attenuation by precipitation | dB $km^{-1}$ |
| $l_m$ | Lower limit of the inertial domain | |
| $L_0$ | Upper limit of the inertial domain | |
| $L^2$ | Attenuation due to the propagation for two paths | |
| $L_r$ | Attenuation due to the finite width of the receiver band | |
| $m$ | Complex refractive index ($= n - ik$) | |
| $m_n$ | Statistical moment of order $n$ | |
| $M$ | Liquid water content | g $m^{-3}$ |
| $n$ | Ordinary refractive index | |
| $N$ | Brunt-Väisälä oscillation frequency | |
| $N$ | Refractivity ($= (n - 1)\ 10^6$) | $N$ unit |
| $N(D)$ | Distribution function of particles with diameter $D$ | $cm^{-4}$ |
| $N_e$ and $N_0$ | Average power of the noise at the input ($i$) and output ($o$) of the receiver | |
| $p$ | Pressure | hPa |
| $P_{av}$ | Transmitted power (average value) | W |
| $\bar{P}_r$ | Average received power | mW |
| $P_t$ | Transmitted power (peak value) | kW |
| $Q_a$ | Absorption cross section | $cm^2$ |
| $Q_s$ | Total scattering cross section | $cm^2$ |
| $Q_t$ | Total attenuation cross section | $cm^2$ |
| $r$ | Radar-target range | km |
| $r_{max}$ | Maximum unambiguous radar-target range | km |
| $R$ | Precipitation intensity | mm $h^{-1}$ |
| $R_f$ | Richardson flux number | |
| $R_i$ | Richardson gradient number | |
| $R(\tau)$ | Complex autocorrelation function | |
| $R$ | Radius of the earth | km |
| $R'$ | Equivalent radius of the earth | km |
| Re | Real part of | |
| $S'(f)$ | Power spectral density of the doppler frequency | |
| $S(f)$ | Normalized power spectral density | |
| $S_i$ | Incident power flux density | W $S_r^{-1}$ |
| $S_{min}$ | Minimum detectable signal | W |
| $t$ | Time | s |
| $t$ | Ordinary temperature | °C |
| $T$ | Absolute temperature | K |
| $T_e$ | Equivalent noise temperature at the input of the receiver | K |
| $T_s$ | Sampling time | |
| $u$ | Doppler velocity along $0x$ | m $s^{-1}$ |
| $v$ | Doppler velocity along $vy$ | m $s^{-1}$ |
| $V$ | Pulse (or sampling) resolution volume | $m^3$ or $km^3$ |
| $V_t$ | Terminal fall velocity in still air | m $s^{-1}$ |

List of Symbols

| Symbol | Meaning | Usual Units |
|---|---|---|
| $V_r$ | Radial component of the doppler velocity | $m\ s^{-1}$ |
| $V_h$ | Horizontal component of the doppler velocity | $m\ s^{-1}$ |
| $w$ | Doppler velocity along $0z$ | $m\ s^{-1}$ |
| $w_d$ | Air velocity along $0z$ | $m\ s^{-1}$ |
| $X(t)$ | Complex amplitude associated with the real signal | |
| $Y(f)$ | Spectral density function of the doppler frequency | |
| $z$ | Vertical coordinate | |
| $Z$ | Radar reflectivity factor | $mm^6\ m^{-3}$ |
| $Z_e$ | Equivalent radar reflectivity factor | $mm^6\ m^{-3}$ |
| $\alpha$ | Elevation angle | deg |
| $\beta$ | Azimuthal angle | deg |
| $\varepsilon$ | Dielectric permittivity | $F\ m^{-1}$ |
| $\varepsilon$ | Rate of dissipation of turbulent kinetic energy | $m^2\ s^{-3}$ |
| $\eta$ | Radar reflectivity | $cm^{-1}$ |
| $\eta_l$ | Antenna efficiency | |
| $\theta$ | Potential temperature | K |
| $\theta$ | Angular coordinate with respect to the beam axis in the horizontal plane | deg |
| $\theta_0$ | 3 dB beamwidth | deg |
| $\lambda$ | Wave length ($= c/f$) | cm |
| $\mu$ | Magnetic permeability | $H\ m^{-1}$ |
| $\rho$ | Density | |
| $\rho(\tau)$ | Autocorrelation function of the signal intensity as a function of the autocorrelation interval | |
| $\sigma$ | Radar backscattering cross section | $m^2$ or $cm^2$ |
| $\sigma_f^2$ | Variance of the frequency spectrum | $Hz^2$ |
| $\sigma_r^2$ | Contribution to the variance of the doppler spectrum due to the fall velocities of the precipitation | $Hz^2$ or $m^2\ s^{-2}$ |
| $\sigma_t^2$ | Contribution to the variance of the doppler spectrum due to turbulence | $Hz^2$ or $m^2\ s^{-2}$ |
| $\sigma_V^2$ | Variance of the velocity spectrum | $m^2\ s^{-2}$ |
| $\tau$ | Duration of the pulse | $\mu s$ |
| $\tau$ | Cross-correlation time interval | |
| $\phi$ | Angular coordinate with respect to the beam axis in the vertical plane | deg |
| $\omega$ | Angular frequency ($= 2\pi f$) | $rad\ s^{-1}$ or $deg\ s^{-1}$ |
| $\Omega$ | Solid angle | Sr |
| $*$ | Complex conjugate | |
| $\langle\ \rangle$ | Average of states in space | |
| $\hat{x}$ | Estimated value of $x$ | |
| $\bar{x}$ | Average value of $x$ | |
| $x'$ | Deviation of $x$ with respect to its average value | |

# *Appendix 3*
# *List of Constants*

| | |
|---|---|
| Boltzmann's constant $k$ | $1.381 \times 10^{-23}$ J K$^{-1}$ |
| Planck's constant $h$ | $6.6256 \times 10^{-34}$ J s |
| Stefan-Boltzmann's constant $\sigma$ | $5.670 \times 10^{-8}$ J m$^{-2}$ K$^{-4}$ s$^{-1}$ |
| Velocity of light $c$ | $2.998 \times 10^{8}$ m s$^{-1}$ |
| Earth mean radius | 6371 km |
| Standard surface gravity | 9.80665 m s$^{-2}$ |
| Standard pressure $p_0$ | $1.01325 \times 10^{5}$ Pa (= 1013.25 mb) |
| Latent heat of vaporization (288 K) | $2.47 \times 10^{6}$ J kg$^{-1}$ |
| Latent heat of fusion | $3.34 \times 10^{5}$ J kg$^{-1}$ |
| Dry adiabatic lapse rate ($\Gamma$) | 9.8 K km$^{-1}$ |
| Density of dry air (273 K; 1013 hPa) | 1.293 kg m$^{-3}$ |
| Density of ice (273 K) | 917 kg m$^{-3}$ |
| Impedance of free space $\eta_0 = (\mu_0/\varepsilon_0)^{1/2}$ | 376.7 Q |
| Joule's constant | 4.187 cal$^{-1}$ |

$\log x \equiv \log_{10} x = 0.4343 \ln x$

$\ln x \equiv \log_e x = 2.3026 \log x$

# *Appendix 4*
# *Definitions and Various Numerical Values*

**Table A4.1**
Designation of the Radar Frequency Bands

| *Band* | *Frequency (MHz)* | *Wavelength (cm)* | *Band* | *Frequency (GHz)* | *Wavelength (cm)* |
|---|---|---|---|---|---|
| VHF | 30–300 | 1000–100 | *X* | 8–12.5 | 3.75–2.40 |
| UHF | 300–1000 | 100–30 | $K_u$ | 12.5–18 | 2.40–1.67 |
| *L* | 1000–2000 | 30–15 | *K* | 18–26.5 | 1.67–1.13 |
| *S* | 2000–4000 | 15–7.50 | $K_a$ | 26.5–40 | 1.13–0.75 |
| *C* | 4000–8000 | 7.50–3.75 | | | |

**Table A4.2(a)**
Variation of Refractive Index of Water as a Function of the Wavelength and Temperature (from [Gunn and East, 1954])

| | *Temperature (°C)* | *Wavelength (cm)* 10 | 3.21 | 1.24 | 0.62 |
|---|---|---|---|---|---|
| $n$ | 20 | 8.88 | 8.14 | 6.15 | 4.44 |
| | 10 | 9.02 | 7.80 | 5.45 | 3.94 |
| | 0 | 8.99 | 7.14 | 4.75 | 3.45 |
| | −8 | | 6.48 | 4.15 | 3.10 |
| $k$ | 20 | 0.63 | 2.00 | 2.86 | 2.59 |
| | 10 | 0.90 | 2.44 | 2.90 | 2.37 |
| | 0 | 1.47 | 2.89 | 2.77 | 2.04 |
| | −8 | | | 2.55 | 1.77 |
| $\|K\|^2$ | 20 | 0.928 | 0.9275 | 0.9193 | 0.8926 |
| | 10 | 0.9313 | 0.9282 | 0.9152 | 0.8726 |
| | 0 | 0.9340 | 0.9300 | 0.9055 | 0.8312 |
| | −8 | | | 0.8902 | 0.7921 |
| $\mathrm{Im}(-K)$ | 20 | 0.00474 | 0.01883 | 0.0471 | 0.0915 |
| | 10 | 0.00688 | 0.0247 | 0.0615 | 0.1142 |
| | 0 | 0.01102 | 0.0335 | 0.0807 | 0.1441 |
| | −8 | | | 0.1036 | 0.1713 |

**Table A4.2(b)**
Variation of Refractive Index of Ice as a Function of Temperature (from [Gunn and East, 1954])

| *Quantity* | *Temperature (°C)* | *Value* |
|---|---|---|
| $n$ | At all temperatures for $\rho = 0.92$ g cm$^{-3}$ | 1.78 |
| $k$ | 0 | 2.4 $10^{-3}$ |
| | −10 | 7.9 $10^{-4}$ |
| | −20 | 5.5 $10^{-4}$ |
| $\|K\|^2$ | at all temperatures | 0.176* |
| | | 0.208** |
| Im($-K$) | 0 | 9.6 $10^{-4}$ |
| | −10 | 3.2 $10^{-4}$ |
| | −20 | 2.2 $10^{-4}$ |

*Notes:*
*For $\rho = 0.92$ g cm$^{-3}$.
**For $\rho = 1$g cm$^{-3}$.

**Table A4.3(a)**
Attenuation Rate* for Water Vapor (in dB km$^{-1}$) (from [Gunn and East, 1954])

| $\lambda$ *(cm)* | *Temperature* 20°C | | 0°C | | −20°C | | −40°C | |
|---|---|---|---|---|---|---|---|---|
| 10 | 7 | $PW \times 10^{-5}$ | 8 | $PW \times 10^{-5}$ | 9 | $PW \times 10^{-5}$ | 10 | $PW \times 10^{-5}$ |
| 5.7 | 2.4 | $PW \times 10^{-4}$ | 2.7 | $PW \times 10^{-4}$ | 3.0 | $PW \times 10^{-4}$ | 3.4 | $PW \times 10^{-4}$ |
| 3.2 | 7 | $PW \times 10^{-4}$ | 8 | $PW \times 10^{-4}$ | 9 | $PW \times 10^{-4}$ | 10 | $PW \times 10^{-4}$ |
| 1.8** | 4.3 | $PW \times 10^{-3}$ | 4.8 | $PW \times 10^{-3}$ | 5.0 | $PW \times 10^{-3}$ | 5.4 | $PW \times 10^{-3}$ |
| 1.24** | 2.2 | $P^{-1}W \times 10^{-2}$ | 2.33 | $P^{-1}W \times 10^{-2}$ | 2.46 | $P^{-1}W \times 10^{-2}$ | 2.61 | $P^{-1}W \times 10^{-2}$ |
| 0.9 | 9.5 | $PW \times 10^{-3}$ | 1.04 | $PW \times 10^{-2}$ | 1.14 | $PW \times 10^{-2}$ | 1.26 | $PW \times 10^{-2}$ |

*Notes:*
$\lambda$ is the wavelength, $P$ is the pressure in atmosphere, and $W$ the water vapor content in g m$^{-3}$.
*For one way.
**In the neighborhood of 1.35 cm water vapor absorption line, no simple power law is accurate.

**Table A4.3(b)**
Attenuation* by Molecular Oxygen at 20°C and under One Atmosphere

| *λ (cm)* | *Attenuation (dB km$^{-1}$)* |
|---|---|
| 10 | $6.5 \times 10^{-3}$ |
| 7.5 | $7.0 \times 10^{-3}$ |
| 3.2 | $7.2 \times 10^{-3}$ |
| 1.8 | $7.5 \times 10^{-3}$ |
| 1.5 | $8.5 \times 10^{-3}$ |
| 1.25 | $1.4 \times 10^{-2}$ |
| 0.8 | $7.5 \times 10^{-2}$ |
| 0.7 | $1.9 \times 10^{0}$ |

*For one way.

**Table A4.3(c)**
Correction Factor for the Attenuation by Oxygen

| *Temperature (°C)* | *Correction Factor* |
|---|---|
| 20 | $1.00\ P^2$ |
| 0 | $1.19\ P^2$ |
| −20 | $1.45\ P^2$ |
| −40 | $1.78\ P^2$ |

*Note:*
$P$ is the pressure in atmosphere.

**Table A4.4(a)**

Attenuation Rate* for Clouds or Precipitation (in dB $km^{-1}$) (from [Gunn and East, 1954])

| | $T$ (°C) | $\lambda = 10$ cm | $\lambda = 5.7$ cm | $\lambda = 3.2$ cm | $\lambda = 1.8$ cm | $\lambda = 1.24$ cm | $\lambda = 0.9$ cm |
|---|---|---|---|---|---|---|---|
| Rain** | 18 | $0.0003\ R^{1.00}$ | $0.0022\ R^{1.17}$ | $0.0074\ R^{1.31}$ | $0.045\ R^{1.14}$ | $0.12\ R^{1.05}$ | $0.22\ R^{1.00}$ |
| $R$ in mm $h^{-1}$ | | | | | | | |
| Snow*** | 0 | $0.035 \times 10^{-5} R^{1.6}$ | $0.33 \times 10^{-5} R^{1.6}$ | $3.3 \times 10^{-5} R^{1.6}$ | $33.2 \times 10^{-5} R^{1.6}$ | | |
| $R$ in mm $h^{-1}$ | | $+22.0 \times 10^{-5} R$ | $+38.5 \times 10^{-5} R$ | $+68.6 \times 10^{-5} R$ | $+122 \times 10^{-5} R$ | | |
| of melted | −10 | $0.035 \times 10^{-5} R^{1.6}$ | $0.33 \times 10^{-5} R^{1.6}$ | $3.3 \times 10^{-5} R^{1.6}$ | $33.2 \times 10^{-5} R^{1.6}$ | | |
| water | | $+7.3 \times 10^{-5} R$ | $+12.9 \times 10^{-5} R$ | $+22.9 \times 10^{-5} R$ | $+40.6 \times 10^{-5} R$ | | |
| | −20 | $0.035 \times 10^{-5} R^{1.6}$ | $0.33 \times 10^{-5} R^{1.6}$ | $3.3 \times 10^{-5} R^{1.6}$ | $33.2 \times 10^{-5} R^{1.6}$ | | |
| | | $+5.0 \times 10^{-5} R$ | $+8.8 \times 10^{-5} R$ | $+15.7 \times 10^{-5} R$ | $+28.0 \times 10^{-5} R$ | | |
| Water cloud | 20 | $0.39 \times 10^{-2} M$ | $1.36 \times 10^{-2} M$ | $4.83 \times 10^{-2} M$ | $12.8 \times 10^{-2} M$ | $31.1 \times 10^{-2} M$ | $64.7 \times 10^{-2} M$ |
| $M$ in g $m^{-3}$ | 10 | $0.56 \times 10^{-2} M$ | $1.96 \times 10^{-2} M$ | $6.30 \times 10^{-2} M$ | $17.9 \times 10^{-2} M$ | $40.6 \times 10^{-2} M$ | $68.1 \times 10^{-2} M$ |
| | 0 | $0.90 \times 10^{-2} M$ | $2.72 \times 10^{-2} M$ | $8.58 \times 10^{-2} M$ | $26.7 \times 10^{-2} M$ | $53.2 \times 10^{-2} M$ | $99 \times 10^{-2} M$ |
| | −8 | | $3.4 \times 10^{-2} M$ | $11.2 \times 10^{-2} M$ | $34 \times 10^{-2} M$ | $68 \times 10^{-2} M$ | $125 \times 10^{-2} M$ |
| | | | (Extrapolated) | (Extrapolated) | (Extrapolated) | | |
| Ice cloud | 0 | $7.87 \times 10^{-4} M$ | $13.8 \times 10^{-4} M$ | $24.6 \times 10^{-4} M$ | $43.6 \times 10^{-4} M$ | $63.5 \times 10^{-4} M$ | $87.4 \times 10^{-4} M$ |
| $M$ in g $m^{-3}$ | −10 | $2.62 \times 10^{-4} M$ | $4.60 \times 10^{-4} M$ | $8.19 \times 10^{-4} M$ | $14.6 \times 10^{-4} M$ | $21.1 \times 10^{-4} M$ | $29.3 \times 10^{-4} M$ |
| | −20 | $1.80 \times 10^{-4} M$ | $3.16 \times 10^{-4} M$ | $5.63 \times 10^{-4} M$ | $10.0 \times 10^{-4} M$ | $14.5 \times 10^{-4} M$ | $20.0 \times 10^{-4} M$ |

*Notes:*

*For one way.

**Empirical relationships.

***These values have been obtained by using the Rayleigh approximation, which loses its validity at wavelengths less than approximately 1.5 cm. The value $R = 10$ or $R^{1.6} = 40$ can be taken as an upper limit for the intensity of snow precipitation.

**Table A4.4(b)**
Parameters of the Relationship $a_p = kR^{\gamma}$ (equation (2.31a)) for Horizontal Polarization ($H$) and Vertical Polarization ($V$) for Rain at 18°C
(according to the [Commission of the European Communities, 1978])

| *Frequency (GHz)* | $k$ | | $\gamma$ | |
|---|---|---|---|---|
| | $H$ | $V$ | $H$ | $V$ |
| 10 | 0.010102 | 0.008870 | 1.2762 | 1.2637 |
| 15 | 0.036684 | 0.034680 | 1.1538 | 1.1279 |
| 20 | 0.075082 | 0.069061 | 1.0992 | 1.0649 |
| 25 | 0.1244 | 0.1130 | 1.0608 | 1.0300 |
| 30 | 0.1871 | 0.1674 | 1.0205 | 0.9997 |
| 35 | 0.2629 | 0.2334 | 0.9789 | 0.9633 |
| 40 | 0.3495 | 0.3098 | 0.9391 | 0.9287 |
| 45 | 0.4424 | 0.3932 | 0.9032 | 0.8965 |
| 50 | 0.5362 | 0.4793 | 0.8725 | 0.8683 |
| 55 | 0.6253 | 0.5632 | 0.8471 | 0.8443 |
| 60 | 0.7069 | 0.6419 | 0.8261 | 0.8243 |
| 65 | 0.7816 | 0.7149 | 0.8083 | 0.8073 |
| 70 | 0.8514 | 0.7836 | 0.7930 | 0.7925 |
| 75 | 0.9167 | 0.8477 | 0.7798 | 0.7799 |
| 80 | 0.9753 | 0.9063 | 0.7687 | 0.7693 |
| 85 | 1.0248 | 0.9572 | 0.7598 | 0.7606 |
| 90 | 1.0640 | 0.9992 | 0.7529 | 0.7537 |
| 95 | 1.0938 | 1.0326 | 0.7476 | 0.7483 |
| 100 | 1.1164 | 1.0590 | 0.7434 | 0.7441 |

**Table A4.5**
Characteristics of Meteorological Radars

| *Designation* | $\lambda$ *(cm)* | $f_r$ *(Hz)* | $\tau$ *(μs)* | $P_t$ *(kW)* | *Diameter of Antenna (m)* | $\theta$ *(3 dB) (deg)* | *P min* (dBm)* |
|---|---|---|---|---|---|---|---|
| MELODI | 10.7 | 1000 | 0.5 | 700 | 4 | 1.8 | −106 |
| (Météorologie Française-André, 1977) | | 250 | 2 | | | | |
| ANATOL | 10.3 | 500 | 0.5 | 250 | 4 | 1.8 | −97 |
| (Observatoire du Puy de Dôme; Waldteufel et al. 1978) | | | | | | | |
| RONSARD | | | | | | | |
| (doppler) | 5.3 | 2928 | 0.34 | 250 | 4 | 0.9 | −115 |
| (CNET—Paris; Nutten | | 1464 | 0.68 | | | | |
| et al., 1979) | | 732 | 1.36 | | | | |
| AN/CPS 9 | 3.2 | 186 | 5 | 250 | 2.3 | 1 | −104 |
| (USA—Speed, 1965) | | 931 | 0.5 | | | | |
| RABELAIS | | | | | | | |
| (doppler) | 0.86 | 1172 | 0.5 | 100 | 1.4 | 0.4 | −100 |
| (Observatoire du Puy de Dôme; Sauvageot et al., 1979) | | | | | | | |
| FM-CW | 10 | 0.15 | | | 3 | 2.3 | −150 |
| (USA—Richter, 1969) | | | | | | | |
| JAFNA | 71.5 | 320 | 0.2 | 6000 | 18.4 | 2.9 | −105 |
| (USA—Wallops Island; | | 960 | 3.0 | | | | |
| Glover et al. 1966) | 10.7 | 320 | 0.1 | 3000 | 18.4 | 0.48 | −110 |
| | | 960 | 1.3 | | | | |
| | 3.2 | 320 | 2 | 900 | 10.4 | 0.21 | −101 |

**Table A4.6**
Characteristics of Various Meteorological Radars

| | $\lambda$ (cm) | $P_t$ (W) | $\tau$ ($\mu s$) | $f_r$ (Hz) | Antenna Size (m) | Beamwidth (deg) | $P_{min}$ (dBm) | $Z_{min}$ at 10 km ($mm^6\ m^{-3}$) | Doppler Capability | Polarization Capability |
|---|---|---|---|---|---|---|---|---|---|---|
| NOAA—Wind profiler (50 MHz) (USA) | 600 | $0.3 \times 10^5$ | 3.9 | 238 | 50 × 50* | 5** | | | Y | N |
| | | | | 672 | | | | | | |
| NOAA—Wind profiler (405 MHz) (USA) | 74 | $0.3 \times 10^5$ | 1 | 10000 | 9 × 9* | 4.3** | | | Y | N |
| | | | 3 | 6600 | | | | | | |
| | | | 9 | 3300 | | | | | | |
| NOAA—Wind profiler (915 MHz) (USA) | 32 | $5.6 \times 10^3$ | 1 | 20000 | 10 × 10* | 1.7** | | | Y | N |
| | | | 3 | 15600 | | | | | | |
| | | | 9 | 9000 | | | | | | |
| CHATANIKA radar (USA) | 23.2 | $3.2 \times 10^6$ | 10 | 294 | 26.8 | 0.6 | −128 | | Y | N |
| WSR 57 (USA) | 10.3 | $4.5 \times 10^5$ | 4 | 164 | 3.65 | 2 | −108 | $1.2 \times 10^{-2}$ | N | N |
| WSR 88 (USA) | 10.3 | $7.5 \times 10^5$ | 1.6 | 318–1304 | 8.54 | 0.95 | −110 | | Y | N |
| | | | 4.5 | 318–452 | | | | | Y | N |
| NSSL (USA) | 10.7 | $5 \times 10^5$ | 1 | 1300 | 9.1 | 0.9 | −110 | $4.5 \times 10^{-3}$ | Y | Y |

| | | | | | | | | | | |
|---|---|---|---|---|---|---|---|---|---|---|
| WALLOPS | | | | | | | | | | |
| *S* band (USA) | 10.7 | $3 \times 10^6$ | 0.1 | 320 | 18.4 | 0.48 | −110 | | N | N |
| | | | 1.3 | 960 | | | | | | |
| *X* band | 3.2 | $9 \times 10^5$ | 2 | 320 | 10.4 | 0.21 | −101 | | N | N |
| CHILBOLTON (UK) | 9.75 | $5.0 \times 10^5$ | 0.5 | 625 | 25 | 0.25 | | | Y | Y |
| NCAR CP | | | | | | | | | | |
| *S* band (USA) | 10.7 | $1.2 \times 10^6$ | 0.15 | <1700 | 8.5 | 0.93 | | $2.5 \times 10^{-3}$ | Y | Y |
| *X* band | 3.2 | $2 \times 10^5$ | 1 | <1000 | 2.4 | 0.94 | | | N | Y |
| FM-CW-WPL (USA) | 10.2 | 200(AV) | | | 2.4 | 2.7 | −150 | $6.2 \times 10^{-3}$ | Y | N |
| RODIN (FR) | 5.6 | $2.5 \times 10^5$ | 2 | 330 | 3 | 1.3 | −107 | | N | N |
| RONSARD (FR) | 5.3 | $2.5 \times 10^5$ | 0.34 | 2928 | 4 | 0.9 | −115 | | Y | N |
| | | | 0.68 | 1464 | | | | | | |
| | | | 1.36 | 732 | | | | | | |
| DFVLR (Germany) | 5.3 | $5 \times 10^5$ | 0.5 | 490 | 5 | 1 | | | Y | Y |
| | | | 1.2 | 1200 | | | | | | |
| | | | | 2400 | | | | | | |
| NOAA ERL/WPL (USA) | 3.2 | $0.9 \times 10^5$ | 0.5 | 2000 | 3.1 | 0.86 | −106 | $1.0 \times 10^{-2}$ | Y | Y |
| SHAPE (Netherlands) | 3.1 | $1 \times 10^5$ | 0.4 | 1–10,000 | 1.8 | 1.2 | | | Y | Y |
| RABELAIS (FR) | 0.86 | $0.7 \times 10^5$ | 0.3 | 2688/3125 | 1.4 | 0.4 | −100 | $5 \times 10^{-4}$ | Y | Y |
| | | | 0.5 | | | | | | | |
| NOAA ERL/WPL (USA) | 0.86 | $1.5 \times 10^5$ | 0.25 | 2000 | 1.12 | 0.5 | −106 | $4.2 \times 10^{-4}$ | Y | Y |

*Notes:*
*In $m^2$.
**For two ways.

# References


Aden, A.L., and M. Kerker, 1951, "Scattering of Electromagnetic Waves by Two Concentric Spheres," *J. Applied Physics,* Vol. 22, pp. 1242–1246.

Adler, R.F., and A.J. Negri, 1988, "A Satellite Infrared Technique to Estimate Tropical Convective and Stratiform Rainfall," *J. Applied Meteorology,* Vol. 27, pp. 30–51.

Anthes, R.A., 1982, *Tropical Cyclones: Their Evolution, Structure and Effects.* Meteorology Monographs, No. 19(41), American Meteorological Society, Boston.

Arakawa, H., 1967, "A Hook-Shaped Echo Accompanied by a Mesocyclone Associated with the Catastrophic Rainstorm of 25 July 1957 in Kyushu, Japan," *J. Applied Meteorology,* Vol. 6, pp. 439–441.

Armijo, L., 1969, "A Theory for the Determination of Wind and Precipitation Velocities with Doppler Radars," *J. Atmospheric Science,* Vol. 26, pp. 570–573.

Armstrong, G.M. and R.J. Donaldson, Jr., 1969, "Plan Shear Indicator for Real-Time Doppler Radar Identification of Hazardous Storm Winds," *J. Applied Meteorology,* Vol. 8, pp. 376–383.

Aso, T., S. Kato, and R.M. Harper, 1977, "Arecibo Middle Atmosphere Experiment," *Geophysics Research Letters,* Vol. 4, pp. 10–12.

Atlas, D., 1954, "The Estimation of Cloud Parameters by Radar," *J. Meteorology,* Vol. 11, pp. 309–317.

Atlas, D., 1963, "Radar Analysis of Severe Storms. In Severe Local Storms," *Meteorology Monographs,* Vol. 5, No. 27, pp. 177–220.

Atlas, D., 1964, "Advances in Radar Meteorology," *Advances in Geophysics,* Vol. 10, pp. 318–478.

Atlas, D., ed., 1990, *Radar in Meteorology,* American Meteorological Society, Boston.

Atlas, D., J. Eckerman, R. Meneghini, and R.K. Moore, 1982, "The Outlook of Precipitation Measurements from Space," *Atmos. Ocean.,* Vol. 20, pp. 50–61.

Atlas, D., K.R. Hardy, and J. Joss, 1964, "Radar Reflectivities of Storms Containing Spongy Hail," *J. Geophysics Research,* Vol. 69, pp. 1955–1961.

Atlas, D., K.R. Hardy, and K. Naito, 1966, "Optimizing the Radar Detection of Clear-Air Turbulence," *J. Applied Meteorology,* Vol. 5, pp. 450–460.

Atlas, D., W.G. Harper, F.H. Ludlam, and W.C. Macklin, 1960, "Radar Scatter by Large Hail," *Quarterly J. Royal Meteorological Society,* Vol. 86, pp. 468–482.

Atlas, D., M. Kerker, and W. Hitschfeld, 1953, "Scattering and Attenuation by Non-Spherical Atmospheric Particles," *J. Atmospheric and Terrestrial Physics,* Vol. 3, pp. 108–119.

Atlas, D., and F.H. Ludlam, 1961, "Multi-Wavelength Radar Reflectivity of Hailstorms," *Quarterly J. Royal Meteorological Society,* Vol. 87, pp. 523–534.

Atlas, D., J.I. Metcalf, J.H. Richter, and E.E. Gossard, 1970, "The Birth of 'CAT' and Microscale Turbulence," *J. Atmospheric Science,* Vol. 27, pp. 903–913.

Atlas, D., D. Rosenfeld, and D.A. Short, 1990, "The Estimation of Convective Rainfall by Area Integrals. 1. The Theoretical and Empirical Basis," *J. of Geophysical Research,* Vol. 95, No. D3, pp. 2153–2160.

Atlas, D., and R.C. Srivastava, 1971, "A Method of Radar Turbulence Detection," *IEEE Trans. on Aerospace and Electronic Systems,* Vol. AES-7, pp. 179–187.

Atlas, D., R.C. Srivastava, and R.S. Sekhon, 1973, "Doppler Radar Characteristics of Precipitation at Vertical Incidence," *Rev. of Geophysics and Space Physics,* Vol. 11, No. 1, pp. 1–35.

Atlas, D., R.C. Srivastava, and P.W. Sloss, 1969, "Wind Shear and Reflectivity Gradient Effects on Doppler Spectra," *J. Applied Meteorology,* Vol. 8, pp. 384–388.

Atlas, D., R. Tatehira, R.C. Srivastava, W. Marker, and R.E. Carbone, 1969, "Precipitation Induced Mesoscale Wind Perturbations in the Melting Layer," *Quarterly J. Royal Meteorological Society,* Vol. 95, pp. 544–560.

Atlas, D., and C.W. Ulbrich, 1974, "The Physical Basis for Attenuation Rainfall Relationships and the Measurement of Rainfall Parameters by Combined Attenuation and Radar Methods," *J. Rech. Atmos.,* Vol. 8, pp. 275–298.

Atlas, D., and C.W. Ulbrich, 1977, "Path and Area Integrated Rainfall Measurement by Microwave Attenuation in the 1–3 cm Band," *J. Applied Meteorology,* Vol. 16, pp. 1322–1331.

Auer, A.H., 1974, "Distribution of Graupel and Hail with Size," *Monthly Weather Rev.,* Vol. 100, pp. 325–328.

Austin, G.L., and A. Bellon, 1982, "Very-Short-Range Forecasting of Precipitation by the Objective Extrapolation of Radar and Satellite Data," K.A. Browning, ed., *Nowcasting,* Academic Press London, pp. 177–190.

Austin, P.M. and R.A. Houze, 1972, "Analysis of the Structure of Precipitation Patterns in New England," *J. Applied Meteorology,* Vol. 11, pp. 926–935.

Aydin, K., T.A. Seliga, and V. Balaji, 1986, "Remote Sensing of Hail with Dual Linear Polarization Radar," *J. Climate and Applied Meteorology,* Vol. 25, pp. 1475–1484.

Balsley, B.B., N. Cianos, D.T. Farley, and M.J. Baron, 1977, "Winds Derived from Radar Measurements in the Arctic Troposphere and Stratosphere," *J. Applied Meteorology,* Vol. 16, pp. 1235–1239.

Balsley, B.B., and D.T. Farley, 1976, "Auroral Zone Winds Detected near the Tropopause with the Chatanika UHF Doppler Radar," *Geophysics Research Letters,* Vol. 3, pp. 525–528.

Balsley, B.B., and K.S. Gage, 1982, "On the Use of Radars for Operational Wind Profiling," *Bull. American Meteorological Society,* Vol. 63, 1009–1018.

Barat, J., and F. Bertin, 1984, "Simultaneous Measurements of Temperature and Velocity Fluctuations within Clear Air Turbulence Layers: Analysis of the Estimate of Dissipation Rate by Remote Sensing Techniques," *J. Atmospheric Science,* Vol. 41, pp. 1613–1619.

Barge, B.L., 1974, "Polarization Measurements of Precipitation Backscatter in Alberta," *J. Rech. Atmos.,* Vol. 8, nos. 1–2, pp. 164–173.

Barnes, A.A., 1972, "Using VHF Radars to Probe the Atmosphere," Environmental Research Paper 426, Air Force Cambridge Research Lab.

Barnes, G.M., E.J. Zipser, D.R. Jorgensen, and F. Marks, Jr., 1983, "Mesoscale and Convective Structure of Hurricane Rainband," *J. Atmospheric Science,* Vol. 40, pp. 2125–2137.

Barrett, E.C., and D.W. Martin, 1981, *The Use of Satellite Data in Rainfall Monitoring*. Academic Press, New York.

Barrick, D.E., 1973, "FM-CW Radar Signals and Digital Processing," NOAA Tech. Rep. ERL 283-WPL-26.

Battan, L.J., 1964, "Some Observations of Vertical Velocities and Precipitation Sizes in a Thunderstorm," *J. Applied Meteorology,* Vol. 3, pp. 415–420.

Battan, L.J., 1971, "Radar Attenuation by Wet Ice Spheres," *J. Applied Meteorology,* Vol. 10, pp. 247–252.

Battan, L.J., 1973, *Radar Observation of the Atmosphere,* The University of Chicago Press, Chicago.

Battan, L.J., 1975, "Doppler Radar Observations of a Hailstorm," *J. Applied Meteorology,* Vol. 14, pp. 98–108.

Battan, L.J., S.R. Browning, and B.M. Herman, 1970a, "Tables of the Radar Cross Sections of Dry and Wet Ice Spheres," Tech. Rep. 21, Tucson, Institute of Atmospheric Physics, University of Arizona.

Battan, L.J., S.R. Browning, and B.M. Herman, 1970b, "Attenuation of Microwaves by Wet Ice Spheres," *J. Applied Meteorology,* Vol. 9, pp. 832–834.

Battan, L.J., and B.M. Herman, 1962, "The Radar Cross Sections of 'Spongy' Ice Spheres," *J. Geophysics Research,* Vol. 67, pp. 5139–5145.

Battan, L.J., and J.B. Theiss, 1970, "Measurements of Vertical Velocities in Convective Clouds by Means of Pulsed-Doppler Radar," *J. Atmospheric Sciences,* Vol. 27, pp. 293–298.

Bean, B.R., and E.J. Dutton, 1968, *Radio Meteorology,* Dover Books, New York.

Beard, K.V., 1976, "Terminal Velocity and Shape of Cloud and Precipitation Drops Aloft," *J. Atmospheric Science,* Vol. 33, pp. 851–864.

Beard, K.V. and D. Chuang, 1987, "A New Model for the Equilibrium Shape of Raindrops," *J. Atmospheric Sciences,* Vol. 44, pp. 1509–1524.

Beard, K.V., and H.R. Pruppacher, 1969, "A Determination of the Terminal Velocity and Drag of Small Water Drops by Means of a Wind Tunnel," *J. Atmospheric Sciences,* Vol. 26, pp. 1066–1072.

Bechtel, M.E., and H.W. Prinsen, 1967, "Errors in *R*-Meter Measurement of Velocity Spread of Meteorological Targets Resulting from Radar Frequency Instabilities," *J. Applied Meteorology,* Vol. 6, pp. 57–60.

Bellon, A., and G.L. Austin, 1986, "On the Relative Accuracy of Satellite and Raingauge Rainfall Measurements over Middle Latitude during Daylight Hours," *J. Climate and Applied Meteorology,* Vol. 25, pp. 1712–1724.

Bendat, J.S., and A.G. Piersol, 1971, *Random Data—Analysis and Measurement Procedures,* Wiley-Interscience, New York.

Bertin, F., A. Cremieux, M. Glass, M. Massebeuf, and M. Petit-Didier, 1985, "The Proust Radar," *Handbook for MAP,* 20, p. 343.

Black, P.G., H.V. Senn, and C.L. Courtright, 1972, "Airborne Radar Observations of Eye Configuration Changes, Bright Band Distribution, and Precipitation Tilt during the 1969 Multiple Seeding Experiments in Hurricane Debbie," *Monthly Weather Rev.*, Vol. 100, pp. 208–217.

Blake, L.V., 1968, "Radio Ray (Radar) Range-Height-Angle Charts," Naval Research Lab. Rep. 6650.

Bluestein, H.B., 1986, "Fronts and Jet Streaks—A Theoretical Perspective," P.S. Ray, ed., *Mesoscale Meteorology,* American Meteorological Society, Boston, pp. 173–215.

Blevis, B.C., 1974, "Microwave Attenuation and Scatter—Theory and Measurement," *J. Rech. Atmos.*, Vol. 8, pp. 5–21.

Bolgiano, R., Jr., 1963, "Evidence of Anisotropy in Tropospheric Microstructure," *J. Geophysics Research,* Vol. 68, pp. 4873–4874.

Booker, H.G., and W.E. Gordon, 1950, "Theory of Radio Scattering in the Troposphere," *Proc. IEEE,* Vol. 38, pp. 401–412.

Borresen, J.A., 1971, "Doppler Radar Study of Shear Zones and Turbulence in a Snowstorm," *J. Applied Meteorology,* Vol. 10, pp. 433–442.

Boucher, R.J., and H. Ottersten, 1971, "Doppler Radar Observation of Wind Structure in Snow," *J. Applied Meteorology,* Vol. 10, pp. 228–233.

Braham, R.R., Jr., ed., 1986, *Precipitation Enhancement. A Scientific Challenge*. Meteorology Monographs No. 43, American Meteorological Society, Boston.

Brandes, E.A., 1975, "Optimizing Rainfall Estimates with the Aid of Radar," *J. Applied Meteorology,* Vol. 14, pp. 1339–1345.

Brandes, E.A., 1977, "Flow in Severe Thunderstorms Observed by Dual-Doppler Radar," *Monthly Weather Rev.,* Vol. 105, pp. 113–120.

Brantley, J.Q., and D.A. Barczys, 1957, "Some Weather Observations with Continuous Wave Doppler Radar," *Proc. 6th Weather Radar Conf.,* American Meteorological Society, Boston, pp. 297–306.

Brewster, K.A., and D.D. Zrnic, 1986, "Comparison of Eddy Dissipation Rates from Spatial Spectra of Doppler Velocities and Doppler Spectrum Widths," *J. Atmospheric and Oceanic Technology,* Vol. 3, pp. 440–452.

Bringi, V.N., and A. Hendry, 1990, "Technology of Polarization Diversity Radars for Meteorology," D. Atlas, ed., *Radar in Meteorology,* American Meteorological Society, pp. 153–190.

Bringi, V.N., T.A. Seliga, and K. Aydin, 1984, "Hail Detection with Differential Reflectivity Radar," *Science,* Vol. 225, pp. 1145–1157.

Bringi, V.N., J. Vivekanandan, and J.D. Tuttle, 1986, "Multiparameter Radar Measurements in Colorado Convective Storms. Part II: Hail Detection Studies," *J. Atmospheric Sciences,* Vol. 43, pp. 2564–2577.

Brown, R.A., L.R. Lemon, and D.W. Burgess, 1978, "Tornado Detection by Pulsed Doppler Radar," *Monthly Weather Rev.,* Vol. 106, pp. 29–38.

Browne, I.C., 1951, "A Radar Echo from Lightning," *Nature,* Vol. 167, p. 438.

Browning, K.A., 1964, "Airflow and Precipitation Trajectories within Severe Local Storms which Travel to the Right of the Winds," *J. Atmospheric Sciences,* Vol. 21, pp. 634–639.

Browning, K.A., 1971, "Structure of the Atmosphere in the Vicinity of Large Amplitude Kelvin-Helmholtz Billows," *Quarterly J. Royal Meteorological Society,* Vol. 97, pp. 283–299.

Browning, K.A., C.G. Collier, P.R. Larke, P. Menmuis, G.A. Monk, and R.G. Owens, 1982, "On the Forecasting of Frontal Rain Using a Weather Radar Network," *Monthly Weather Rev.,* Vol. 110, pp. 534–552.

Browning, K.A., G.W. Bryant, J.R. Starr, and D.N. Axford, 1973, "Air Motion within Kelvin-Helmholtz Billows Determined from Simultaneous Doppler Radar and Aircraft Measurements," *Quarterly J. Royal Meteorological Society,* Vol. 99, pp. 608–618.

Browning, K.A., J.C. Fankhauser, J.P. Chalon, P.J. Eccles, R.G. Strauch, F.H. Merrem, D.J. Musil, E.L. May and W.H. Sand, 1976, "Structure of an Evolving Hailstorm. Part V: Synthesis and Implications for Hail Growth and Hail Suppression," *Monthly Weather Rev.,* Vol. 104, pp. 603–610.

Browning, K.A. and G.B. Foote, 1976, "Airflow and Hail Growth in Supercell Storms and Some Implications for Hail Suppression," *Quarterly J. Royal Meteorological Society,* Vol. 102, pp. 499–533.

Browning, K.A., M.E. Hardman, T.W. Harrold, and C.W. Pardoe, 1973, "The Structure of Rainbands within a Mi-Latitude Depression," *Quarterly J. Royal Meteorological Society,* Vol. 99, pp. 215–231.

Browning, K.A., and T.W. Harrold, 1969, "Air Motion and Precipitation Growth in a Wave Depression," *Quarterly J. Royal Meteorological Society,* Vol. 95, pp. 288–309.

Browning, K.A., and T.W. Harrold, 1970, "Air Motion and Precipitation Growth in a Cold Front," *Quarterly J. Royal Meteorological Society,* Vol. 96, pp. 369–389.

Browning, K.A., T.W. Harrold, and J.R. Starr, 1970, "Richardson Number Limited Shear Zones in the Free Atmosphere," *Quarterly J. Royal Meteorological Society,* Vol. 96, pp. 40–49.

Browning, K.A., and F.H. Ludlam, 1962, "Airflow in Convective Storms," *Quarterly J. Meteorological Society,* Vol. 88, pp. 117–135.

Browning, K.A. and C.D. Watkins, 1970, "Observations of Clear Air Turbulence by High Power Radar," *Nature,* Vol. 227, pp. 260–263.

Browning, K.A., and R. Wexler, 1968, "A Determination of Kinematic Properties of a Wind Field Using Doppler Radar," *J. Applied Meteorology,* Vol. 7, pp. 105–113.

Bryant, G.W., and K.A. Browning, 1975, "Multi-Level Measurements of Turbulence over the Sea during the Passage of a Frontal Zone," *Quarterly J. Royal Meteorological Society,* Vol. 101, pp. 35–54.

Businger, S.A., and P.V. Hobbs, 1987, "Mesoscale Structure of Two Comma Cloud Systems over the Pacific Ocean," *Monthly Weather Rev.,* Vol. 115, pp. 1908–1928.

Byers, H.R., and R.R. Braham, Jr., 1949, *The Thunderstorm,* US Department of Commerce, Washington, DC.

Campistron, B., 1975, "Characteristic Distributions of Angel Echoes in the Lower Atmosphere and Their Meteorological Implications," *Boundary-Layer Meteorology,* Vol. 9, pp. 411–426.

Campistron, B., G. Despaux, and J.P. Lacaux, 1987, "A Microcomputer Data-Acquisition System for Real-Time Processing of Raindrop Size Distribution Measured with the RD69 Disdrometer," *J. Atmospheric and Oceanic Technology,* Vol. 4, pp. 536–540.

Campistron, B. and H. Sauvageot, 1974, "Sur l'évolution matinale de la distribution dans l'espace des échos radar en air clair," *C.R. Ac. Sc. Paris,* Série B, pp. 479–482.

Campistron, B., and H. Sauvageot, 1981, "Dual Doppler Radar Analysis of a Convection Line," *Proc. 20th Conf. on Radar Meteorology,* American Meteorological Society, Boston, pp. 71–76.

Carbone, R.E., 1977, "Tetrahedral Scan for Tri-Doppler Radar and Other Remote Sensing Applications," *J. Applied Meteorology,* Vol. 16, pp. 844–850.

Carbone, R.E., D. Atlas, P. Eccles, R. Fetter, and E. Mueller, 1973, "Dual Wavelength Radar Hail Detection," *Bull. American Meteorological Society,* Vol. 54, pp. 921–924.

Carbone, R.E., and A.R. Bohne, 1975, "Cellular Snow Generation—A Doppler Radar Study," *J. Atmospheric Science,* Vol. 32, pp. 1384–1394.

Carbone, R.E., M.J. Carpenter, and D.C. Burghart, 1985, "Doppler Radar Sampling Limitations in Convective Storms," *J. Atmospheric and Oceanic Technology,* Vol. 2, pp. 357–361.

Carbone, R.E., and L.D. Nelson, 1978, "The Evolution of Raindrop Spectra in Warm-Based Convective Storms as Observed and Numerically Modeled," *J. Atmospheric Science,* Vol. 12, pp. 2302–2314.

du Castel, F., 1966, *Tropospheric Radio Wave Propagation beyond the Horizon,* Pergamon Press, Oxford.

Cataneo, R.A., and G.E. Stout, 1968, "Raindrop-Size Distributions in Humid Continental Climatic and Associated Rainfall Rate-Radar Reflectivity Relationships," *J. Applied Meteorology,* Vol. 7, pp. 901–907.

Caylor, I.J., and A.J. Illingworth, 1987, "Radar Observations and Modelling of Warm Rain Initiation," *Quarterly J. Royal Meteorological Society,* Vol. 113, pp. 1171–1191.

Caylor, I.J., and A.J. Illingworth, 1989, "Identification of the Bright Band and Hydrometeors Using Co-Polar Dual Polarization Radar," *Proc. 24th Radar Meteorology Conf.,* American Meteorological Society, Boston, pp. 9–12.

Cerni, T.A., 1976, "Experimental Investigation of the Radar Cross Section of Cloud-to-Ground Lightning," *J. Applied Meteorology,* Vol. 15, pp. 795–798.

Chadwick, R.B., K.P. Moran, R.G. Strauch, G.E. Morrison, and W.C. Campbell, 1976, "A New Radar for Measuring Winds," *Bull. American Meteorological Society,* Vol. 57, pp. 1120–1125.

Chalon, J.P., G. Jaubert, F. Roux, and J.P. Lafore, 1988, "The West African Squall Line Observed on 23 June 1981 during COPT 81: Mesoscale Structure and Transports," *J. Atmospheric Science,* Vol. 45, pp. 2744–2763.

Chandrasekar, V., and V.N. Bringi, 1988, "Error Structure of Multiparameter Radar and Surface Measurements of Rainfall. Part I: Differential Reflectivity," *J. Atmospheric and Oceanic Technology,* Vol. 5, pp. 783–802.

Changnon, S.A., 1970, "Hailstreaks," *J. Atmospheric Science,* Vol. 27, pp. 109–125.

Changnon, S.A., 1977, "The Scales of Hail," *J. Applied Meteorology,* Vol. 16, pp. 626–648.

Cheng, L., and M. English, 1983, "A Relationship between Hailstone Concentration and Size," *J. Atmospheric Science,* Vol. 40, pp. 204–213.

Chernikov, A.A., Y.V. Mel'Nichuk, N.Z. Pinus, S.M. Shmeter, and N.K. Vinnichenko, 1969, "Investigations of the Turbulence in Convective Atmosphere Using Radar and Aircraft," *Radio Science,* Vol. 4, pp. 1257–1259.

Chiu, L.S., 1988, "Estimating Area Rainfall from Rain Area," J.S. Theon and N. Fugono, eds., *Tropical Rainfall Measurements,* A. Deepak, Hampton, VA, pp. 361–367.

Chong, M., and J. Testud, 1983, "Three-Dimensional Wind Field Analysis from Dual Doppler Radar. Part 3: The Boundary Condition: An Optimum Determination Based on Variational Concept," *J. Climate and Applied Meteorology,* Vol. 22, pp. 1227–1241.

Chong, M., J. Testud, and F. Roux, 1983, "Three Dimensional Wind Field Analysis from Dual Doppler Radar. Part 2: Minimizing the Error due to Temporal Variation," *J. Climate and Applied Meteorology,* Vol. 22, pp. 1216–1226.

Collier, C.G., C.A. Fair, and D.H. Newsome, 1988, "International Weather Radar Networking in Western Europe," *Bull. American Meteorological Society,* Vol. 69, pp. 16–21.

Collier, C.G., and P.R. Larke, 1978, "A Case Study of the Measurement of Snowfall by Radar—An Assessment of Accuracy," *Quarterly J. Royal Meteorological Society,* Vol. 104, pp. 615–621.

Cooley, J.W., and J. Tukey, 1965, "An Algorithm for the Machine Calculation of Complex Fourier Series," *Mathematics of Computation,* Vol. 19, pp. 297–301.

Creutin, J.D., G. Delrieu, and T. Lebel, 1988, "Rain Measurement by Raingauge-Radar Combination: A Geostatistical Approach," *J. Atmospheric and Oceanic Technology,* Vol. 5, pp. 102–115.

Crochet, M., 1985, "The Provence ST Radar," *Handbook for MAP,* Vol. 20, p. 346.

Czechowsky, P., J. Klostermeyer, J. Röttger, R. Ruster, G. Schmidt, and R.F. Woodman, 1976, "The Sousy-VHF Radar for Tropo, Strato and Mesospheric Sounding," *Proc. 17th Radar Meteorology Conf.,* American Meteorological Society, Boston, pp. 349–353.

Davies-Jones, R.P., 1983, "Tornado Dynamics," E. Kessler, ed., *Thunderstorm Morphology and Dynamics,* University of Oklahoma Press, Norman, pp. 197–257.

Dawson, G.A., 1972, "Radar as a Diagnostic Tool for Lightning," *J. Geophysics Research,* Vol. 77, pp. 4518–4528.

Debye, P., 1929, *Polar Molecules,* Chemical Catalogue Co, New York.

Deirmendjian, D., 1969, *Electromagnetic Scattering on Spherical Polydispersions,* Elsevier, New York.

Dennis, A.S., 1980, *Weather Modification by Cloud Seeding,* International Geophysics Series, No. 24, Academic Press, New York.

Dessens, H., 1960, *Severe Hailstorms Are Associated with Very Strong Winds between 6000 and 12000 Meters,* Geophysics Monographs, No. 5, American Geophysics Union, pp. 333–338.

Dickison, R.B.B., M.J. Haggis, R.C. Rainey, and L.M.D. Burns, 1986, "Spruce Budworm Moth Flight and Storms: Further Studies Using Aircraft and Radar," Vol. 25, pp. 1600–1608.

Donaldson, R.J., Jr., 1964, "A Demonstration of Antenna Beam Errors in Radar Reflectivity Patterns," *J. Applied Meteorology,* Vol. 3, pp. 611–623.

Donaldson, R.J., Jr., 1965, "Resolution of a Radar Antenna for Distributed Targets," *J. Applied Meteorology,* Vol. 4, pp. 727–740.

Donaldson, R.J., Jr., 1970, "Vortex Signature Recognition by a Doppler Radar," *J. Applied Meteorology,* Vol. 9, pp. 661–670.

Donaldson, R.J., Jr., 1978, "Observations of the Union City Tornadic Storm by Plan Shear Indicator," *Monthly Weather Rev.,* Vol. 106, pp. 39–47.

Doneaud, A.A., S.I. Niscov, D.L. Priegnitz, and P.L. Smith, 1984, "The Area-Time Integral as an Indicator for Convective Rain Volumes," *J. Climate and Applied Meteorology,* Vol. 23, pp. 555–561.

Doswell, C.A. III, 1986, "Short-Range Forecasting," P.S. Ray, ed., *Mesoscale Meteorology and Forecasting,* American Meteorological Society, Boston, pp. 689–719.

Douglas, R.H., K.L.S. Gunn, and J.S. Marshall, 1957, "Pattern in the Vertical of Snow Generation," *J. Meteorology,* Vol. 14, pp. 95–114.

Doviak, R.J., and M. Berger, 1980, "Turbulence and Waves in the Optically Clear Planetary Boundary Layer Resolved by Dual Doppler Radar," *Radio Sciences,* Vol. 15, pp. 297–317.

Doviak, R.J., P.S. Ray, R.G. Strauch, and L.J. Miller, 1976, "Error Estimation in Wind Fields Derived from Dual-Doppler Radar Measurement," *J. Applied Meteorology,* Vol. 15, pp. 868–878.

Doviak, R.J., D. Sirmans, D. Zrnic', and G.B. Walker, 1978, "Considerations for Pulse-Doppler Radar Observations of Severe Thunderstorms," *J. Applied Meteorology,* Vol. 17, pp. 189–205.

Doviak, R.J., and D.S. Zrnic', 1984, *Doppler Radar and Weather Observations,* Academic Press.

Drake, V.A., and R.A. Farrow, 1983, "The Nocturnal Migration of the Australian Plague Locust, Chortoicetes terminifera (Walker) (Orthoptera-Acrididae): Quantitative Radar Observations of a Series of Northward Flights," *Bull. Entemology Research,* Vol. 73, pp. 567–585.

Drake, V.A., K.F. Helm, J.L. Readshaw, and D.G. Reid, 1981, "Insect Migration across Bass Strait during Spring: A Radar Study," *Bull. Entemological Research,* Vol. 71, pp. 449–466.

du Toit, P.A., 1967, "Doppler Radar Observations of Drop Sizes in Continuous Rain," *J. Applied Meteorology,* Vol. 6, pp. 1082–1087.

Dyer, R.M., 1968, "Doppler Measurements in Stratiform Rain," *Proc. 13th Radar Meteorology Conf.,* American Meteorological Society, Boston, pp. 144–147.

Eagleman, J.R., and W.C. Lin, 1977, "Severe Thunderstorm Internal Structure from Dual-Doppler Radar Measurements," *J. Applied Meteorology,* Vol. 16, pp. 1036–1048.

Easterbrook, C.C., 1967, "Some Doppler Radar Measurements of Circulation Patterns in Convective Storms," *J. Applied Meteorology,* Vol. 6, pp. 882–888.

Eastwood, E., 1967, *Radar Ornithology,* Methuen, London.

Eccles, P.J., and D. Atlas, 1973, "A Dual Wavelength Radar Hail Detector," *J. Applied Meteorology,* Vol. 12, pp. 847–854.

Eccles, P.J., and E.A. Mueller, 1971, "X-Band Attenuation and Liquid Water Content Estimation by Dual-Wavelength Radar," *J. Applied Meteorology,* Vol. 10, pp. 1252–1259.

Eklund, W.L., D.A. Carter, and B.B. Balsley, 1988, "A UHF Wind Profile for the Boundary Layer: Brief Description and Initial Results," *J. Atmospheric and Oceanic Technology,* Vol. 5, pp. 432–441.

Eklund, W.L., D.A. Carter, and K.S. Gage, 1977, "Sounding of the Lower Atmosphere with a Portable 50 MHz Coherent Radar," *J. Geophysics Research,* Vol. 82, No. 31, pp. 4969–4971.

Elliott, R.D., and E.L. Hovind, 1964, "On the Convection Bands within Pacific Coast Storms and Their Relation to Storm Structure," *J. Applied Meteorology,* Vol. 3, pp. 143–154.

Eymard, L., and A. Weill, 1982, "Investigation of Clear Air Convective Structures in the PBL Using a Dual Doppler Radar and Doppler Sodar," *J. Applied Meteorology,* Vol. 21, pp. 1891–1906.

Eymard, L., and A. Weill, 1988, "Dual Doppler Radar Investigation of the Tropical Convective Boundary Layer," *J. Atmospheric Science,* Vol. 45, pp. 853–864.

Fankhauser, S.C., 1971, "Thunderstorm-Environment Interactions Determined from Aircraft and Radar Observations," *Monthly Weather Rev.,* Vol. 99, pp. 171–192.

Feingold, G., and Z. Levin, 1986, "The Log Normal Fit to Raindrop Spectra from Convective Clouds in Israël," *J. Climate and Applied Meteorology,* Vol. 25, pp. 1346–1363.

Feingold, G., and Z. Levin, 1987, "Application of the Log Normal Raindrop Distribution to Differential Reflectivity Radar Measurement," *J. Atmospheric and Oceanic Technology,* Vol. 4, pp. 377–382.

Flitcroft, I.D., J.R. Milford, and G. Dugdale, 1989, "Relating Point to Area Average Rainfall in Semiarid West Africa and the Implications for Rainfall Estimates Derived from Satellite Data," *J. Applied Meteorology,* Vol. 28, pp. 252–266.

Foote, G.B., and H.W. Franck, 1983, "Case Study of a Hailstorm in Colorado. Part III: Airflow from Triple Doppler Measurements," *J. Atmospheric Sciences,* Vol. 40, p. 686–707.

Foote, G.B., and C.A. Knight, 1977, *Hail: A Review of Hail Science and Hail Suppression,* Meteorology Monographs, No. 38, American Meteorological Society, Boston.

Foote, G.B., and P.S. du Toit, 1969, "Terminal Velocity of Raindrops Aloft," *J. Applied Meteorology,* Vol. 8, pp. 249–253.

Frankel, M.S., N.J.F. Chang, and M.J. Sanders, 1977, "A High-Frequency Radio Acoustic Sounder for Remote Measurement of Atmospheric Winds and Temperature," *Bull. American Meteorological Society,* Vol. 58, pp. 928–934.

Friend, A.W., 1940, "Developments in Meteorological Soundings by Radio Waves," *J. Aeronautical Sciences,* Vol. 7, pp. 347–352.

Friend, A.W., 1949, "Theory and Practice of Tropospheric Sounding by Radar," *Proc. IEEE,* Vol. 37, pp. 116–138.

Frisch, A.S., R.B. Chadwick, W.R. Moninger, and J.M. Young, 1976, "Observations of Boundary-Layer Convection Cells Measured by Dual-Doppler Radar and Echosonde and by Microbarograph Array," *Boundary-Layer Meteorology,* Vol. 10, pp. 55–68.

Frisch, A.S., and S.F. Clifford, 1974, "A Study of Convection Capped by a Stable Layer Using Doppler Radar and Acoustic Echo Sounders," *J. Atmospheric Science,* Vol. 31, pp. 1622–1628.

Frisch, A.S., B.E. Martner, and J.S. Gibson, 1989, "Measurement of the Vertical Flux of Turbulent Kinetic Energy with a Single Doppler Radar," *Boundary-Layer Meteorology,* Vol. 49, pp. 331–337.

Frisch, A.S., B.L. Weber, R.G. Strauch, D.A. Merritt, and K.P. Moran, 1986, "The Altitude Coverage of the Colorado Wind Profilers at 50, 405 and 915 MHz."

Fujita, T.T., 1985, "The Downburst: Microburst and Macroburst," Satellite and Mesometeorology Research Project, Research Paper No. 210, University of Chicago.

Fujita, T.T., D.L. Bradbury, and C.F. van Thullenar, 1970, "Palm Sunday Tornadoes of April 11, 1965," *Monthly Weather Rev.,* Vol. 98, pp. 29–69.

Fujita, T.T., and H.R. Byers, 1977, "Spearhead Echo and Downburst in the Crash of an Airliner," *Monthly Weather Rev.,* Vol. 105, pp. 129–146.

Fujita, T.T., and H. Grandoso, 1968, "Split of a Thunderstorm into Anticyclonic Storms and Their Motion from Numerical Model Experiments," *J. Atmospheric Science,* Vol. 25, pp. 416–439.

Fujiwara, M., 1965, "Raindrop-Size Distribution from Individual Storms," *J. Atmospheric Science,* Vol. 22, pp. 585–591.

Gage, K.S., and W.L. Clark, 1978, "Mesoscale Variability of Jet Stream Winds Observed by the Sunset VHF Doppler Radar," *J. Applied Meteorology,* Vol. 17, pp. 1412–1416.

Gage, K.S., and J.L. Green, 1979, "Evidence for Specular Reflection from Monostatic VHF Radar Observation of the Stratosphere," *Radio Science,* Vol. 13, pp. 991–1001.

Gage, K.S., and J.L. Green, 1982, "A Technique for Determining the Temperature Profile from VHF Radar Observations," *J. Applied Meteorology,* Vol. 21, pp. 1146–1149.

Gal-Chen, T., 1978, "A Method for Initialization of the Anclastic Equations: Implications for Matching Models with Observation," *Monthly Weather Rev.,* Vol. 106, pp. 587–606.

Gans, R., 1912, "Uber die Form ultramikroskopischer Goldteilchen [Sur la forme des particules d'or ultra-microscopiques]", *Annals of Physics,* Vol. 37, pp. 881–900.

Garand, L., 1989, "Two Automated Methods to Derive Probability of Precipitation Fields over Oceanic Areas from Satellite Imagery," *J. Applied Meteorology,* Vol. 28, pp. 913–924.

Gentry, R.C., 1983, "Genesis of Tornadoes Associated with Hurricanes," *Monthly Weather Rev.,* Vol. 111, pp. 1793–1805.

Geotis, S.G., and W.M. Silver, 1976, "An Evaluation of Techniques for Automatic Ground-Echo Rejection," *Proc. 17th Radar Meteorological Conf.,* American Meteorological Society, Boston, pp. 448–452.

Gertzman, H.S., and D. Atlas, 1977, "Sampling Errors in the Measurement of Rain and Hail Parameters," *J. Geophysics Research,* Vol. 82, pp. 4955–4966.

Gilet, M., 1984, "Un système intégral pour le prévision immédiate: ARAMIS," *La Météorologie,* Vol. 7, pp. 25–34.

Gilet, M., S. Nicoloff, V. Klaus, and C. Gaillard, 1980, "Dual Doppler Radar Measurements of Winds and Turbulence ahead of a Cold Front," *Proc. 19th Conf. on Radar Meteorology,* American Meteorological Society, Boston, pp. 30–37.

Glover, K.M., J.R. Branche, J.H. Turner, and H.L. Groginsky, 1976, "Precise Calibration of Coherent and Non-Coherent Weather Radars by Means of a Radar Transponder," *Proc. 17th Radar Meteorology Conf.,* American Meteorological Society, Boston, pp. 49–52.

Glover, K.M., K.R. Hardy, T.G. Konrad, W.N. Sullivan, and A.S. Michaels, 1966, "Radar Observations of Insects in Free Flight," *Science,* Vol. 154, pp. 967–972.

Godard, S., 1965, "Propriétés de l'atténuation par la pluie des ondes radioélectriques dans la bande 0,86 cm," *J. Rech. Atmos.,* Vol. 2, pp. 121–167.

Godske, C.L., T. Bergeron, J. Bjerknes, and R.C. Bundgaard, 1957, *Dynamic Meteorology and Weather Forecasting,* American Meteorological Society, Boston.

Goldhirsh, J., 1988, "Analysis of Algorithms for Retrieval of Rain-Rate Profiles from a Spaceborne Dual Wavelength Radar," *IEEE Trans. on Geoscience and Remote Sensing,* Vol. GE 26, pp. 98–114.

Goldhirsh, J., and I. Katz, 1974, "Estimation of Raindrop Size Distribution Using Multiple Wavelength Radar Systems," *Radio Sciences,* Vol. 9, pp. 439–446.

Gorelik, A.G., 1968, "Wind Structure Investigations of Boundary Layer by Radar 'Clear Air' Returns," *Proc. 13th Radar Meteorology Conf.,* American Meteorological Society, Boston, pp. 248–251.

Gorelik, A.C., and V. Mel'Nichuk, 1966, "Radar Measurements of Turbulent Parameters in Clouds and Precipitation," *Proc. 12th Weather Radar Conf.,* American Meteorological Society, Boston, p. 104.

Gossard, E.E., 1988, "Measuring Drop-Size Distributions in Clouds with a Clear-Air-Sensing Doppler Radar," *J. Atmospheric and Oceanic Technology,* Vol. 5, pp. 640–649.

Gossard, E.E., D.R. Jensen, and J.H. Richter, 1971, "An Analytical Study of Tropospheric Structure as Seen by High-Resolution Radar," *J. Atmospheric Sciences,* Vol. 28, pp. 794–807.

Gossard, E.E., and J.H. Richter, 1970, "The Shape of Internal Waves of Finite Amplitude from High Resolution Radar Sounding of the Lower Atmosphere," *J. Atmospheric Sciences,* Vol. 27, pp. 971–973.

Gossard, E.E., J.H. Richter, and D. Atlas, 1970, "Internal Waves in the Atmosphere from High-Resolution Radar Measurements," *J. Geophysics Research,* Vol. 75, pp. 3523–3526.

Gossard, E.E., and R.G. Strauch, 1983, *Radar Observation of Clear Air and Clouds,* Elsevier, Developments in Atmospheric Science, No. 14.

Green, J.L., J.M. Warnock, R.H. Winkler, and T.E. Vanzandt, 1975, "Studies of Winds in the Upper Troposphere with a Sensitive VHF Radar," *Geophysics Research Letters,* Vol. 2, pp. 19–21.

Griffith, C.G., W.L. Woodley, P.G. Grube, D.W. Martin, J. Stont, and D.N. Sikdar, 1978, "Rain Estimation from Geosynchronous Satellite Imagery. Visible and Infrared Studies," *Monthly Weather Rev.,* Vol. 106, pp. 1151–1171.

Groginsky, H.L., 1972, "Pulse Pair Estimation of Doppler Spectrum Parameters," *Proc. 15th Conf. on Radar Meteorology,* American Meteorological Society, Boston, pp. 233–236.

Gunn, K.L.S., and T.W.R. East, 1954, "The Microwave Properties of Precipitation Particles," *Quarterly J. Royal Meteorological Society,* Vol. 80, pp. 522–545.

Gunn, R., and G.D. Kinzer, 1949, "The Terminal Velocity of Fall for Water Droplets in Stagnant Air," *J. Meteorology,* Vol. 6, pp. 243–248.

Gunn, K.L.S., and J.S. Marshall, 1955, "Effect of Wind Shear on Falling Precipitation," *J. Meteorology,* Vol. 12, pp. 339–349.

Gunn, K.L.S., and J.S. Marshall, 1958, "The Distribution with Size of Aggregate Snowflakes," *J. Meteorology,* Vol. 15, pp. 452–466.

Hall, M.P.M., 1979, *Effects of the Troposphere on Radio Communication,* IEE Electromagnetics Waves Series 8, Peter Peregrinus, London.

Hane, C.E., R.B. Wilhelmson, and T. Gal-Chen, 1981, "Retrieval of Thermodynamic Variables within Deep Convective Clouds: Experiments in Three Dimensions," *Monthly Weather Rev.*, Vol. 109, pp. 564–576.

Hardy, K.R., D. Atlas, and K.M. Glover, 1966, "Multiwavelength Backscatter from the Clear Atmosphere," *J. Geophysical Research,* Vol. 71, pp. 1537–1552.

Hardy, K.R., and H. Ottersten, 1969, "Radar Investigations of Convective Patterns in the Clear Atmosphere," *J. Atmospheric Science,* Vol. 26, pp. 666–672.

Hardy, K.R., R.J. Reed, and G.K. Mather, 1973, "Observation of Kelvin-Helmholtz Billows and Their Mesoscale Environment by Radar, Instrumented Aircraft and a Dense Radiosonde Network," *Quarterly J. Royal Meteorological Society,* Vol. 99, pp. 279–293.

Harris, F.I., 1977, The Effects of Evaporation at the Base of Ice Precipitation Layers—Theory and Radar Observations," *J. Atmospheric Science,* Vol. 34, pp. 651–672.

Harrold, T.W., and P.M. Austin, 1974, "The Structure of Precipitation Systems—A Review," *J. Rech. Atmos.,* Vol. 8, pp. 41–57.

Harrold, T.W., and K.A. Browning, 1971, "Identification of Preferred Areas of Shower Development by Means of High Powered Radar," *Quarterly J. Royal Meteorological Society,* Vol. 97, pp. 330–339.

Harrold, T.W., E.J. English, and C.A. Nicholass, 1974, "The Accuracy of Radar-Derived Rainfall Measurements in Hilly Terrain," *Quarterly J. Royal Meteorological Society,* Vol. 100, pp. 331–350.

Hauser, D., and P. Amayenc, 1981, "A New Method for Deducing Hydrometeor-Size Distribution and Vertical Air Motions from Doppler Radar at Vertical Incidence," *J. Applied Meteorology,* Vol. 20, pp. 547–555.

Hauser, D., and P. Amayenc, 1986, Retrieval of Cloud Water and Water Vapor Contents from Doppler Radar Data in a Tropical Squall Line," *J. Atmospheric Science,* Vol. 43, pp. 823–838.

Hawkins, H.F., and S.M. Imbembo, 1976, "The Structure of a Small, Intense Hurricane Inez 1966," *Monthly Weather Rev.*, Vol. 104, pp. 418–442.

Hendry, A., and Y.M.M. Antar, 1984, "Precipitation Particle Identification with Centimeter Wavelength Dual-Polarization Radars," *Radio Science,* Vol. 19, pp. 115–122.

Hendry, A., and Y.M.M. Antar, 1987, "Rain Rate Estimates from Differential Polarization Measurements," *J. Atmospheric and Oceanic Technology,* Vol. 4, pp. 588–598.

Hendry, A., and G.C. McCormick, 1976, "Radar Observations of the Alignment of Precipitation Particles by Electrostatic Fields in Thunderstorms," *J. Geophysical Research,* Vol. 81, pp. 5353–5357.

Henrion, X., and H. Sauvageot, 1977, "Spectral Analysis of Stratiform Cloud Radar Observation," *Geophysics Research Letters,* Vol. 4, pp. 360–362.

Henrion, X., H. Sauvageot, and D. Ramond, 1978, "Finestructure of Precipitation and Temperature in a Stratocumulus Cloud," *J. Atmospheric Science,* Vol. 35, pp. 2315–2324.

Herman, B.M., and L.J. Battan, 1961a, "Calculations of Mie Backscattering of Microwaves from Ice Spheres," *Quarterly J. Royal Meteorological Society,* Vol. 87, pp. 223–230.

Herman, B.M., and L.J. Battan, 1961b, "Calculations of Mie Backscattering from Melting Ice Spheres," *J. Meteorology,* Vol. 18, pp. 468–478.

Herman, B.M., S.R. Browning, and L.J. Battan, 1961, "Tables of the Radar Cross Sections of Water Spheres," Tech. Rep. 9, Tucson, Institute of Atmospheric Physics, University of Arizona.

Heymsfield, G.M., 1978, "Kinematic and Dynamic Aspects of the Harrah Tornadic Storm Analysed from Dual-Doppler Radar Data," *Monthly Weather Rev.,* Vol. 106, pp. 233–254.

Heymsfield, G.M., 1979, "Doppler Radar Study of a Warm Frontal Region," *J. Atmospheric Science,* Vol. 36, pp. 2093–2107.

Hicks, J.J., 1969, "Radar Observations of a Gravitational Wave in Clear Air near the Tropopause Associated with CAT," *J. Applied Meteorology,* Vol. 8, pp. 627–633.

Hicks, J.J., and J.K. Angell, 1968, "Radar Observations of Breaking Gravitational Waves in the Visually Clear Atmosphere," *J. Applied Meteorology,* Vol. 7, pp. 114–121.

Hildebrand, P.H., 1977, "A Radar Study of Turbulent Diffusion in the Lower Atmosphere," *J. Atmospheric Science,* Vol. 16, pp. 493–510.

Hildebrand, P.H., 1978, "Iterative Correction for Attenuation of 5 cm Radar in Rain," *J. Applied Meteorology,* Vol. 17, pp. 508–514.

Hildebrand, P.H., and C.K. Mueller, 1985, "Evaluation of Meteorological Airborne Radar. Part I: Dual Doppler Analyses of Air Motions," *J. Atmospheric and Oceanic Technology,* Vol. 2, pp. 362–1380.

Hildebrand, P.H., and R.S. Sekhon, 1974, "Objective Determination of Noise Level in Doppler Spectra," *J. Applied Meteorology,* Vol. 13, pp. 808–811.

Hildebrand, P.H., N. Towery, and M.R. Snell, 1979, "Measurement of Convective Mean Rainfall over Small Areas Using High-Density Rain Gages and Radar," *J. Applied Meteorology,* Vol. 18, pp. 1316–1326.

Hill, R.J., 1978, "Spectra of Fluctuations in Refractivity, Temperature, Humidity and the Temperature-Humidity Cospectrum in the Inertial and Dissipation Ranges," *Radio Science,* Vol. 13, pp. 953–961.

Hjelmfelt, M.R., 1988, "Structure and Life Cycle of Microburst Outflows Observed in Colorado," *J. Applied Meteorology,* Vol. 27, pp. 900–927.

Hiser, H.W., 1973, "Sferics and Radar Studies of South Florida Thunderstorms," *J. Applied Meteorology,* Vol. 12, pp. 479–1483.

Hobbs, P.V., 1978, "Organization and Structure of Clouds and Precipitation on the Mesoscale and Microscale in Cyclonic Storms," *Rev. Geophysics and Space Physics,* Vol. 16, pp. 741–755.

Hobbs, P.V., and J.D. Locatelli, 1978, "Rainbands, Precipitation Cores and Generating Cells in a Cyclonic Storm," *J. Atmospheric Science,* Vol. 35, pp. 230–241.

Hobbs, P.V., T.J. Matejka, P.H. Herzegh, I.D. Locatelli, and R.A. Houze Jr., 1980, "The Mesoscale and Microscale Structure and Organization of Clouds and Precipitation in Midlatitude Cyclones. I. A Case Study of a Cold Front," *J. Atmospheric Science,* Vol. 37, pp. 568–596.

Hogg, D.C., *et al.*, 1983, "An Automatic Profile of the Temperature, Wind and Humidity in the Troposphere," *J. Climate and Applied Meteorology,* Vol. 22, pp. 807–831.

Holitza, F.J., and H.W. Kasemir, 1974, "Accelerated Decay of Thunderstorm Electric Fields by Chaff Seeding," *J. Geophysics Research,* Vol. 79, pp. 425–429.

Holt, A.R., 1982, "The Scattering of Electromagnetic Waves by Single Hydrometeors," *Radio Science,* Vol. 17, pp. 929–945.

Holton, J.R., 1972, *Introduction to Dynamic Meteorology,* International Geophysics Series. Academic Press, New York.

Hoose, J.M., and J.A. Colon, 1970, "Some Aspects of the Radar Structure of Hurricane Beulah on September 9, 1967," *Monthly Weather Rev.,* Vol. 98, pp. 529–533.

Houze, R.A., 1977, "Structure and Dynamics of a Tropical Squall-Line System," *Monthly Weather Rev.* Vol. 105, pp. 1540–1567.

Huff, F.A., 1970, "Sampling Errors in Measurement of Mean Precipitation," *J. Applied Meteorology,* Vol. 9, pp. 35–44.

Huff, F.A., 1971, "Evaluation of Precipitation Records in Weather Modification Experiments," *Advances in Geophysics,* Academic Press, New York, Vol. 15, pp. 59–134.

Huschke, R.E., ed., 1959, *Glossary of Meteorology,* American Meteorological Society, Boston.

Illingworth, A.J., and I.J. Caylor, 1989, "Polarization Radar Estimates of Raindrop Size Spectra and Rainfall Rates," *J. Atmospheric and Oceanic Technology,* Vol. 6, pp. 939–949.

Illingworth, A.J., J.W.F. Goddard, and S.M. Cherry, 1986, "Detection of Hail by Dual Polarization Radar," *Nature,* Vol. 320, pp. 431–433.

Illingworth, A.J., J.W.F. Goddard, and S.M. Cherry, 1987, "Polarization Radar Studies of Precipitation Development in Convective Storm," *Quarterly Royal Meteorological Society,* Vol. 113, pp. 469–489.

Istok, M.J., and R.J. Doviak, 1986, "Analysis of the Relation between Doppler Spectral Width and Thunderstorm Turbulence," *J. Atmospheric Sciences,* Vol. 43, pp. 2199–2214.

Jacobson, E.A., and E.P. Krider, 1976, "Electrostatic Field Changes Produced by Florida Lightning," *J. Atmospheric Sciences,* Vol. 33, pp. 103–117.

Jameson, A.R., 1983a, "Microphysical Interpretation of Multi-Parameter Radar Measurements in Rain. Part I: Interpretation of Polarization Measurements and Estimation of Raindrop Shapes," *J. Atmospheric Sciences,* Vol. 40, pp. 1792–1802.

Jameson, A.R., 1983b, "Microphysical Interpretation of Multi-Parameter Radar Measurements in Rain. Part II: Estimation of Raindrop Distribution Parameters by Combined Dual-Wavelength and Polarization Measurements," *J. Atmospheric Sciences,* Vol. 40, pp. 1803–1813.

Jameson, A.R., 1985, "Deducing the Microphysical Character of Precipitation from Multiple-Parameter Radar Polarization Measurements," *J. Climate and Applied Meteorology,* Vol. 24, pp. 1037–1047.

Jameson, A.R., and R.C. Srivastava, 1978, "Dual-Wavelength Doppler Radar Observations of a Hail at Vertical Incidence," *J. Applied Meteorology,* Vol. 17, pp. 1694–1703.

Johnson, C.G., 1969, *Migration and Dispersal of Insects by Flight,* Chaucer Press.

Johnson, D.B., 1984, "The Effect of Antenna Sidelobes on Multiple-Parameter Radar Measurements," *J. Atmospheric and Oceanic Technology,* Vol. 1, pp. 287–290.

Jorgensen, D.P., and F.D. Marks, Jr., 1984, "Airborne Doppler Radar Study of the Structure and Three-Dimensional Airflow within a Hurricane Rainbands," *Proc. 22nd Conf. on Radar Meteorology,* American Meteorological Society, Boston, pp. 572–577.

Joss, J., and A. Waldvogel, 1967, "Ein Spektrograph für niederschlagstropfen mit automatischer Auswertung," *Pure and Applied Geophysics,* Vol. 68, pp. 240–246.

Joss, J., and A. Waldvogel, 1969, "Raindrop Size Distribution and Sampling Size Errors," *J. Atmospheric Science,* Vol. 26, pp. 566–569.

Joss, J., and A. Waldvogel, 1970, "Raindrop Size Distributions and Doppler Velocities," *Proc. 14th Radar Meteorological Conf.,* American Meteorological Society, Boston, pp. 153–156.

Kasemir, H.W., F.J. Holitza, W.E. Cobb, and W.D. Rust, 1976, "Lightning Suppression by Chaff Seeding at the Base of Thunderstorms," *J. Geophysics Research,* Vol. 81, pp. 1965–1970.

Kerker, M., 1969, *The Scattering of Light and Other Electromagnetic Radiation,* Academic Press, New York.

Kessler, E., ed., 1983, *Thunderstorms: A Social, Scientific and Technological Documentary,* Vols. 1, 2, and 3. University of Oklahoma Press, Norman.

Keyser, D., 1986, "Atmospheric Fronts: An Observational Perspective," P.S. Ray, ed., *Mesoscale Meteorology and Forecasting,* American Meteorological Society, Boston, pp. 216–258.

Konrad, T.G., 1970, "The Dynamics of the Convective Process in the Clear Air as Seen by Radar," *J. Atmospheric Science,* Vol. 27, pp. 1138–1147.

Konrad, T.G., and F.L. Robison, 1972, "Simultaneous Measurements of Radar Reflectivity and Refractive Index Spectra in Clear Air Convection," *J. Applied Meteorology,* Vol. 11, pp. 1114–1119.

Koscielny, A.J., R.J. Doviak, and R. Rabin, 1982, "Statistical Considerations in the Estimation of Divergence from Single Doppler Radar and Application to Prestorm Boundary Layer Observations," *J. Applied Meteorology,* Vol. 21, pp. 197–210.

Krajewski, W.F., 1987, "Cokriging Radar-Rainfall and Rain Gauge Data." *J. Geophysical Research,* Vol. 92, pp. 9571–9580.

Krehhiel, P.R., and M. Brook, 1979, "A Broad-Band Noise Technique for Fast-Scanning Radar Observations of Clouds and Clutter Targets. *IEEE Trans. on Geoscience and Electronics Spec.*, Issue on Radio Meteorology, Vol. GE-17, No. 4, pp. 196–204.

Krehbiel, P., M. Brook, R. McCrory, and D. Tarbox, 1976, "Lightning Charge Center Locations Relative to Precipitation in a Thunderstorm," *Proc. Int. Cloud Physics Conf.*, American Meteorological Society, Boulder, CO, pp. 642–643.

Kreitzberg, C.W., and H.A. Brown, 1970, "Mesoscale Weather Systems within an Occlusion," *J. Applied Meteorology,* Vol. 9, pp. 417–432.

Kropfli, R.A., 1971, "Simultaneous Radar and Instrumented Aircraft Observations in a Clear Air Turbulent Layer," *J. Applied Meteorology,* Vol. 10, pp. 796–802.

Kropfli, R.A., 1986, "Single Doppler Radar Measurements of Turbulence Profiles in the Convective Boundary Layer," *J. Atmospheric and Oceanic Technology,* Vol. 3, pp. 305–314.

Kropfli, R.A., and N.M. Kohn, 1978, "Persistent Rolls in the Urban Mixed Layer as Revealed by Dual-Doppler Radar," *J. Applied Meteorology,* Vol. 17, pp. 669–676.

Kropfli, R.A., and L.J. Miller, 1976, "Kinematic Structure and Flux Quantities in a Convective Storm from Dual-Doppler Radar Observations," *J. Atmospheric Science,* Vol. 33, pp. 520–525.

Kuo, Y.H., and R.A. Anthes, 1985, "Calculation of Geopotential and Temperature Fields from an Array of Nearly Continuous Wind Observations," *J. Atmospheric and Oceanic Technology,* Vol. 2, pp. 22–34.

Labrum, N.R., 1952, "The Scattering of Radio Waves by Meteorological Particles," *J. Applied Physics,* Vol. 23, pp. 1324–1330.

Langleben, M.P., 1954, "The Terminal Velocity of Snow Flakes," *Quarterly J. Royal Meteorological Society,* Vol. 80, pp. 174–181.

Larsen, M.F., and J. Röttger, 1982, "VHF and UHF Doppler Radars as Tools for Synoptic Research," *Bull. American Meteorological Society,* Vol. 63, pp. 996–1008.

Larsen, M.F., and J. Röttger, 1982, "The Spaced Antenna Technique for Radar Wind Profiling," *J. Atmospheric and Oceanic Technology,* Vol. 6, pp. 920–938.

Laws, J.O. and D.A. Parsons, 1943, "The Relation of Raindrop Size to Intensity," *Trans. American Geophysics Union,* Vol. 24, pp. 432–460.

Lemon, L.R., D.W. Burgess, and R.A. Brown, 1978, "Tornadic Storm Airflow and Morphology Derived from Single-Doppler Radar Measurements," *Monthly Weather Rev.*, Vol. 106, pp. 48–61.

Lhermitte, R.M., 1963, "Weather Echoes in Doppler and Conventional Radars," *Proc. 10th Weather Radar Conf.*, American Meteorological Society, Boston, pp. 323–329.

Lhermitte, R.M., 1966, "Probing Air Motion by Doppler Analysis of Radar Clear Air Returns," *J. Atmospheric Science,* Vol. 23, pp. 575–591.

Lhermitte, R.M., 1968a, "New Developments in Doppler Radar Methods," *Proc. 13th Radar Meteorology Conf.*, American Meteorological Society, Boston, pp. 14–17.

Lhermitte, R.M., 1968b, "Turbulent Air Motion as Observed by Doppler Radar," *Proc. 13th Radar Meteorology Conf.*, American Meteorological Society, Boston, pp. 498–503.

Lhermitte, R.M., 1969a, "Atmospheric Probing by Doppler Radar," *Atmospheric Exploration by Remote Probes,* National Academy of Science, Washington, DC, Vol. 2, pp. 253–285.

Lhermitte, R.M., 1969b, "Note on the Observation of Small-Scale Atmospheric Turbulence by Doppler Radar Techniques," *Radio Sciences,* Vol. 4, pp. 1241–1246.

Lhermitte, R.M., 1971, "Probing of Atmospheric Motion by Airborne Pulse-Doppler Radar Techniques," *J. Applied Meteorology,* Vol. 10, pp. 234–246.

Lhermitte, R.M., 1972, "Real Time Processing of Meteorological Doppler Radar Signals," *Proc. 15th Conf. on Radar Meteorology,* American Meteorological Society, Boston, pp. 364–367.

Lhermitte, R.M., 1988, "Observation of the Rain at Vertical Incidence with 94 GHz Doppler Radar: An Insight on Mie Scattering," *Geophysical Research Letters,* Vol. 15, pp. 1125–1128.

Lhermitte, R.M., 1988, "Cloud and Precipitation Remote Sensing at 94 GHz," *IEEE Trans. on Geospace and Remote Sensing,* Vol. 26, pp. 207–216.

Lhermitte, R.M., 1990, "Attenuation and Scattering of Millimeter Wave–Length Radiation by Clouds and Precipitation," *J. Atmospheric and Oceanic Technology,* Vol. 7, pp. 464–479.

Lhermitte, R.M., and D. Atlas, 1961, "Precipitation Motion by Pulse Doppler Radar," *Proc. 9th Weather Radar Conf.,* American Meteorological Society, Boston, pp. 218–223.

Lhermitte, R.M., and M. Gilet, 1975, "Dual-Doppler Radar Observation and Study of Sea Breeze Convective Storm Development," *J. Applied Meteorology,* Vol. 14, pp. 1346–1361.

Lhermitte, R.M., and L.J. Miller, 1970, "Doppler Radar Methodology for the Observation of Convective Storms," *Proc. 14th Radar Meteorology Conf.,* American Meteorological Society, Boston, pp. 133–138.

Ligda, M.G.H., 1956, "The Radar Observation of Lightning," *J. Atmospheric and Terrestrial Physics,* Vol. 9, pp. 329–346.

List, R., T.B. Low, and J.D. McTaggart-Cowan, 1974, "Collisions of Raindrops with Chaffs," *J. Applied Meteorology,* Vol. 13, pp. 796–799.

Little, C.G., 1969, "Acoustic Methods for Remote Probing of the Lower Atmosphere," *Proc. IEEE,* Vol. 57, pp. 571–578.

Lob, W.H., 1968, "R-Meter Correction for Boxcar-ing and Receiver Noise in Incoherent Weather Radars," *J. Applied Meteorology,* Vol. 7, pp. 1018–1025.

Ludlam, F.H., 1980, *Clouds and Storm,* Pennsylvania State University Press, State College.

Maddox, R.A., 1980, "Mesoscale Convective Complexes," *Bull. American Meteorological Society,* Vol. 61, pp. 1374–1387.

Maresca, J.W., and J.R. Barnum, 1979, "Remote Measurement of the Position and Surface Circulation of Hurricane Eloise by HF Skywave Radar," *Monthly Weather Rev.,* Vol. 107, pp. 1648–1652.

Marks, F.D., 1985, "Evolution of the Structure of Precipitation in Hurricane Allen," *Monthly Weather Rev.,* Vol. 113, pp. 909–930.

Marks, F.D., Jr., and R.A. Houze, Jr., 1984, "Airborne Doppler Radar Observations in Hurricane Debby," *Bull. American Meteorological Society,* Vol. 65, pp. 569–582.

Marshall, J.S., 1953, "Precipitation Trajectories and Patterns," *J. Meteorology,* Vol. 10, pp. 25–29.

Marshall, J.S., and K.L.S. Gunn., 1952, "Measurement of Snow Parameters by Radar," *J. Meteorology,* Vol. 9, pp. 322–327.

Marshall, J.S., and W. Hitschfeld, 1953, "Interpretation of the Fluctuating Echo from Randomly Distributed Scatterers—Pt. I," *Canadian J. Physics,* Vol. 31, pp. 962–994.

Marshall, J.S., and W.M.K. Palmer, 1948, "The Distribution of Raindrops with Size," *J. Meteorology,* Vol. 5, pp. 165–166.

Marshall, J.S., and S. Radhakant, 1978, "Radar Precipitation Maps as Lightning Indicators," *J. Applied Meteorology,* Vol. 17, pp. 206–212.

Marwitz, J.D., 1972a, *The structure and motion of severe hailstorms.* Part I. Supercell Storms. J. Appl. Meteor., 11, 166–179.

Marwitz, J.D., 1972b, *The structure and motion of severe hailstorms.* Part II. Multi-Cell Storms. J. Appl. Meteor., 11, 180–188.

Marwitz, J.D., 1972c, *The structure and motion of severe hailstorms,* Part III. Severely Sheared Storms. J. Appl. Meteor., 11, 189–201.

Marwitz, J.D., 1973, "Trajectories within the Weak Echo Regions of Hailstorms," *J. Applied Meteorology,* Vol. 12, pp. 1174–1182.

Masuda, Y., 1988, "Influence of Wind and Temperature on the Height Limit of a Radio Acoustic Sounding System," *Radio Science,* Vol. 23, pp. 647–651.

May, P.T., K.P. Moran, and R.G. Strauch, 1989, "The Accuracy of RASS Temperature Measurements," *J. Applied Meteorology,* Vol. 28, pp. 1329–1335.

Mazur, V., D.S. Zrnic', and W.D. Rust, 1985, "Lightning Channel Properties Determined with a Vertically Pointing Doppler Radar," *J. Geophysics Research,* Vol. 90, No. D, pp. 6165–6174.

McCormick, G.C., and A. Hendry, 1975, "Principles for the Radar Determination of the Polarization Properties of Precipitation," *Radio Science,* Vol. 10, pp. 421–434.

McGinley, 1986, "Nowcasting Mesoscale Phenomena," P.S. Ray, ed., *Mesoscale Meteorology and Forecasting,* American Meteorological Society, Boston, pp. 657–688.

Medhurst, R.G., 1965, "Rainfall Attenuation of Centimeter Waves—Comparison of Theory and Measurement," *IEEE Trans. on Antennas and Propagation,* Vol. AP-13, pp. 550–564.

Megaw, E.C.S., 1950, "Scattering of Electromagnetic Waves by Atmospheric Turbulence," *Nature,* Vol. 166, pp. 1100–1104.

Mel'Nichuk, Y.V., G.A. Smirnova, and A.A. Chernikov, 1968, "Radar Measurements of the Rate of Turbulent Energy Dissipation in Clouds and Precipitation," *Proc. 13th Radar Meteorology Conf.,* American Meteorological Society, Boston, pp. 486–489.

Meneghini, R., J. Eckerman, and D. Atlas, 1983, "Determination of Rain Rate from Spaceborne Radar Using Measurements of Total Attenuation," *IEEE Trans. on Geoscience and Remote Sensing,* Vol. GE-21, pp. 34–43.

Meneghini, R., and D. Atlas, 1986, "Simultaneous Ocean Cross Section and Rainfall Measurements from Space with a Radio-Looking Radar," *J. Atmospheric and Oceanic Technology,* Vol. 3, pp. 400–413.

Meneghini, R., J.A. Jones, and L.H. Gesell, 1987, "Analysis of Dual-Wavelength Surface Radar Reference Technique," *IEEE Trans. on Geosciences and Remote Sensing,* Vol. GE-25, pp. 456–471.

Metcalf, J.I., 1975a, "Gravity Waves in a Low-Level Inversion," *J. Atmospheric Science,* Vol. 32, pp. 351–361.

Metcalf, J.I., 1975b, "Microstructure of Radar Echo Layers in the Clear Atmosphere," *J. Atmospheric Science,* Vol. 32, pp. 362–370.

Metcalf, J.I., 1986, "Interpretation of the Autocovariance and Cross-Covariance from a Polarization Diversity Radar," *J. Atmospheric Sciences,* Vol. 43, pp. 2479–2498.

Metcalf, J.I., and J.D. Echard, 1978, "Coherent Polarization-Diversity Radar Techniques in Meteorology," *J. Atmospheric Sciences,* Vol. 35, pp. 2010–2019.

Metcalf, J.I., and J.S. Ussailis, 1984, "Radar System Errors in Polarization Diversity Measurements," *J. Atmospheric and Oceanic Technology,* Vol. 1, pp. 105–114.

Meyer, J.H., 1971, "Radar Observations of Land Breeze Fronts," *J. Applied Meteorology,* Vol. 10, pp. 1224–1232.

Mie, G., 1908, "Beiträge zur Optik trüber Medien, speziell kolloidaler Metallösungen [Contribution à l'optique des milieux en suspension, particulièrement les suspensions métalliques colloïdales]," *Annals of Physics,* Vol. 25, pp. 377–445.

Mielke, K.B., and D.D. Houghton, 1977, "An Analysis of Echo Radar Systems for the Upper Midwestern United States," *J. Applied Meteorology,* Vol. 16, pp. 833–843.

Miller, K.S., and M.M. Rochwarger, 1972, "A Covariance Approach to Spectral Moment and Estimation," *IEEE Trans. on Information Theory,* Vol. IT-18, pp. 588–596.

Mon, J.P., 1982, "Backward and Forward Scattering of Microwaves by Ice Particles: A Review," *Radio Science,* Vol. 17, pp. 953–971.

Monin, A.S., and A.M. Yaglom, 1975, *Statistical Fluid Mechanics—Mechanics of Turbulence,* 2 vols., MIT Press, Cambridge, MA.

Moninger, W.R., and R.A. Kropfli, 1982, "Radar Observations of a Plume from an Elevated Continuous Point Source," *J. of Applied Meteorology,* Vol. 21, pp. 1685–1697.

Motell, C.E., and B.C. Weare, 1987, "Estimating Tropical Pacific Rainfall Using Digital Satellite Data," *J. Climate and Applied Meteorology,* Vol. 26, pp. 1436–1446.

Mueller, C.K., and P.H. Hildebrand, 1985, "Evaluation of Meteorological Airborne Doppler Radar. Part II: Triple-Doppler Analyses of Air Motions," *J. Atmospheric and Oceanic Technology,* Vol. 2, pp. 381–392.

Mueller, E.A., and R.P. Larkin, 1985, "Insects Observed Using Dual Polarization Radar," *J. Atmospheric and Oceanic Technology,* Vol. 2, pp. 49–54.

Nathanson, F.E., 1969, *Radar Design Principles,* McGraw-Hill, New York.

Nathanson, F.E., and J.P. Reilly, 1968, "Radar Precipitation Echo Experiments on Temporal, Spatial and Frequency Correlation," *IEEE Trans. on Aerospace and Electronic Systems,* Vol. AES-4, pp. 55–104.

Nathanson, F.E., and P.L. Smith, 1972, "A Modified Coefficient for the Weather Radar Equation," Proc. 15th Radar Meteorological Conf., American Meteorological Society, pp. 228–231.

Negri, A.J., and R.F. Adler, 1981, "Relation of Satellite-Based Thunderstorm Intensity to Radar-Estimated Rainfall," *J. Applied Meteorology,* Vol. 20, pp. 288–300.

Nelson, A.R., and N.R. Gokhale, 1972, "Oscillation Frequencies of Freely Suspended Water Drops," *J. Geophysics Research,* Vol. 77, pp. 2724–2727.

Newton, C.W., 1963, "Dynamics of Severe Convective Storms," *Meteorology Monographs,* Vol. 5, No. 27, pp. 33–58.

Newton, C.W., and J.C. Fankhauser, 1975, "Movement and Propagation of Multicellular Convective Storms," *Pure and Applied Geophysics,* Vol. 113, pp. 747–764.

Nozumi, Y., and H. Arakawa, 1968, "Prefrontal Rainbands Located in the Warm Sector of Subtropical Cyclones over the Ocean," *J. Geophysics Research,* Vol. 73, pp. 487–492.

Nutten, B., D. Hauser, F. Roux, and G. Scialom, 1979, "The Ronsard Radars—Internal Calibration Techniques Using Coherent and Noise Sources," *IEEE Trans. for Geoscience Electronics,* Vol. GE-17, No. 4, pp. 288–295.

Oetzel, G.N., and E.T. Pierce, 1968, "The Radio Emissions from Close Lightning," *Planetary Electrodynamics,* Gordon & Breach, New York, Vol. 1, pp. 543–569.

Oguchi, T., 1981, "Scattering from Hydrometeors: A Survey," *Radio Science,* Vol. 16, pp. 691–730.

Oguchi, T., 1983, "Electromagnetic Wave Propagation and Scattering in Rain and Other Hydrometeors," *Proc. IEEE,* Vol. 71, pp. 1029–1078.

Oguchi, T., and Y. Hosoya, 1974a, "Differential Attenuation and Differential Phase Shift of Radio Waves due to Rain: Calculations at Microwave and Millimeter Wave Regions," *J. Rech. Atmos.,* Vol. 8, pp. 121–128.

Oguchi, T., and Y. Hosoya, 1974b, "Scattering Properties of Oblate Raindrops and Cross Polarization of Radio Waves Due to Rain. Part II: Calculations at Microwave and Millimeter Wave Region," *J. Radio Science Labs.,* Vol. 12, pp. 191–259.

Ohtake, T., 1969, "Observations of Size Distribution of Hydrometeors through the Melting Layer," *J. Atmospheric Science,* Vol. 26, pp. 545–557.

Ottersten, H., 1969a, "Atmospheric Structure and Radar Backscattering in Clear Air," *Radio Sciences,* Vol. 4, pp. 1179–1193.

Ottersten, H., 1969b, "Radar Backscattering from the Turbulent Clear Atmosphere," *Radio Sciences,* Vol. 4, pp. 1251–1255.

Ottersten, H., 1969c, "Mean Vertical Gradient of Potential Refractive Index in Turbulent Mixing and Radar Detection of CAT," *Radio Sciences,* Vol. 4, pp. 1247–1249.

Palmen, E., and C.W. Newton, 1969, *Atmospheric Circulation Systems,* Academic Press, New York.

Papoulis, A., 1973, "Minimum Bias Windows for High Resolution Spectral Estimates," *IEEE Trans. on Information Theory,* Vol. IT-19, p. 1.

Papoulis, A., 1977, *Signal Analysis,* McGraw-Hill, New York.

Parrish, J.R., R.W. Burpee, F.D. Marks, Jr., and R. Grebe, 1982, "Rainfall Pattern Observed by Digitized Radar during the Landfall of Hurricane Frederic (1979)," *Monthly Weather Rev.,* Vol. 110, pp. 1933–1944.

Peace, R.L., Jr., R.A. Brown, and H.G. Camnitz, 1969, "Horizontal Motion Field Observations with a Single Pulse Doppler Radar," *J. Atmospheric Science,* Vol. 26, pp. 1096–1103.

Petitdidier, M., A. Desautez, M. Glass, and G. Penazzi, 1985, "A Decoder for a 30 m Height Resolution Radar," *Radio Sciences,* Vol. 20, pp. 1141–1145.

Pointin, Y., D. Ramond, and J. Fournet-Fayard, 1988, "Radar Differential Reflectivity $Z_{DR}$: A Real-Case Evaluation of Errors Induced by Antenna Characteristics," *J. Atmospheric and Oceanic Technology,* Vol. 5, pp. 416–423.

Priestley, M.B., 1981, *Spectral Analysis and Time Series*. Academic Press, New York.

Probert-Jones, J.R., 1962, "The Radar Equation in Meteorology," *Quarterly J. Royal Meteorological Society,* Vol. 88, pp. 485–495.

Pruppacher, H.R., and J.D. Klett, 1978, *Microphysics of Clouds and Precipitation,* D. Reidel, Dordrecht.

Pruppacher, H.R., and R.L. Pitter, 1971, "A Semi-Empirical Determination of the Shape of Cloud and Rain Drops," *J. Atmospheric Science,* Vol. 28, pp. 86–94.

Puhakka, T., 1974, "On the Variability of the Z-R Relationship Related to Radar Echo Pattern," *Geophysica,* Vol. 13, pp. 103–119.

Rabin, R., and D. Zrnic', 1980, "Subsynoptic-Scale Vertical Wind Revealed by Dual Doppler-Radar and VAD Analysis," *J. Atmospheric Science,* Vol. 37, pp. 644–654.

Rankine, W.J.M., 1901, *A Manual of Applied Mechanics,* Charles Criff and Co., pp. 574–578.

Ramond, D., 1978, "Pressure Perturbations in Deep Convection: An Experimental Study," *J. Atmospheric Science,* Vol. 35, pp. 1704–1711.

Ray, P.S., ed., 1986, *Mesoscale Meteorology and Forecasting,* American Meteorological Society, Boston.

Ray, P.S., R.J. Doviak, G.B. Walker, D. Sirmans, J. Carter, and B. Bumgarner, 1975, "Dual-Doppler Observation of a Tornadic Storm," *J. Applied Meteorology,* Vol. 14, pp. 1521–1530.

Ray, P.S., and D.P. Jorgensen, 1988, "Uncertainties Associated with Combining Airborne and Ground-Based Doppler Radar Data," *J. Atmospheric and Oceanic Technology,* Vol. 5, pp. 177–196.

Ray, P.S., D.P. Jorgensen, and S.L. Wang, 1985, "Airborne Doppler Radar Observations of Convective Storm," *J. Climate and Applied Meteorology,* Vol. 24, pp. 687–698.

Ray, P.S., and K.L. Sangren, 1983, "Multiple Doppler Radar Networks Design," *J. Climate and Applied Meteorology,* Vol. 22, pp. 1444–1454.

Ray, P.S., K.K. Wagner, K.W. Johnson, J.J. Stephens, W.C. Bumgarner, and E.A. Mueller, 1978, "Triple-Doppler Observations of a Convective Storm," *J. Applied Meteorology,* Vol. 17, pp. 1201–1212.

Ray, P.S., C.L. Ziegler, W. Bumgarner, and R.J. Serafin, 1980, "Single and Multiple Doppler Radar Observations of Tornadic Storms," *Monthly Weather Rev.,* Vol. 108, pp. 1607–1625.

Reed, R.J., and K.R. Hardy, 1972, "A Case Study of Persistent, Intense Clear Air Turbulence in an Upper Level Frontal Zone," *J. Applied Meteorology,* Vol. 11, 541–549.

Richard, P., and G. Auffray, 1985, "VHF-UHF Interferometric Measurements, Applications to Lightning Discharges Mapping," *Radio Sciences,* Vol. 20, 171–192.

Richter, J.H., 1969, "High Resolution Tropospheric Radar Sounder," *Radio Science,* Vol. 4, pp. 1261–1268.

Richter, J.H., D.R. Jensen, R.A. Pappert, and V.R. Noonkester, 1973, "New Developments in FM-CW Radar Sounding," *Boundary-Layer Meteorology,* Vol. 4, pp. 179–199.

Rihaczek, A.W., 1969, *Principles of High-Resolution Radar,* McGraw-Hill.

Riley, J.R., and D.R. Reynolds, 1983, "A Long-Range Migration of Grasshoppers Observed in the Sahelian Zone of Mali by Two Radars," *J. Animal Ecology,* Vol. 52, pp. 167–183.

Rinehart, R.E., and J.D. Tuttle, 1982, "Antenna Beam Patterns and Dual-Wavelength Processing," *J. Applied Meteorology,* Vol. 21, pp. 1865–1880.

Roberts, R.D., and J.W. Wilson, 1989, "A Proposed Microburst Nowcasting Procedure Using Single-Doppler Radar," *J. Applied Meteorology,* Vol. 28, pp. 285–303.

Rogers, R.R., 1963, "Radar Measurements of Velocities of Meteorological Scatterers," *J. Atmospheric Science,* Vol. 20, pp. 170–174.

Rogers, R.R., 1964, "An Extension of the Z-R Relationship for Doppler Radar," *Proc. 11th Weather Radar Conf.,* American Meteorological Society, Boston, pp. 158–160.

Rogers, R.R., 1976, *A Short Course in Cloud Physics,* Pergamon Press.

Rogers, R.R., and B.R. Tripp, 1964, "Some Radar Measurements of Turbulence in Snow," *J. Applied Meteorology,* Vol. 3, pp. 603–610.

Rosenfeld, D., D. Atlas, and D.A. Short, 1990, "The Estimation of Convective Rainfall by Area Integrals: 2. The Height Area Rainfall Threshold (HART) Method," *J. Geophysical Research,* Vol. 95, pp. 2161–2176.

Röttger, J., 1980, "Reflection and Scattering of VHF Radar Signals from Atmospheric Refractivity Structures," *Radio Science,* Vol. 15, pp. 259–276.

Roux, F., 1985, "Retrieval of Thermodynamic Fields from Multiple Doppler Radar Data Using the Equations of Motion and the Thermodynamic Equation," *Monthly Weather Rev.*, Vol. 113, pp. 2142–2157.

Roux, F., 1988, "The West African Squall Line Observed on 23 June 1981 during COPT 81: Kinematics and Thermodynamics of the Convective Region," *J. Atmospheric Science,* Vol. 45, pp. 406–426.

Rowland, J.R., 1976, "Clear Air Convective Behavior Revealed by Radar Chaff," *J. Applied Meteorology,* Vol. 15, pp. 521–526.

Rummler, W.D., 1968, "Introduction of a New Estimator for Velocity Spectral Parameters," Tech. Memo 68-4121-5, Bell Telephone Labs.

Rust, W.D., and P.R. Krehbiel, 1977, "Microwave radiometric detection of corona from chaff within thunderstorms," *J. Geophys. Res.,* 82, 3945–3950.

Rutkowski, W., and A. Fleisher, 1955, "The R-Meter: An Instrument for Measuring Gustiness," *Proc. 5th Weather Radar Conf.,* American Meteorological Society, Fort Monmouth, NJ, pp. 255–260.

Rutledge, S.A., and P.V. Hobbs, 1984, "The Mesoscale and Microscale Structure and Organization of Clouds and Precipitation in Midlatitude Cyclones. XII: A Diagnostic Modeling Study of Precipitation Development in Narrow Cold-Frontal Rainbands," *J. Atmospheric Science,* Vol. 41, pp. 2949–2952.

Sachidananda, M., and D.S. Zrnic', 1985, "$Z_{DR}$ Measurements Considerations for a Fast Scan Capability Radar," *Radio Science,* Vol. 20, pp. 907–922.

Sachidananda, M., and D.S. Zrnic', 1987, "Rain Rate Estimates from Differential Polarization Measurements," *J. Atmospheric and Oceanic Technology,* Vol. 4, pp. 588–598.

Sassen, K., 1987, "Ice Cloud Content from Radar Reflectivity," *J. Climate and Applied Meteorology,* Vol. 26, pp. 1050–1053.

Sauvageot, H., 1974, "On the Fine Structure of Precipitation Generating Cells," *J. Rech. Atmos.,* Vol. 8, pp. 213–219.

Sauvageot, H., 1976, "Circulation de l'air et formation des précipitations dans les stratocumulus en rouleaux," *J. Rech. Atmos.,* Vol. 10, pp. 17–24.

Sauvageot, H., 1982, *Radarmeteorologie,* Eyrolles.

Sauvageot, H., 1987, "Radar Observation of Snowfall from a Natural-Draft Cooling Tower Plume," *J. Climate and Applied Meteorology,* Vol. 26, pp. 1471–1481.

Sauvageot, H., 1989, "Radar Measurement of Cooling Tower Drift," *J. Applied Meteorology,* Vol. 28, pp. 925–935.

Sauvageot, H., R. Auria, and B. Campistron, 1982, "Long-Lasting Precipitation Cells in Stratiform Clouds."

Sauvageot, H., R. Auria, and B. Campistron, 1984, "Microscale Structure of Convection in Stratiform Clouds: An Observational and Numerical Study," *Proc. 9th Int. Cloud Physics Conf.,* Tallinn, USSR, pp. 363–366.

Sauvageot, H., K. Kouadio, and C.A. Etty, 1986, "The Influence of Temperature and Supersaturation on the Polarization of Radar Signals," *Proc. 23rd Conf. on Radar Meteorology,* American Meteorological Society, Boston, pp. 173–176.

Sauvageot, H., G. Lafon, and M. Oruba, 1982, "Air Motion Measurements in and Around Thermal Plumes Using Radar and Chaff," *J. Applied Meteorology,* Vol. 21, pp. 656–665.

Sauvageot, H., and J. Omar, 1987, "Radar Reflectivity of Cumulus Clouds," *J. Atmospheric and Oceanic Technology,* Vol. 4, pp. 264–272.

Schaefer, G.W., 1976, "Radar Observation of Insect Flight, R.C. Rainey, ed., *Insect Flight,* Blackwell Scientific Publications, Oxford, pp. 157–197.

Scialom, G., and J. Testud, 1986, "Retrieval of Horizontal Wind Field and Mesoscale Vertical Vorticity in Stratiform Precipitation by Conical Scannings with Two Doppler Radars," *J. Atmospheric and Oceanic Technology,* Vol. 3, pp. 693–703.

Scorer, R.S. 1978, *Environmental Aerodynamics,* Ellis Horwood,

Sekhon, R.S., and R. C. Srivastava, 1970, "Snow Size Spectra and Radar Reflectivity," *J. Atmospheric Science,* Vol. 27, pp. 299–307.

Sekhon, R.S., and R.C. Srivastava, 1971, "Doppler Radar Observations of Drop-Size Distribution in a Thunderstorm," *J. Atmospheric Science,* Vol. 28, pp. 983–994.

Seliga, T.A., K. Aydin, and H. Direskeneli, 1986, "Disdrometer Measurements during an Intense Rainfall Event in Central Illinois: Implications for Differential Reflectivity Radar Observations," *J. Climate and Applied Meteorology,* Vol. 25, pp. 835–846.

Seliga, T.A., and V.N. Bringi, 1976, "Potential Use of Radar Differential Reflectivity Measurements at Orthogonal Polarizations for Measuring Precipitation," *J. Applied Meteorology,* Vol. 15, pp. 69–76.

Serafim, R.J., 1975. "Information Extraction from Meteorological Radar," *Atmospheric Technology,* Vol. 6, pp. 46–54.

Serafin, R.J., L.C. Peach, and D. Atlas, 1972, "Radar Measurements of the Atmospheric Turbulence Structure Function," *Proc. 15th Radar Meteorology Conf.,* American Meteorological Society, Boston, pp. 280–285.

Shenk, W.E., and E.B. Rodgers, 1978, "Nimbus 3/ATS 3 Observations of the Evolution of Hurricane Camille," *J. Applied Meteorology,* Vol. 17, pp. 458–476.

Shupiatskii, A.B., 1959, "Radiolokatzionnoe rasseianie nesfericheskimi chastitsami [Diffusion radar par les particules non sphériques]", Trudy Tsentral. Aero. Obs., [Moscow], Vol. 30, pp. 39–52.

Silver, S., 1949, *Microwave Antenna Theory and Design,* McGraw-Hill, New York.

Silverman, R.A., 1956, "Turbulent Mixing Theory Applied to Radio Scattering," *J. Applied Meteorology,* Vol. 27, pp. 690–705.

Simpson, J., R.F. Adler, and G.R. North, 1988, "A Proposed Tropical Rainfall Measuring Mission (TRMM) Satellite," *Bull. American Meteorological Society,* Vol. 69, pp. 278–295.

Simpson, J., G.W. Brier, and R.H. Simpson, 1967, "Stormfury Cumulus Seeding Experiment 1965: Statistical Analysis and Main Results," *J. Atmospheric Science,* Vol. 24, pp. 508–521.

Sirmans, D., and B. Bumgarner, 1975a, "Numerical Comparison of Five Mean Frequency Estimators," *J. Applied Meteorology,* Vol. 14, pp. 991–1003.

Sirmans, D., and B. Bumgarner, 1975b, "Estimation of Spectral Density Mean and Variance by Covariance Argument Technique," *Proc. 16th Radar Meteorology Conf.,* American Meteorological Society, Boston, pp. 6–13.

Skolnik, M.I., 1962, *Introduction to Radar Systems,* McGraw-Hill, New York.

Skolnik, M.I., ed., 1970, *Radar Handbook,* McGraw-Hill, New York.

Sloss, P.W., and D. Atlas, 1968, "Wind shear and reflectivity gradient effects on Doppler radar spectra." *J. Atmos. Sci.,* 25, 1080–1089.

Smagorinsky, J., 1974, "Global Atmosphere Modeling and the Numerical Simulation of Climate," W.N. Hess, ed., *Weather and Climate Modification,* John Wiley and Sons, New York, pp. 633–686.

Smith, R.L., and D.W. Holmes, 1961, "Use of Doppler Radar in Meteorological Observations," *Monthly Weather Rev.,* Vol. 89, pp. 1–7.

Smith, P.L., 1964, "Interpretation of the Fluctuating Echo from Randomly Distributed Scatterers—Part 3," McGill University Rep. MW 39.

Smith, S.D., and R.M. Rabin, 1989, "Estimation of Divergence in the Prestorm Boundary Layer," *J. Atmospheric and Oceanic Technology,* Vol. 6, pp. 446–458.

Smull, F.B. and R.A. Houze, 1985, "A Midlatitude Squall Line with a Trailing Region of Stratiform Rain: Radar and Satellite Observations," *Monthly Weather Rev.*, Vol. 113, pp. 117–133.

Spencer, R.W., 1986, "A Satellite Passive 37 GHz Scattering-Based Method for Measuring Oceanic Rain Rates," *J. Applied Meteorology,* Vol. 25, pp. 754–766.

Srivastava, R.C, and D. Atlas, 1974, "The Effects of Finite Radar Pulse Volume on Turbulence Measurements," *J. Applied Meteorology,* Vol. 13, pp. 472–480.

Srivastava, R.C., A.R. Jameson, and P.H. Hildebrand, 1979, "Time-Domain Computation of Mean and Variance of Doppler Spectra," *J. Applied Meteorology,* Vol. 18, pp. 189–194.

Srivastava, R.C., T.J. Matejka, and T.J. Lorello, 1986, "Doppler Radar Study of the Trailing Anvil Region Associated with a Squall Line," *J. Atmospheric Sciences,* Vol. 43, pp. 356–377.

Starr, J.R., and K.A. Browning, 1972, "Observations of Lee Waves by High-Power Radar," Quarterly *J. Royal Meteorological Society,* Vol. 98, pp. 73–85.

Steiner, M., and A. Waldvogel, 1987, "Peaks in Raindop Size Distributions," *J. Atmospheric Sciences,* Vol. 44, pp. 3127–3133.

Stephens, J.J., 1961, "Radar Cross Sections for Water and Ice Spheres," *J. Meteorology,* Vol. 18, pp. 348–359.

Stratmann, E., D. Atlas, J.H. Richter, and D.R. Jensen, 1971, "Sensitivity Calibration of a Dual-Beam Vertically Pointing FM-CW Radar," *J. Applied Meteorology,* Vol. 10, pp. 1260–1265.

Stratton, J.A., 1941, *Electromagnetic Theory,* McGraw-Hill, New York.

Strauch, R.G., D.A. Merritt, K.P. Moran, K.B. Earnshaw, and D. Van de Kamp, 1984, "The Colorado Wind-Profiling Network," *J. Atmospheric and Oceanic Technology,* Vol. 1, pp. 37–49.

Strauch, R.G., B.L. Weber, A.S. Frisch, C.G. Little, D.A. Merritt, K.P. Moran, and D.C. Welsh, 1987, "The Prevision and Relative Accuracy of Profiler Wind Measurements," *J. Atmospheric and Oceanic Technology,* Vol. 5, pp. 563–571.

Sulakvelidze, G.K., N. Sh. Bibilashvili, and V.F. Lapcheva, 1965, *Formation of Precipitation and Modification of Hail Processes,* Hydromet. Publ. House, Leningrad (Trans. in 1967, Available in English from Clearinghouse for Federal Scientific and Technical Information U.S. Dept. of Commerce, Washington, DC).

Summers, P.W., G.K. Mather, and D.S. Treddenich, 1972, "The Development and Testing of an Airborne Droppable Pyrotechnic Flare System for Seeding Alberta Hailstorms," *J. Applied Meteorology,* Vol. 11, pp. 695–703.

Takeda, T., and S. Murabayashi, 1981, "Observation of Clear-Air Echoes with 3.2 cm Radars," *J. Meteorological Society* [Japan], Vol. 59, pp. 864–874.

Tatarski, V.I., 1961, *Wave Propagation in a Turbulent Medium,* McGraw-Hill, New York.

Tatehira, R., 1968, "A Study of Rainband," *Geophysics Magazine,* Vol. 34, pp. 115–136.

Tatehira, R., and H. Itakura, 1966, "Radar Observation of Typhoon Lucy by Mt Fuji Radar," *Proc. 12th Conf. on Radar Meteorology,* American Meteorological Society, Boston, pp. 432–435.

Tatehira, R., and T. Shimizu, 1978, "Intensity Measurement of Precipitation Echo Superposed on Ground Clutter—A New Automatic Technique for Ground Clutter Rejection," *Proc. 18th Radar Meteorological Conf.,* American Meteorological Society, Boston, pp. 364–369.

Testud, J., and P. Amayenc, 1989, "Stereoradar Meteorology: A Promising Technique for Observation of Precipitation from a Mobile Platform," *J. Atmospheric and Oceanic Technology,* Vol. 6, pp. 89–108.

Testud, J., G. Breger, P. Amayenc, M. Chong, B. Nutten, and A. Sauvageot, 1980, "A Doppler Radar Observation of a Cold Front—Three-Dimensional Air Circulation, Related Precipitation System and Associated Wavelike Motions," *J. Atmospheric Science,* Vol. 37, pp. 78–98.

Testud, J., and M. Chong, 1983, "Three-Dimensional Wind Field Analysis from Dual Doppler Radar. Part 1: Filtering, Interpolating and Differentiating the Raw Data," *J. Climate and Applied Meteorology,* Vol. 22, pp. 1204–1215.

Todd, C.J., and W.E. Howell, 1985, "Weather Modification," D.H. Houghton, ed., *Handbook of Applied Meteorology,* Chapter 38, pp. 1065–1139.

Torlaschi, E., R.G. Humphries, and B.L. Barge, 1984, "Circular Polarization for Precipitation Measurements," *Radio Science,* Vol. 19, pp. 193–200.

Tsonis, A.A., and G.A. Isaac, 1985, "On a New Approach for Instantaneous Rain Area Delineation in the Mid Latitudes Using GOES Data," *J. Climatology and Applied Meteorology,* Vol. 24, pp. 1208–1218.

Tuttle, J.D., and R.E. Rinehart, 1983, "Attenuation Correction in Dual-Wavelength Analyses," *J. Climatology and Applied Meteorology,* Vol. 22, pp. 1914–1921.

Tuttle, J.D., V.M. Bringi, H.D. Orville, and F.J. Kopp, 1989, "Multiparameter Radar Study of a Microburst: Comparison with Model Results," *J. Atmospheric Science,* Vol. 46, pp. 601–620.

Ulaby, F.T., R.K. Moore, and A.K. Fung, 1981, *Microwave Remote Sensing, Vol. 1, Fundamentals and Radiometry,* Addison-Wesley, Reading, MA.

Ulaby, F.T., R.K. Moore, and A.K. Fung, 1982, *Microwave Remote Sensing. Vol. II. Radar Remote Sensing and Surface Scattering and Emission Theory.* Addison-Wesley, Reading, MA.

Ulbrich, C.W., 1978, "Relationships of Equivalent Reflectivity Factor to the Vertical Fluxes of Mass and Kinetic Energy of Hail," *J. Applied Meteorology,* Vol. 17, pp. 1803–1808.

Ulbrich, C.W., 1983, "Natural Variations in the Analytical Form of the Raindrop Size Distribution," *J. Climate and Applied Meteorology,* Vol. 22, 1764–1775.

Ulbrich, C.W., and D. Atlas, 1978, "The Rain Parameter Diagram: Methods and Applications," *J. Geophysics Research,* Vol. 83, No. C3, pp. 1319–1325.

Uman, M.A., 1969, *Lightning,* McGraw-Hill, New York.

Uman, M.A., and R. E. Orville, 1964, "Electron Density Measurement in Lightning from Stark Broadening of H$\alpha$," *J. Geophysics Research,* Vol. 69, pp. 5151–5154.

Van de Hulst, H.C., 1957, *Light Scattering by Small Particles,* John Wiley and Sons, New York.

Van Vleck, J.H., 1947a, "Absorption of Microwaves by Oxygen," *Physics Rev.,* Vol. 71, pp. 413–424.

Van Vleck, J.H., 1947b, "The Absorption of Microwaves by Uncondensed Water Vapor," *Physics Rev.,* Vol. 71, pp. 425–433.

Van Zandt, T.E., J.L. Green, K.S. Gage, and W.L. Clark, 1978, "Vertical Profiles of Refractivity Turbulence Structure Constant—Comparison of Observations by the Sunset Radar with a New Theoretical Model," *Radio Sciences,* Vol. 13, pp. 819–829.

Vincent, R.A., and J. Röttger, 1980, "Spaced Antenna VHF Radar Observations of Tropospheric Velocities and Irregularities," *Radio Sciences,* Vol. 15, pp. 319–335.

Wakasugi, K., A. Mizutani, M. Matsuo, S. Fukao, and S. Kato, 1986, "A Direct Method for Deriving Drop-Size Distribution and Vertical Air Velocities from VHF Doppler Radar Spectra," *J. Atmospheric and Oceanic Technology,* Vol. 3, pp. 623–629.

Waldteufel, P., 1976, "An Analysis of Weather Spectra Variance in a Tornadic Storm," NOAA Tech. Memo. ERL/NSSL-76.

Waldteufel, P., and H. Corbin, 1979, "On the Analysis of Single-Doppler Radar Data," *J. Applied Meteorology,* Vol. 18, pp. 532–542.

Waldvogel, A., L. Klein, D.J. Musil, and P.L. Smith, 1987, "Characteristics of Radar-Identified Big Drop Zones in Swiss Hailstorm," *J. Climate and Applied Meteorology,* Vol. 26, pp. 861–877.

Waldvogel, A., W. Schmid, and B. Federer, 1978a, "The Kinetic Energy of Hailfalls. Part I: Hailstorm spectra," *J. Applied Meteorology,* Vol. 17, pp. 515–520.

Waldvogel, A., W. Schmid, and J.F. Mezeix, 1978b, "The Kinetic Energy of Hailfalls. Part II: Radar and hailpads," *J. Applied Meteorology,* Vol. 17, pp. 1680–1693.

Walker, G.B., and W.T. Cronenwett, 1974, "Experimental Studies of Modified R-Meter Theory," *J. Applied Meteorology,* Vol. 13, pp. 558–562.

Walker, G.B., P.S. Ray, D. Zrnic', and R. Doviak, 1980, "Time, Angle and Range Averaging of Radar Echoes from Distributed Targets," *J. Applied Meteorology,* Vol. 19, pp. 315–323.

Wallace, P.R., 1953, "Interpretation of the Fluctuating Echo from Randomly Distributed Scatterers—II," *Canadian J. Physics,* Vol. 31, pp. 995–1009.

Weill, A., C. Klapisz, B. Strauss, F. Baudin, C. Jaupart, P. Van Grunderbeeck, and J.P. Goutorbe, 1980, "Measuring Heat Flux and Structure Functions of Temperature Fluctuations with an Acoustic Doppler Sodar," *J. Applied Meteorology,* Vol. 19, pp. 199–205.

Weinstein, A.I., and P.B. MacCready, Jr., 1969, "An Isolated Cumulus Cloud Modification Project," *J. Applied Meteorology,* Vol. 8, pp. 936–947.

Wexler, R., and D. Atlas, 1959, "Precipitation Generation Cells," *J. Meteorology,* Vol. 16, pp. 327–332.

Wexler, R., and D. Atlas, 1963, "Radar Reflectivity and Attenuation of Rain," *J. Applied Meteorology,* Vol. 2, pp. 276–280.

Whitten, R.C., P.L. Smith, and A.C. Harbuck, 1976, "Calibration of Weather Radar Systems Using the Sun as a Radio Source," *Proc. 17th Radar Meteorology Conf.,* American Meteorological Society, Boston, pp. 60–65.

Wilheit, T.T., 1986, "Some Comments on Passive Microwave Measurement of Rain," *Bull. American Meteorological Society,* Vol. 67, pp. 1226–1232.

Wilheit, T.T., A.T.C. Chang, J.L. King, E.B. Rodgers, R.A. Nieman, B.M. Krupp, A.S. Milman, J.S. Stratigos, and H. Siddalingaiah, 1982, "Microwave Radiometric Observations near 19.35, 92 and 183 GHz of Precipitation in Tropical Storm Cora," *J. Applied Meteorology,* Vol. 21, pp. 1137–1145.

Wilheit, T.T., A.T.C. Chang, M.S.V. Rao, E.B. Rodgers, and J.S. Theon, 1977, "A Satellite Technique for Quantitatively Mapping Rainfall Rates over the Oceans," *J. Applied Meteorology,* Vol. 16, pp. 551–560.

Willis, P.T., and J.F. Spahn, 1977, "*Z-R* Relations from Drop Size Measurements in Hurricane Eloise on 22 September 1975," *11th Tech. Conf. on Hurricanes and Tropical Meteorology,* American Meteorological Society, Boston, pp. 227–231.

Wilson, D.A., 1970, "Doppler Radar Studies of Boundary Layer Wind Profile and Turbulence in Snow Conditions," *Proc. 14th Radar Meteorology Conf.,* American Meteorological Society, Boston, pp. 191–196.

Wilson, J.W., 1970, "Integration of Radar and Raingauge Data for Improved Rainfall Measurement," *J. Applied Meteorology,* Vol. 9, pp. 489–497.

Wilson, J.W., and D. Reum, 1988, "The Flare Echo: Reflectivity and Velocity Signature," *J. Atmospheric and Oceanic Technology,* Vol. 5, pp. 197–205.

Wobus, H.B., F.W. Murray, and L.R. Koenig, 1971, "Calculation of the Terminal Velocity of Water Drops," *J. Applied Meteorology,* Vol. 10, pp. 751–754.

Wood, V.T., and R.A. Brown, 1983, "Single Doppler Velocity Signatures: An Atlas of Pattern in Clear Air/Widespread Precipitation and Convective Storms," NOAA Tech. Memo. ERL NSSL-95. Environmental Research Laboratories (NTIS # PB84-155779).

Woodley, W., A. Olsen, A. Herndon, and V. Wiggert, 1975, "Comparison of Gage and Radar Methods of Convective Rain Measurement," *J. Applied Meteorology,* Vol. 14, pp. 909–928.

Woodman, R.F., and A. Guillen, 1974, "Radar Observations of Winds and Turbulence in the Stratosphere and Mesosphere," *J. Atmospheric Science,* Vol. 31, pp. 493–505.

Wu, R., J.A. Weinman, and R.T. Chin, 1985, "Determination of Rainfall Rates from GOES Satellite Images by Pattern Recognition Technique," *J. Atmospheric and Oceanic Technology,* Vol. 2, pp. 314–330.

Zawadski, I.I., 1975, "On Radar-Raingage Comparison," *J. Applied Meteorology,* Vol. 14, pp. 1430–1436.

Ziegler, C.L., 1985, "Retrieval of Thermal and Microphysical Variables in Observed Convective Storms. Part I: Model Development and Preliminary Testing," *J. Atmospheric Science,* Vol. 42, pp. 1487–1509.

Zrnic', D.S., 1975, "Moments of Estimated Input Power for Finite Sample Averages of Radar Receiver Outputs," *IEEE Trans. on Aerospace and Electronic Systems,* Vol. AES-11, pp. 109–113.

Zrnic', D.S., 1977, "Spectral Moment Estimates from Correlated Pulse Pairs," *IEEE Trans. on Aerospace and Electronics Systems,* Vol. AES-13, No. 4, pp. 344–354.

Zrnic', D.S., 1987, "Three-Body Scattering Produces Precipitation Signature of Special Diagnostic Value," *Radio Sciences,* Vol. 22, pp. 76–78.

Zrnic', D.S., D.W. Burgess, and L.D. Hernington, 1985, "Automatic Detection of Mesocyclonic Shear with Doppler Radar," *J. Atmospheric and Oceanic Technology,* Vol. 2, pp. 425–438.

Zrnic', D.S., and R.J. Doviak, 1975, "Velocity Spectra of Vortices Scanned with a Pulse-Doppler Radar," *J. Applied Meteorology,* Vol. 14, pp. 1531–1539.

Zrnic', D.S., W.D. Rust, and W.L. Taylor, 1982, "Doppler Radar Echoes of Lightning and Precipitation at Vertical Incidence," *J. Geophysics Research,* Vol. 87, pp. 7179–7191.

# INDEX

## *The Artech House Radar Library*

David K. Barton, *Series Editor*

*Active Radar Electronic Countermeasures* by Edward J. Chrzanowski

*Airborne Pulsed Doppler Radar* by Guy V. Morris

*AIRCOVER: Airborne Radar Vertical Coverage Calculation Software and User's Manual* by William A. Skillman

*Analog Automatic Control Loops in Radar and EW* by Richard S. Hughes

*Aspects of Modern Radar* by Eli Brookner, *et al.*

*Aspects of Radar Signal Processing* by Bernard Lewis, Frank Kretschmer, and Wesley Shelton

*Bistatic Radar* by Nicholas J. Willis

*Detectability of Spread-Spectrum Signals* by Robin A. Dillard and George M. Dillard

*Electronic Homing Systems* by M.V. Maksimov and G.I. Gorgonov

*Electronic Intelligence: The Analysis of Radar Signals* by Richard G. Wiley

*Electronic Intelligence: The Interception of Radar Signals* by Richard G. Wiley

*EREPS: Engineer's Refractive Effects Prediction System Software and User's Manual,* developed by NOSC

*Handbook of Radar Measurement* by David K. Barton and Harold R. Ward

*High Resolution Radar* by Donald R. Wehner

*High Resolution Radar Cross-Section Imaging* by Dean L. Mensa

*Interference Suppression Techniques for Microwave Antennas and Transmitters* by Ernest R. Freeman

*Introduction to Electronic Defence Systems* by F. Neri

*Introduction to Electronic Warfare* by D. Curtis Schleher

*Introduction to Sensor Systems* by S.A. Hovanessian

*Logarithmic Amplification* by Richard Smith Hughes

*Modern Radar System Analysis* by David K. Barton

*Modern Radar System Analysis Software and User's Manual* by David K. Barton and William F. Barton

*Monopulse Principles and Techniques* by Samuel M. Sherman

*Monopulse Radar* by A.I. Leonov and K.I. Fomichev

*MTI and Pulsed Doppler Radar* by D. Curtis Schleher

*Multifunction Array Radar Design* by Dale R. Billetter

*Multisensor Data Fusion* by Edward L. Waltz and James Llinas

*Multiple-Target Tracking with Radar Applications* by Samuel S. Blackman

*Multitarget-Multisensor Tracking: Advanced Applications,* Yaakov Bar-Shalom, ed.

*Over-The-Horizon Radar* by A.A. Kolosov, et al.

*Principles and Applications of Millimeter-Wave Radar,* Charles E. Brown and Nicholas C. Currie, eds.

*Principles of Modern Radar Systems* by Michel H. Carpentier

*Pulse Train Analysis Using Personal Computers* by Richard G. Wiley and Michael B. Szymanski

*Radar and the Atmosphere* by Alfred J. Bogush, Jr.

*Radar Anti-Jamming Techniques* by M.V. Maksimov, *et al.*

*Radar Cross Section Analysis and Control* by A.K. Bhattacharyya and D.L. Sengupta

*Radar Cross Section* by Eugene F. Knott, *et al.*

*Radar Detection* by J.V. DiFranco and W.L. Rubin

*Radar Electronic Countermeasures System Design* by Richard J. Wiegand

*Radar Evaluation Handbook* by David K. Barton, *et al.*

*Radar Evaluation Software* by David K. Barton and William F. Barton

*Radar Propagation at Low Altitudes* by M.L. Meeks

*Radar Range-Performance Analysis* by Lamont V. Blake

*Radar Reflectivity Measurement: Techniques and Applications,* Nicholas C. Currie, ed.

*Radar Reflectivity of Land and Sea* by Maurice W. Long

*Radar System Design and Analysis* by S.A. Hovanessian

*Radar Technology,* Eli Brookner, ed.

*Receiving Systems Design* by Stephen J. Erst

*Radar Vulnerability to Jamming* by Robert N. Lothes, Michael B. Szymanski, and Richard G. Wiley

*RGCALC: Radar Range Detection Software and User's Manual* by John E. Fielding and Gary D. Reynolds

*SACALC: Signal Analysis Software and User's Guide* by William T. Hardy

*Secondary Surveillance Radar* by Michael C. Stevens

*SIGCLUT: Surface and Volumetric Clutter-to-Noise, Jammer and Target Signal-to-Noise Radar Calculation Software and User's Manual* by William A. Skillman

*Signal Theory and Random Processes* by Harry Urkowitz

*Solid-State Radar Transmitters* by Edward D. Ostroff, *et al.*

*Space-Based Radar Handbook,* Leopold J. Cantafio, ed.

*Spaceborne Weather Radar* by Robert M. Meneghini and Toshiaki Kozu

*Statistical Theory of Extended Radar Targets* by R.V. Ostrovityanov and F.A. Basalov

*The Scattering of Electromagnetic Waves from Rough Surfaces* by Peter Beckmann and Andre Spizzichino

*VCCALC: Vertical Coverage Plotting Software and User's Manual* by John E. Fielding and Gary D. Reynolds